Soil Strength and Slope Stability

Soil Strength and Slope Stability

J. Michael Duncan
Stephen G. Wright

WILEY

JOHN WILEY & SONS, INC.

Copyright © 2005 by John Wiley & Sons, Inc. All rights reserved

Published by John Wiley & Sons, Inc., Hoboken, New Jersey
Published simultaneously in Canada

For general information on our other products and services or for technical support, please contact our Customer Care Department within the United States at (800) 762-2974, outside the United States at (317) 572-3993 or fax (317) 572-4002.

Wiley also publishes its books in a variety of electronic formats. Some content that appears in print may not be available in electronic books. For more information about Wiley products, visit our web site at www.wiley.com.

Library of Congress Cataloging-in-Publication Data:
Duncan, J. M. (James Michael)
    Soil strength and slope stability / J. Michael Duncan, Stephen G. Wright.
        p. cm.
    Includes index.
    ISBN 0-471-69163-1 (cloth : alk. paper)
    1. Slopes (Soil mechanics) I. Wright, Stephen G. (Stephen Gailord), 1943–
II. Title.
  TA710.D868 2005
  624.1′51363--dc22

2004019535

Printed in the United States of America

10 9 8 7 6 5 4

CONTENTS

PREFACE

This book has as its purpose to draw together lessons of the past 35 years regarding soil strength and slope stability. During this period, techniques and equipment for laboratory and in situ tests have improved, our ability to perform analyses of slope stability has been revolutionized by the widespread use of computers, new methods of reinforcing and stabilizing slopes have been developed, and a large number of investigations into slope failures have provided well-documented and valuable case histories to guide engineering practice. It is therefore an appropriate time to bring these elements together in a book that can serve geotechnical graduate students and professionals.

Development of this book would not have been possible without the assistance of many colleagues, whose contributions to our understanding we gratefully acknowledge. Foremost among these is Professor Harry Seed, who taught both of us and was the inspiration for our lifelong interest in soil strength and slope stability. We are also grateful for the opportunity to work with Nilmar Janbu, who during his sabbatical at Berkeley in 1969 taught us many valuable lessons regarding analysis of slope stability and the shear strength of soils. Our university colleagues Jim Mitchell, Roy Olson, Clarence Chan, Ken Lee, Peter Dunlop, Guy LeFebvre, Fred Kulhawy, Suphon Chirapuntu, Tarciso Celestino, Dean Marachi, Ed Becker, Kai Wong, Norman Jones, Poul Lade, Pat Lucia, Tim D'Orazio, Jey Jeyapalan, Sam Bryant, Erik Loehr, Loraine Fleming, Bak Kong Low, Bob Gilbert, Vern Schaefer, Tim Stark, Mohamad Kayyal, Marius DeWet, Clark Morrison, Tom Brandon, Ellen Rathje, George Filz, Mike Pockoski, and Jaco Esterhuizen have also contributed greatly to our understanding of solid strength and stability. Our experiences working with professional colleagues Al Buchignani, Laurits Bjerrum, Jim Sherard, Tom Leps, Norbert Morgenstern, George Sowers, Robert Schuster, Ed Luttrell, Larry Franks, Steve Collins, Dave Hammer, Larry Cooley, John Wolosick, Luis Alfaro, Max DePuy and his group at the Panama Canal Authority, and Fernando Bolinaga have helped us to see the useful relationships among teaching, research, and professional practice. Special thanks goes to Chris Meehan for his invaluable assistance with figures, references, proofing, and indexing. Finally, we express our deepest appreciation and love to our wives, Ann and Ouida, for their support, understanding, and constant encouragement throughout our careers and during the countless hours we have spent working together on this book.

J. Michael Duncan
Stephen G. Wright

Soil Strength and Slope Stability

CHAPTER 1

Introduction

Evaluating the stability of slopes in soil is an important, interesting, and challenging aspect of civil engineering. Concerns with slope stability have driven some of the most important advances in our understanding of the complex behavior of soils. Extensive engineering and research studies performed over the past 70 years provide a sound set of soil mechanics principles with which to attack practical problems of slope stability.

Over the past decades, experience with the behavior of slopes, and often with their failure, has led to development of improved understanding of the changes in soil properties that can occur over time, recognition of the requirements and the limitations of laboratory and in situ testing for evaluating soil strengths, development of new and more effective types of instrumentation to observe the behavior of slopes, improved understanding of the principles of soil mechanics that connect soil behavior to slope stability, and improved analytical procedures augmented by extensive examination of the mechanics of slope stability analyses, detailed comparisons with field behavior, and use of computers to perform thorough analyses. Through these advances, the art of slope stability evaluation has entered a more mature phase, where experience and judgment, which continue to be of prime importance, have been combined with improved understanding and rational methods to improve the level of confidence that is achievable through systematic observation, testing, and analysis. This seems an appropriate stage in the development of the state of the art to summarize some of these experiences and advances in a form that will be useful for students learning about the subject and for geotechnical engineers putting these techniques into practice. This is the objective that this book seeks to fill.

Despite the advances that have been made, evaluating the stability of slopes remains a challenge. Even when geology and soil conditions have been evaluated in keeping with the standards of good practice, and stability has been evaluated using procedures that have been effective in previous projects, it is possible that surprises are in store. As an example, consider the case of the Waco Dam embankment.

In October 1961, the construction of Waco Dam was interrupted by the occurrence of a slide along a 1500-ft section of the embankment resting on the Pepper shale formation, a heavily overconsolidated, stiff-fissured clay. A photograph of the 85-ft-high embankment section, taken shortly after the slide occurred, is shown in Figure 1.1. In the slide region, the Pepper shale had been geologically uplifted to the surface and was bounded laterally by two faults crossing the axis of the embankment. The slide was confined to the length of the embankment founded on Pepper shale, and no significant movements were observed beyond the fault boundaries.

The section of the embankment involved in the slide was degraded to a height of approximately 40 ft, and an extensive investigation was carried out by the U.S. Army Corps of Engineers to determine the cause of the failure and to develop a method for repairing the slide. The investigation showed that the slide extended for several hundred feet downstream from the embankment, within the Pepper shale foundation. A surprising finding of the studies conducted after the failure was the highly anisotropic nature of the Pepper shale, which contained pervasive horizontal slickensided fissures spaced about $\frac{1}{8}$ in. (3 mm) apart. The strength along horizontal planes was found to be only about 40% as large as the strength measured in conventional tests on vertical specimens. Although conventional testing and analysis indicated that the embankment would be stable throughout construction, analyses performed using the lower strengths on horizontal planes

Figure 1.1   Slide in the downstream slope of the Waco Dam embankment.

produced results that were in agreement with the failure observed (Wright and Duncan, 1972).

This experience shows that the conventional practice of testing only vertical samples can be misleading, particularly for stiff fissured clays with a single dominant fissure orientation. With the lesson of the Waco Dam experience in mind, geotechnical engineers are better prepared to avoid similar pitfalls.

The procedures we use to measure soil strengths and evaluate the stability of slopes are for the most part rational and may appear to be rooted solidly in engineering science. The fact that they have a profound empirical basis is illustrated by the case of an underwater slope in San Francisco Bay. In August 1970, during construction of a new shipping terminal at the Port of San Francisco, a 250-ft (75-m)-long portion of an underwater slope about 90 ft (30 m) high failed, with the soil on one side sliding into the trench, as shown in Figure 1.2. The failure took place entirely within the San Francisco Bay mud, a much-studied highly plastic marine clay.

Considerable experience in the San Francisco Bay area had led to the widely followed practice of excavating underwater slopes in Bay mud at 1 (horizontal) : 1 (vertical). At this new shipping terminal, however, it was desired to make the slopes steeper, if possible, to reduce the volume of cut and fill and the cost of the project. Thorough investigations, testing, and analyses were undertaken to study this question.

Laboratory tests on the best obtainable samples, and extensive analyses of stability, led to the conclusion

that it would be possible to excavate the slopes at 0.875 : 1. At this inclination, the factor of safety computed for the slopes would be 1.17. Although such a low factor of safety was certainly unusual, the conditions involved were judged to be exceptionally well known and understood, and the slopes were excavated at the steep angle. The result was the failure depicted in Figure 1.2. An investigation after the failure led to the conclusion that the strength of the Bay mud that could be mobilized in the field over a period of several weeks was lower than the strength measured in laboratory tests in which the Bay mud was loaded to failure in a few minutes, and that the cause of the difference was creep strength loss (Duncan and Buchignani, 1973).

The lesson to be derived from this experience is that our methods may not be as scientifically well founded as they sometimes appear. If we alter our conventional methods by "improving" one aspect, such as the quality of samples used to measure the undrained strength of Bay mud, we do not necessarily achieve a more accurate result. In the case of excavated slopes in Bay mud, conventional sample quality and conventional test procedures, combined with conventional values of factor of safety, had been successful many times. When the procedures were changed by "refining" the sampling and strength testing procedures, the result was higher values of undrained shear strength than would have been measured if conventional procedures had been used. When, in addition, the value of the safety factor was reduced, the result was a decision to use an

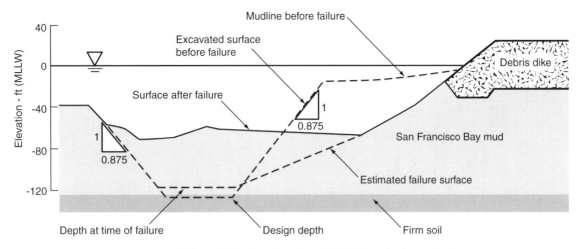

Figure 1.2 Failure of the San Francisco LASH Terminal trench slope.

excessively steep slope, which failed. Altering conventional practice and reducing the factor of safety led to use of a procedure that was not supported by experience.

SUMMARY

The broader messages from these and similar cases are clear:

1. We learn our most important lessons from experience, often from experience involving failures. The state of the art is advanced through these failures and the lessons they teach. As a result, the methods we use depend strongly on experience. Despite the fact that our methods may have a logical background in mechanics and our understanding of the behavior of soils and rocks, it is important to remember that these methods are semi-empirical. We depend as much on the fact that the methods have worked in the past as we do on their logical basis. We cannot count on improving these methods by altering only one part of the process that we use.

2. We should not expect that we have no more lessons to learn. As conditions arise that are different from the conditions on which our experience is based, even in ways that may at first seem subtle, we may find that our semi-empirical methods are inadequate and need to be changed or expanded. The slide in Waco Dam served clear notice that conventional methods were not sufficient for evaluating the shear strength of Pepper shale and the stability of embankments founded on it. The lesson learned from that experience is now part of the state of the art, but it would be imprudent to think that the current state of knowledge is complete. We need to keep abreast of advances in the state of the art as they develop and practice our profession with humility, in recognition that the next lesson to be learned may be lurking in tomorrow's project.

The objective of this book is to draw together some of the lessons that have been learned about measuring soil strengths and performing limit equilibrium analyses of stability into a consistent, clear, and convenient reference for students and practicing engineers.

CHAPTER 2

Examples and Causes of Slope Failure

Experience is the best teacher but not the kindest. Failures demand attention and always hold lessons about what not to do again. Learning from failures—hopefully from other people's failures—provides the most reliable basis for anticipating what might go wrong in other cases. In this chapter we describe 10 cases of slope failure and recount briefly the circumstances under which they occurred, their causes, and their consequences. The examples are followed by an examination of the factors that influence the stability of slopes, and the causes of instability, as illustrated by these examples.

EXAMPLES OF SLOPE FAILURE

The London Road and Highway 24 Landslides

The London Road landslide in Oakland, California, occurred in January 1970 during a period of heavy rainfall. Front-page headlines in the January 14, 1970, *Oakland Tribune* exclaimed "Storm Hammers State—Slide Menaces 14 Homes—The Helpless Feeling of Watching Ruin Approach." Figure 2.1 shows houses in the slide area that were destroyed by the slide and had to be abandoned by their owners. The slide covered an area of about 15 acres, and the sliding surface was estimated to be as deep as 60 ft (20 m) beneath the surface of the ground. As evident from the height of the headscarp in comparison with the houses on the right in Figure 2.1, the slide movements were very large. Some 14 houses were destroyed, and a jet fuel pipeline at the bottom of the hill was never used again because of the danger that it would be ruptured by slide movements. Because the cost of stabilizing the massive slide was greater than the economic benefit, it was not repaired, and an entire neighborhood was lost permanently.

Not only was there heavy rainfall during January 1970, the entire preceding year had been unusually wet, with about 140% of the average rainfall recorded at the nearest rain gage station. As discussed later, these prolonged wet conditions played a significant role in the occurrence of the massive landslide.

The Highway 24 landslide shown in Figure 2.2 occurred in January 1982, when a storm blew in from the Pacific Ocean and stalled over the San Francisco Bay area. In a 24-hour period in early January, the storm dumped nearly 10 in. of rain on the area, where the normal yearly rainfall is about 25 in. The sudden enormous deluge resulted in literally thousands of landslides in the San Francisco Bay area. Typically, these slides were shallow. The intense rainfall saturated the upper few feet of the ground on the hillsides, which came sliding and flowing down in many places, knocking down trees, destroying houses, and blocking roads. Figure 2.2 shows Highway 24 near Orinda, California, partially blocked by a flow of sloppy, saturated soil that flowed onto the roadway, just one of the thousands of slides that occurred during the storm.

The London Road landslide and the flow slide on Highway 24, less than 10 miles apart, illustrate two very different types of slope failures that occur in the same area. The London Road landslide was very deep-seated and followed two successive years of above-normal rainfall. Although detailed soil exploration was not carried out at the London Road site, some interesting facts may be surmised based on what could be seen at the ground surface. The slide movement exposed serpentine rock in one area. Serpentine is a metamorphic rock that can be hard and strong but is subject to rapid deterioration to a weak powdery mass when exposed to air and water. Although the exposed serpentine retained its rocklike appearance, it could be penetrated several inches with a bare hand and had

Figure 2.1   London Road landslide, Oakland, California. (J. M. Duncan photos.)

essentially no strength or stiffness. It can be surmised that the strength of the serpentine, and other soils and rocks underlying the London Road area, had been deteriorating slowly over a period of many tens, hundreds, even thousands of years since the hillside was formed. Such deterioration results from chemical and physical processes that can gradually change the properties of earth materials. Eventually, this reduction in strength, combined with two years of heavy rainfall and resulting high groundwater levels, led to the very deep-seated landslide.

In contrast, the slide that blocked Highway 24 was very shallow, probably no more than 3 ft (1 m) deep. This type of slide develops very quickly as a result of relatively brief, extremely intense rainfall. Infiltration within a brief period affects only the upper few feet of soil. Within this depth, however, the soil may become saturated and lose much of its strength. In the area east of San Francisco Bay, the hillsides are blanketed by silty and sandy clays of low plasticity over the top of less weathered and stronger rock. The thickness of the soil cover ranges from zero to 15 ft (5 m). The soil

Figure 2.2 Flow slide on Highway 24 near Orinda, California. (J. M. Duncan photo.)

has formed from the underlying rocks and has reached its present condition through processes of weathering, erosion, shallow sliding, and deposition farther downhill. When dry, the soils that blanket the hillsides are stiff and strong, and the slopes they form are stable. During intense rains, however, water infiltrates the ground rapidly because the ground contains many cracks that provide secondary permeabilty. Although the processes leading to this type of slide are still the subject of research study, it is clear that conditions can change very quickly, and that the transition from stable ground to a fluid mass in rapid motion can take place within minutes. The high velocities with which these slides move makes them very dangerous, and many lives have been lost when they flowed down and crushed houses without warning.

The London Road and the Highway 24 slides illustrate a relationship between rainfall and landslides that has been observed in many places: Long periods of higher-than-average rainfall cause deep-seated, slow-moving slides, with shear surfaces that can extend tens of feet below the ground surface. One or two days of very intense rainfall, in contrast, tend to cause shallow slides involving only a few feet of soil, which move with high velocity once they are in motion.

The Landslide at Tuve, Sweden

In December 1977 a large landslide occurred on a gentle hillside in the town of Tuve, Sweden, a suburb north of Göteborg. A photograph and three cross sections through the slide are shown in Figure 2.3. The soil at the site was a layer of quick clay overlying a thin layer of permeable granular material on top of rock. The slide covered an area of 15 ha (40 acres), destroyed about 50 houses, and took 11 lives. Quick clays of the type involved in this slide are noted for their great sensitivity and extremely brittle behavior. When they fail, they lose practically all shear strength and flow like a viscous liquid.

The slide is believed to have started as a small slope failure in the side of a road embankment. The small slide left unstable the slope it slid away from, and a slightly larger failure of that slope took place. The process was repeated as the slide grew in the uphill direction, covering a larger and larger area. Houses in the area were undermined by the retrogressing slides and cruised downhill on the weakened slippery clay, crashing into other houses. As soil and houses from the failed area moved downhill, the soil in the area into which it moved was loaded and disturbed, and it began to fail also. The slide thus grew uphill and downhill from the original small slope failure in the middle.

Slides that grow uphill by increments are called *retrogressive;* slides that grow downhill by increments are called *progressive*. The Tuve slide was both. The small embankment failure that started the Tuve slide is believed to have been caused by erosion steepening the slope of the roadway embankment at the location of a small drainage culvert. This small slope failure was the trigger for a slide of immense proportions because of the metastable structure of the Swedish quick clay. It is likely that the strength of the bottom part of the quick clay layer was unusually low as a result of ar-

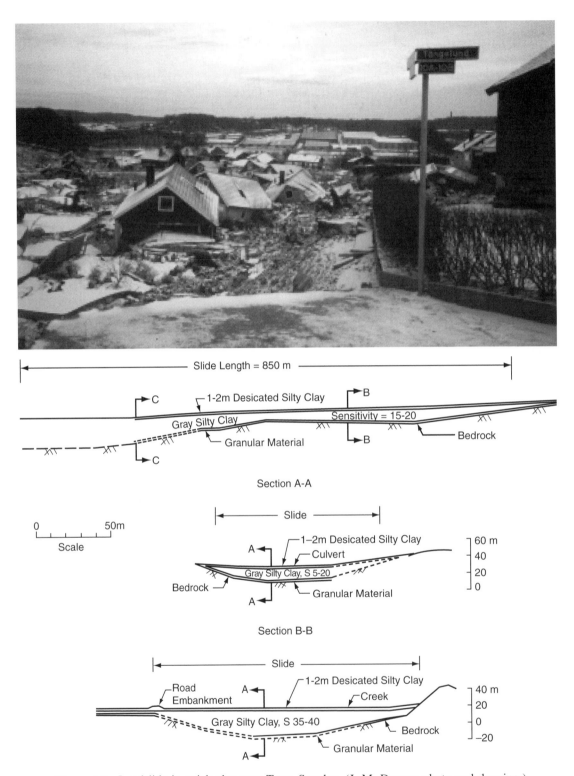

Figure 2.3 Landslide in quick clay near Tuve, Sweden. (J. M. Duncan photo and drawing.)

tesian pressures in the permeable granular material beneath the clay. Similar conditions in Norway have been found to be especially treacherous.

Slope Failures in Highway and Dam Embankments

Pinole, California, slide. Figure 2.4 shows a slope failure that occurred on a section of Interstate 80 near Pinole, California, where the road was supported on an embankment of well-compacted clayey soil. It can be seen that the back scarp of the failure is very steep. This is an indication that the embankment material was very strong, or it would not have remained stable in this nearly vertical slope, which was about 30 ft (10 m) high. The weak link was the foundation, which contained organic soil that had not been removed when the highway was constructed.

The natural ground sloped upward away from the south side of the embankment (the left side in Figure 2.4) and downward away from the right side. During rains, water tended to pond against the embankment because there was no underdrainage, and water seeped from south to north through the foundation of the embankment. The slide occurred after a period of heavy rain in the winter of 1969.

Houston, Texas, slide. A slide in a highway embankment near Houston, Texas, is shown in Figure 2.5. The embankment was constructed of compacted highly plastic clay and was built with 2 (horizontal):1 (vertical) side slopes. The fill was well compacted. The embankment was stable when it was built and remained stable for many years afterward. However, as time went by and the fill was wetted and dried repeat-

edly during alternating rainy and dry periods, it gradually swelled and grew softer and weaker. Finally, about 20 years after the embankment was built, the failure shown in Figure 2.5 occurred.

San Luis Dam, California, slide. On September 4, 1981, a massive slide occurred in the upstream slope of San Luis Dam, about 100 miles southeast of San Francisco, California. A photograph of the slide is shown in Figure 2.6. At the left end of the slide the movements were about 35 ft (about 10 m). The amount of displacement decreased to the right, diminishing to zero in a length of about 1100 ft (350 m). In the area where the slide occurred, the embankment was 200 ft (60 m) high and was constructed on a layer of highly plastic clay slope wash overlying the rock that formed the hillside. This material was formed from the underlying rocks by the same processes of weathering, erosion, shallow sliding, and deposition farther downhill that formed the soils on the hills east of San Francisco Bay.

When the San Luis Dam embankment was constructed in 1969, the highly plastic clay slope-wash that covered the foundation was dry and very strong. However, when it was wetted by the water stored in the reservoir, it became much weaker. Furthermore, over the period of 12 years between construction of the dam and the slide shown in Figure 2.6, the water level in the reservoir moved up and down several times as the pumped-storage reservoir was filled in the wet season and as water was withdrawn in the dry season. Stark and Duncan (1991) performed tests on the slope-wash and analyses of the slide which indicated that the

Figure 2.4   Interstate 80 embankment slope failure.

Figure 2.5 Slide in highway embankment near Houston, Texas. Embankment constructed of highly plastic clay.

Figure 2.6 San Luis Dam upstream slope failure.

strength of the slope-wash was gradually reduced to a low residual value due to the wetting and the cyclic variations in shear stress caused by the changes in the water level in the reservoir. Finally, in September 1981, the slide shown in Figure 2.6 occurred following the largest and fastest drawdown of the reservoir. The slide was stabilized by rebuilding the failed part of the dam,

adding a 60-ft (18-m)-high buttress at the base (ENR, 1982).

The Olmsted Landslide

The Olmsted Locks and Dam project was built on the Ohio River about 50 miles (80 km) above the confluence of the Ohio with the Mississippi. To satisfy navigational requirements, the project had to be built at a location where there was a massive active landslide on the Illinois bank of the river. The slide extends for about 3300 ft (1000 m) along the river bank. Evidence of instability on the Illinois shore at Olmsted was first discovered in 1987 during the foundation investigation for the proposed locks and dam. The extent of the unstable ground was mapped on the basis of slide scarps, cracks, leaning trees, and hummocky terrain. The difference in elevation from the toe of the slide to the scarp shown in Figure 2.7 is about 70 ft. Slope inclinometers were installed to determine the location of the shear surface, and piezometers were installed to determine water levels.

In late May and early June 1988, a drop of 7 ft (2.1 m) in the river level took place over a period of 10 days. During this period the groundwater levels within the slope dropped 1 to 3 ft (0.3 to 0.9 m). When the river level dropped, the slide moved, and nearly-vertical scarps about 3 ft (0.9 m) high developed at the head of the slide. The location of the sliding surface in the 1988 movement was determined based on data from 12 slope inclinometers, and the elevations of crimping in the riser tubes of five standpipe piezometers.

It was found that the shear surface was located within the McNairy I formation, which consists of interbedded layers of clay, silt, and sand. The thicknesses of these layers vary from fractions of an inch to as much as a foot. The layering within the McNairy I formation makes determination of shear strengths very difficult. First, the shear strength varies with the direction of the shear plane, being much higher when the shear plane crosses silt and sand layers than when it passes entirely through clay. Second, it is difficult to obtain representative undisturbed samples for laboratory testing. However, because the conditions at the time of sliding were well defined, it was possible to determine the shearing resistance of the McNairy I by *back analysis.* The strength was estimated, slope stability analyses were performed, the estimated strengths were adjusted, and the analyses were repeated until the calculated factor of safety was 1.0. The strengths back-calculated from the observed failure were used in subsequent analyses to design flatter slopes and buttresses that ensured long-term stability of the slope.

Panama Canal Landslides

The Panama Canal has been plagued by slope failures ever since the beginning of construction by the French (McCullough, 1999). To achieve even marginal stabil-

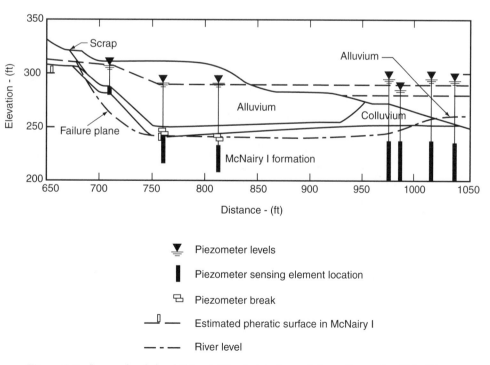

Figure 2.7 Lower bank landslide at Olmsted locks and dam site on the Ohio River.

ity, it was necessary to excavate much gentler slopes than anticipated when the first optimistic estimates of the volume of excavation were made. Unfortunately, the clay shales in which most of the slopes were cut are subject to serious deterioration over time, and many slopes that stood when first excavated failed later.

Construction of the canal was completed in 1914, but slope failures continued for many years. In October 1986 a large landslide occurred on the Cucaracha reach of the canal, where the slopes had failed many times before. The 1986 landslide, shown in Figure 2.8, closed the canal for 12 hours and impeded traffic until December 1986, when the slide mass was cleared from the navigation channel by dredging.

For some years preceding the 1986 Cucaracha slide, the budget devoted to landslide problems in the canal had gradually decreased, because no large slides had occurred. The 1986 slide engendered renewed appreciation of the important effect that landslides could have on the canal, and the Panama Canal Commission immediately devoted more resources to detection and control of landslides. The commission increased the size of the geotechnical engineering staff and instituted a landslide control program to reduce the hazard that landslides pose to the operation of the canal. The landslide control program included investigation of the causes of the Cucaracha slide and measures to stabilize it, a program of precise and essentially continuous measurements of surface movements on slopes, sys-

tematic inspections of slopes for indications of instability, improvement of surface drainage, installation of horizontal drains for subsurface drainage, and excavation to flatten and unload slopes. This approach, which treats landslides along the canal as a hazard that requires continuing attention and active management, has been highly successful.

The Rio Mantaro Landslide

On April 25, 1974, one of the largest landslides in recorded history occurred on a slope in the valley of the Rio Montaro in Peru (Lee and Duncan, 1975). A photograph of the landslide is shown in Figure 2.9, and a cross section through the slope is shown in Figure 2.10. As shown in Figure 2.10, the slope on which the landslide occurred was about 3.7 miles (6 km) long and 1.2 miles (2 km) high. It was approximately the same height and steepness as the south rim of the Grand Canyon in Arizona, also shown in Figure 2.10. The volume of earth involved in the slide was about 2 billion cubic yards (about 1.5 billion cubic meters). It was estimated that the sliding mass achieved a velocity of 120 miles/hr (190 km/h) as it moved down the slope. When it slammed into the opposite side of the valley, it splashed up to a height of 600 ft (200 m) above the bottom of the valley and then slumped back to form a landslide dam about 550 ft (170 m) high. The impact as the sliding mass hit the opposite valley wall was recorded at a seismographic station 30 miles

Figure 2.8 1986 Cucaracha landslide at the Panama Canal.

Figure 2.9 Rio Mantaro landslide in Peru, 1974.

(50 km) away as an event comparable to a magnitude 4 earthquake.

Prior to the slide, a town of about 450 inhabitants, Mayunmarca, was situated on the slope where the landslide occurred. After the landslide, no trace was found of the town or any of its inhabitants.

Kettleman Hills Landfill Slope Failure

On March 19, 1988, a slide occurred in a 90-ft (27-m)-high slope of a hazardous-waste landfill at Kettle-

man Hills, California (Mitchell et al., 1990; Seed et al., 1990). The failure involved about 580,000 cubic yards of waste (Golder Associates, 1991). A plan view and cross section through the fill are shown in Figure 2.11. The failure occurred by sliding on interfaces within the composite liner beneath the waste. The liner included three geomembranes, six geotextiles, three layers of granular fill, and two layers of compacted clay. Mitchell et al. (1990) found that some of the interfaces within the liner system had interface friction angles

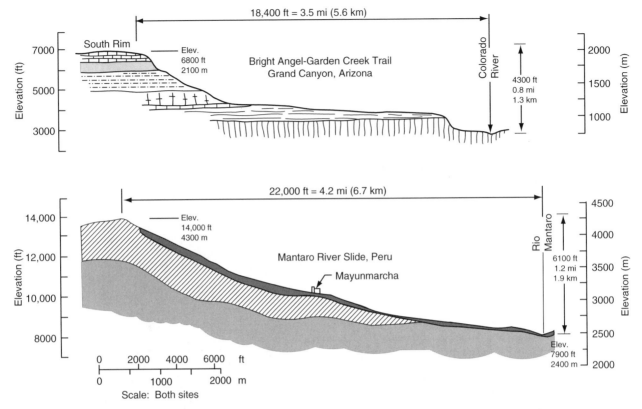

Figure 2.10 Comparison of cross sections: Grand Canyon, Arizona, and Mantaro River landslide.

as low as 8°. The lowest friction angles were found for interfaces between high-density polyethylene (HDPE) geomembranes and geotextiles, between geomembranes and geonets, and between geomembranes and compacted clay that was saturated after compaction. Seed et al. (1990) showed that factors of safety calculated for the conditions at failure were near 1.0 if the effect of wetting of the lower, nearly flat portion of the liner was taken into account and if consideration was given to three-dimensional effects. One particularly interesting aspect of the failure is that the maximum section (C1–C2 in Figure 2.11) did not have the lowest factor of safety. A shallower section near the top of Figure 2.11 had a considerably smaller factor of safety (Seed et al., 1990).

Fs wrt strength

CAUSES OF SLOPE FAILURE

It is important to understand the agents of instability in slopes for two reasons. First, for purposes of designing and constructing new slopes, it is important to be able to anticipate the changes in the properties of the soil within the slope that may occur over time and the various loading and seepage conditions to which the slope will be subjected over the course of its life. Second, for purposes of repairing failed slopes, it is important to understand the essential elements of the situation that lead to its failure, so that repetition of the failure can be avoided. Experience is the best teacher—from experiences with failures of slopes come the important lessons regarding what steps are necessary to design, construct, and repair slopes so that they will remain safe and stable.

In discussing the various causes of slope failures, it is useful to begin by considering the fundamental requirement for stability of slopes: that *the shear strength of the soil must be greater than the shear stress required for equilibrium.* Given this basic requirement, it follows that the most fundamental cause of instability is that for some reason, the shear strength of the soil is *less than* the shear strength required for equilibrium. This condition can be reached in two ways:

- Through a decrease in the shear strength of the soil

  • disturbance
  • anisotropy
  • strain rate effects

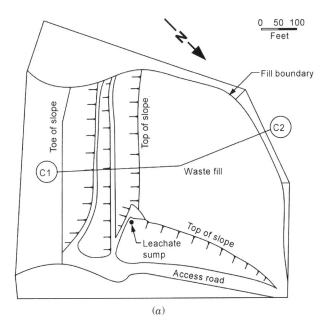

(a)

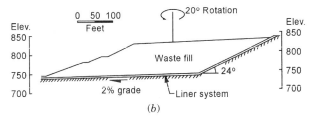

(b)

Figure 2.11 Kettleman Hills, California, landfill slope failure: (a) surface topography, March 15, 1988; (b) cross section C1–C2. (After Mitchell et al., 1990.)

- Through an increase in the shear stress required for equilibrium

The slope failures discussed in Chapters 1 and 2 include examples of both of these causes of instability.

Decrease in Shear Strength

Several different processes can lead to reduction in the shear strengths of soils. Experience has shown that the following processes are of particular importance with regard to slope stability:

1. *Increased pore pressure (reduced effective stress).* Rise in groundwater levels and more adverse seepage, frequently during periods of heavy rainfall, are the most frequent reasons for increased pore pressures and associated decrease in effective stresses within slopes. All types of soils are affected. The length of time required for the pore pressures to change depends on the permeability of the soil. In soils with high permeability, changes can occur rapidly, and in soils with low permeability, changes can be slow. Although the matrix permeability of clayey soils is usually very low, clay masses can have surprisingly high secondary permeability, due to cracks, fissures, and lenses of more permeable materials. As a result, pore pressures within clay deposits can change with surprising rapidity.

2. *Cracking.* Slope failures are frequently preceded by development of cracks through the soil near the crest of the slope. These cracks develop as a result of tension in the soil at the ground surface that exceeds the tensile strength of the soil. Cracks are possible only in soils that have some tensile strength. Quite clearly, once the soil is cracked, all strength on the plane of the crack is lost.

3. *Swelling (increase in void ratio).* Clays, especially highly plastic and heavily overconsolidated clays, are subject to swell when in contact with water. Low confining pressures and long periods of access to water promote swell. It has generally not been possible to achieve the same amount of swell in laboratory tests as occurs in the field. Kayyal (1991) studied highway embankments near Houston, Texas, constructed of highly plastic compacted clays, which failed 10 to 20 years after construction as a result of swell and strength loss. Similar shallow slides in highly plastic clays have occurred in many areas. Skempton (1964) showed three cases of slides in the overconsolidated London clay, where zones of higher water content extended for about an inch on either side of the shear surfaces, indicating that the shear stresses within the developing rupture zone led to localized dilation of the heavily overconsolidated clay.

4. *Development of slickensides.* Slickensided surfaces develop in clays, especially highly plastic clays, as a result of shear on distinct planes of slip. As shear displacements occur on a distinct plane, platelike clay particles tend to be realigned parallel to the plane of slip. The result is a smooth surface that exhibits a dull luster, comparable in appearance to the lustrous surface of a new bar of soap. The clay separates readily across these surfaces, and they can be found by breaking hand samples in tension or by picking at the walls of trenches. Slickensided surfaces are weaker than the surrounding clay where par-

ticles are randomly oriented. The friction angle on slickensided surfaces is called the *residual friction angle*. In highly plastic clays this may be only 5 or 6°, compared with peak friction angles of 20 or 30° in the same clay. Slickensides develop most prominently in clays that consist predominantly of clay-size particles; significant silt or sand content inhibits their formation. In some deposits randomly oriented slickensides develop as a result of tectonic movements. These have less significance for slope stability than a single set of slickensides with an adverse orientation.

5. *Decomposition of clayey rock fills.* Clay shales and claystones excavated for use as fill may break into pieces of temporarily sound rock that can be compacted into a seemingly stable rock fill. Over time, however, as the fill is wetted by infiltration or by groundwater seepage, the pieces of rock may slake and revert to chunks of disaggregated clay particles. As the clay swells into the open voids within the fill, it can lose a great deal of its strength, and the fill can become unstable.

6. *Creep under sustained loads.* Clays, especially highly plastic clays, deform continuously when subjected to sustained loading. These clays may eventually fail under these sustained loads, even at shear stresses that are significantly smaller than the short-term strength. Creep is exacerbated by cyclic variations in conditions, such as freeze–thaw and wet–dry. When the cyclically varying conditions are at their adverse extremes, movements occur in the downhill direction. These movements are permanent—they are not recovered when conditions are less adverse. The long-term result is ratcheting downslope movement that gradually increases from year to year, and this may eventually result in sliding on a continuous failure plane.

7. *Leaching.* Leaching involves changes in the chemical composition of pore water as water seeps through the voids. Leaching of salt from the pore water of marine clays contributes to the development of quick clays, which virtually no strength when disturbed.

8. *Strain softening.* Brittle soils are subject to strain softening. After the peak of the stress–strain curve has been reached, the shearing resistances of brittle soils decrease with increasing strain. This type of stress–strain behavior makes progressive failure possible and makes it impossible to count on mobilizing the peak strength simultaneously at all points around a shear surface (see Chapter 3).

9. *Weathering.* Rocks and indurated soils are subject to strength loss as a result of weathering, which involves various physical, chemical, and biological processes (Mitchell, 1993). Physical processes break the strong soil or rock into smaller pieces, and the chemical and biological process change it into material with fundamentally different properties. Weaker soils are also subject to weathering effects, but may become stronger, rather than weaker, as a result (Mitchell, 1993).

10. *Cyclic loading.* Under the influence of cyclic loads, bonds between soil particles may be broken and pore pressures may increase. The soils most subject to loss of strength due to cyclic loads are loose soils and soils with particles that are weakly bonded into loose structures. Loose sands may liquefy under cyclic loading, lose virtually all strength, and flow like a liquid.

Water plays a role in many of the processes that reduce strength, and as discussed in the following section, water is also involved in many types of loads on slopes that increase shear stresses. It is not surprising, therefore, that virtually every slope failure involves the destabilizing effects of water in some way, and often in more than one way.

Another factor involved in most slope failures is the presence of soils that contain clay minerals. The behavior of clayey soils is much more complicated than the behavior of sands, gravels, and nonplastic silts, which consist of chemically inert particles. The mechanical behavior of clays is affected by the physicochemical interaction between clay particles, the water that fills the voids between the particles, and the ions in the water. The larger the content of clay minerals, and the more active the clay mineral, the greater is its potential for swelling, creep, strain softening, and changes in behavior due to physicochemical effects. The percentage of clay in a soil and the activity of clay minerals are reflected qualitatively by the value of the *plasticity index* (PI). For that reason PI affords a useful first indication of the potential for problems that a clayey soil poses: The higher the PI, the greater the potential for problems. See Mitchell (1993) for a thorough discussion of the physicochemical behavior of clays.

It is safe to say that except for the effects of water and clayey soils, slope failures would be extremely rare. The truth of this statement is illustrated by the slopes on the surface of the moon, where there is neither water nor clay. In that environment, slopes remain stable for eons, failing only under the influence of violent meteor impacts. On Earth, however, both water and clays are common, and slope failures occur frequently, sometimes with little or no warning.

Increase in Shear Stress

Even if the strength of the soil does not change, slopes can fail if the loads on them change, resulting in increased shear stresses within the soil. Mechanisms through which shear stresses can increase include:

1. *Loads at the top of the slope.* If the ground at the top of a slope is loaded, the shear stress required for equilibrium of the slope will increase. Common occurrences that load the ground are placement of fill and construction of buildings supported on shallow foundations. To avoid significantly increasing the shear stresses in the slope, such loads should be kept away from the top of the slope. An acceptable distance can be determined by slope stability analysis.

2. *Water pressure in cracks at the top of the slope.* If cracks at the top of a slope are filled with water (or partially filled), the hydrostatic water pressure in the cracks loads the soil within the slope, increasing shear stresses and destabilizing the slope. If the cracks remain filled with water long enough for seepage toward the slope face to develop, the pore pressures in the soil increase, leading to an even worse condition.

3. *Increase in soil weight due to increased water content.* Infiltration and seepage into the soil within a slope can increase the water content of the soil, thereby increasing its weight. This increase in weight is appreciable, especially in combination with the other effects that accompany increased water content.

4. *Excavation at the bottom of the slope.* Excavation that makes a slope steeper or higher will increase the shear stresses in the soil within the slope and reduce stability. Similarly, erosion of soil by a stream at the base of a slope has the same effect.

5. *Drop in water level at the base of a slope.* External water pressure acting on the lower part of a slope provides a stabilizing effect. (This is perhaps the only good thing that water can do to a slope.) If the water level drops, the stabilizing influence is reduced and the shear stresses within the soil increase. When this occurs rapidly, and the pore pressures within the slope do not decrease in concert with the drop in outside water level, the slope is made less stable. This condition, called *rapid drawdown* or *sudden drawdown,* is an important design condition for the upstream slopes of dams and for other slopes that are partially submerged.

6. *Earthquake shaking.* Earthquakes subject slopes to horizontal and vertical accelerations that result in cyclic variations in stresses within the slope, increasing them above their static values for brief periods, typically fractions of a second. Even if the shaking causes no change in the strength of the soil, the stability of the slope is reduced for those brief instants when the dynamic forces act in adverse directions. If the cyclic loading causes reduction in soil strength, the effects are even more severe.

SUMMARY

When a slope fails, it is usually not possible to pinpoint a single cause that acted alone and resulted in instability. For example, water influences the stability of slopes in so many ways that it is frequently impossible to isolate one effect of water and identify it as the single cause of failure. Similarly, the behavior of clayey soils is complex, and it might not be possible to determine in some particular instance whether softening, progressive failure, or a combination of the two was responsible for failure of a slope. Sowers (1979) expressed the difficulties in attempting to isolate *the* cause of failure in these words: "In most cases, several 'causes' exist simultaneously; therefore, attempting to decide which one finally produced failure is not only difficult but also technically incorrect. Often the final factor is nothing more than a trigger that sets a body of earth in motion that was already on the verge of failure. Calling the final factor *the cause* is like calling the match that lit the fuse that detonated the dynamite that destroyed the building *the* cause of the disaster."

The fact that it is so difficult to isolate a single cause of failure highlights the importance of considering and evaluating all potential causes of failure in order to develop an effective means of repairing and stabilizing slopes that have failed. Similarly, in designing and constructing new slopes, it is important to attempt to anticipate all of the changes in properties and conditions that may affect a slope during its life and to be sure that the slope is designed and constructed so that it will remain stable despite these changes.

CHAPTER 3

Soil Mechanics Principles

For slope stability analyses to be useful, they must represent the *correct problem, correctly formulated*. This requires (1) mastery of the principles of soil mechanics, (2) knowledge of geology and site conditions, and (3) knowledge of the properties of the soils at the site. In this chapter we deal with the principles of soil mechanics that are needed to understand and to formulate analyses of slope stability problems correctly.

DRAINED AND UNDRAINED CONDITIONS

The concepts of drained and undrained conditions are of fundamental importance in the mechanical behavior of soils, and it is worthwhile to review these concepts at the beginning of this examination of soil mechanics principles. The lay definitions of *drained* and *undrained* (drained = dry or emptied, undrained = not dry or not emptied) do not describe the way these words are used in soil mechanics. The definitions used in soil mechanics are related to the ease and speed with which water moves in or out of soil in comparison with the length of time that the soil is subjected to some change in load. The crux of the issue is whether or not changes in changes in load cause changes in pore pressure:

- *Drained* is the condition under which water is able to flow into or out of a mass of soil in the length of time that the soil is subjected to some change in load. Under drained conditions, changes in the loads on the soil do not cause changes in the water pressure in the voids in the soil, because the water can move in or out of the soil freely when the volume of voids increases or decreases in response to the changing loads.
- *Undrained* is the condition under which there is no flow of water into or out of a mass of soil in the length of time that the soil is subjected to some change in load. Changes in the loads on the soil cause changes in the water pressure in the voids, because the water cannot move in or out in response to the tendency for the volume of voids to change.

An example that illustrates these conditions is shown in Figure 3.1, which shows a clay test specimen in a direct shear test apparatus. The permeability of the clay is low, and its compressibility is high. When the normal load P and the shear load T are increased, there is a tendency for the volume of the clay to decrease. This decrease in volume of the clay would take place entirely by reduction of the volume of the voids because the clay particles themselves are virtually incompressible. However, for the volume of the voids in the clay to decrease, water would have to run out of the clay because water is also virtually incompressible.

If the loads P and T were increased quickly, say in 1 second, the clay specimen would be in an undrained state for some period of time. Within the period of 1 second involved in increasing P and T, there would not be enough time for any significant amount of water to flow out of the clay. It is true that even in a period of 1 second there would be some small amount of flow, but this would be insignificant. For all practical purposes the clay would be undrained immediately after the loads were changed.

If the loads P and T were held constant for a longer period, say one day, the state of the clay specimen would change from undrained to drained. This is because within a period of one day, there would be sufficient time for water to flow out of the clay. Within this time the volume of the voids would decrease and come essentially to equilibrium. It is true that equilibrium would be approached asymptotically, and strictly

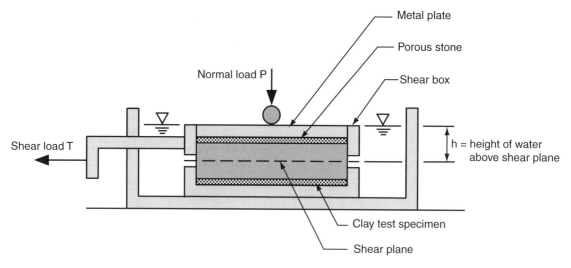

Figure 3.1   Direct shear test apparatus.

speaking, equilibrium would be approached closely but never be reached. However, for all practical purposes the clay would be drained after the loads were held constant for one day.

It is clear from this example that the difference between undrained and drained, as these words are used in soil mechanics, is *time*. Every mass of soil has characteristics that determine how long is required for transition from an undrained to a drained condition. A practical measure of this time is t_{99}, the time required to achieve 99% of the equilibrium volume change, which for practical purposes, we consider to be equilibrium. Using Terzaghi's theory of consolidation, we can estimate the value of t_{99}:

$$t_{99} = 4\,\frac{D^2}{c_v} \qquad (3.1)$$

where t_{99} is the time required for 99% of the equilibrium volume change, D the greatest distance that water must travel to flow out of the soil mass (length units), and c_v the coefficient of consolidation (length squared per unit of time). For the test specimen in Figure 3.1, D would be half the specimen thickness, about 1.0 cm, and c_v would be about 2 cm^2/h (19 ft^2/yr). Using these numbers, we would estimate that t_{99} would be 2.0 h. One second after the new loads were applied, the test specimen would be undrained. After 2 hours or longer, the test specimen would be drained.

Parenthetically, it should be noted that the use of the direct shear test as an example of drained and undrained conditions is not meant to indicate that the direct shear apparatus is suitable for both drained and undrained shear tests on soils. Direct shear tests are suitable for drained shear tests on soils, but not for undrained tests. Drained direct tests are performed using thin specimens so that D is small, and using a slow rate of shearing so that the specimen is drained throughout the test. Direct shear tests are not good for undrained tests, because the only way to prevent drainage is to apply the loads very quickly, which can result in higher measured strength due to strain rate effects. Triaxial tests are better suited to undrained testing in the laboratory, because drainage can be prevented completely by sealing the test specimens in impermeable membranes. Undrained triaxial tests can therefore be performed slowly enough to eliminate undesirable rate effects, and still be undrained.

Recapitulation

- The difference between undrained and drained conditions is time.
- *Undrained* signifies a condition where changes in loads occur more rapidly than water can flow in or out of the soil. The pore pressures increase or decrease in response to the changes in loads.
- *Drained* signifies a condition where changes in load are slow enough, or remain in place long enough, so that water is able to flow in or out of the soil, permitting the soil to reach a state of equilibrium with regard to water flow. The pore pressures in the drained condition are controlled by the hydraulic boundary conditions, and are unaffected by the changes in loads.

TOTAL AND EFFECTIVE STRESSES

Stress is defined as force per unit area. *Total stress* is the sum of all forces, including those transmitted through interparticle contacts and those transmitted through water pressures, divided by the total area. *Total area* includes both the area of voids and the area of solid.

Effective stress includes only the forces that are transmitted through particle contacts. It is equal to the total stress minus the water pressure. The *total normal stress* on the potential shear plane in the test specimen in Figure 3.1 is equal to

$$\sigma = \frac{W + P}{A} \qquad (3.2)$$

where σ is the total stress (force per unit of area); W the weight of the upper half of the specimen, porous stone, metal plate, and the steel ball through which the load is applied; P the applied normal load (F); and A the total area (L^2). For a typical direct shear apparatus, with a 102-mm^2 (4-in^2) shear box, W would be about 12.4 N (2.8 lb).

Before any load is applied to the specimen (when $P = 0$), the normal stress on the horizontal plane is

$$\sigma_0 = \frac{12.4 \text{ N}}{0.0103 \text{ m}^2} = 1.2 \text{ kPa} \qquad (3.3)$$

The effective stress is equal to the total stress minus the water pressure. Consider the condition before any load is applied to the specimen (when $P = 0$): If the specimen has had enough time to come to a drained condition, the water pressure would be hydrostatic and its value would be governed by the depth of water in the reservoir around the shear box. For a typical direct shear apparatus the depth of water (h in Figure 3.1) would be about 2 in. (about 0.051 m). The corresponding hydrostatic water pressure at the level of the horizontal plane would be

$$u_0 = \gamma_w h = (9.81 \text{ kN/m}^3)(0.051 \text{ m}) = 0.5 \text{ kPa} \qquad (3.4)$$

where u_0 is the initial water pressure in the specimen, γ_w the unit weight of water = 9.81 kN/m^3, and h the height of water above the horizontal plane = 0.051 m.

With $\sigma = 1.2$ kPa and $u_0 = 0.5$ kPa, the effective stress is equal to 0.7 kPa:

$$\sigma_0' = \sigma_0 - u_0 = 1.2 \text{ kPa} - 0.5 \text{ kPa} = 0.7 \text{ kPa} \qquad (3.5)$$

where σ_0' is the initial effective stress. If a load $P = 200$ N is applied to the specimen, the change in normal stress would be

$$\Delta\sigma = \frac{200 \text{ N}}{0.0103 \text{ m}^2} = 19.4 \text{ kPa} \qquad (3.6)$$

and the total stress after the load is applied would be

$$\sigma = \sigma_0 + \Delta\sigma = 1.2 \text{ kPa} + 19.4 \text{ kPa} = 20.6 \text{ kPa} \qquad (3.7)$$

The values of total stress are defined without reference to how much of the force might be carried by contacts between particles or to how much is transmitted through water pressure. Total stress is the same for the undrained and drained conditions. The value of total stress depends only on equilibrium; it is equal to the total of all normal forces divided by the total area.

When the load P is applied rapidly and the specimen is undrained, the pore pressure changes. The specimen is confined within the shear box and cannot deform. The clay is saturated (the voids are filled with water), so the volume of the specimen cannot change until water flows out. In this condition, the added load is carried entirely by increased water pressure. The soil skeleton (the assemblage of particles in contact with one another) does not change shape, does not change volume, and carries none of the new applied load.

Under these conditions the increase in water pressure is equal to the change in total stress:

$$\Delta u = \Delta\sigma = 19.4 \text{ kPa} \qquad (3.8)$$

where Δu is the increase in water pressure due to the change in load in the undrained condition. The water pressure after the load is applied is equal to the initial water pressure plus this change in pressure:

$$u = u_0 + \Delta u = 0.5 \text{ kPa} + 19.4 \text{ kPa} = 19.9 \text{ kPa} \qquad (3.9)$$

The effective stress is equal to the total stress [Eq. (3.7)] minus the water pressure [Eq. (3.9)]:

$$\sigma' = 20.6 \text{ kPa} - 19.9 \text{ kPa} = 0.7 \text{ kPa} \qquad (3.10)$$

Because the increase in water pressure caused by the 200-N load is equal to the increase in total stress, the effective stress does not change.

The effective stress after the load is applied [Eq. (3.10)] is the same as the effective stress before the load is applied [Eq. (3.5)]. This is because the specimen is undrained. Water does not have time to drain as the load is applied, so there is no volume change in the saturated specimen. As a result, the soil skeleton does not strain. The load carried by the soil skeleton, which is measured by the value of effective stress, does not change.

If the load is maintained over a period of time, drainage will occur, and eventually the specimen will be drained. The drained condition is achieved when there is no difference between the water pressures inside the specimen (the pore pressure) and the water pressure outside, governed by the water level in the reservoir around the direct shear apparatus. This condition will be achieved (for practical purposes) in about 2 hours and will persist until the load is changed again. After 2 hours the specimen will have achieved 99% equilibrium, the volume change will be essentially complete, and the pore pressure on the horizontal plane will be equal to the hydrostatic head at that level, $u = 0.5$ kPa.

In this drained condition the effective stress is

$$\sigma' = 20.6 \text{ kPa} - 0.5 \text{ kPa} = 20.1 \text{ kPa} \quad (3.11)$$

and all of the 200-N load is carried by the soil skeleton.

Recapitulation

- *Total stress* is the sum of all forces, including those transmitted through particle contacts and those transmitted through water pressures, divided by total area.
- *Effective stress* is equal to the total stress minus the water pressure. It is the force transmitted through particle contacts, divided by total area.

DRAINED AND UNDRAINED SHEAR STRENGTHS

Shear strength is defined as the maximum value of shear stress that the soil can withstand. The *shear stress* on the horizontal plane in the direct shear test specimen in Figure 3.1 is equal to the shear force divided by the area:

$$\tau = \frac{T}{A} \quad (3.12)$$

The shear strength of soils is controlled by effective stress, whether failure occurs under drained or undrained conditions. The relationship between shear strength and effective stress can be represented by a Mohr–Coulomb strength envelope, as shown in Figure 3.2. The relationship between τ and σ' shown in Figure 3.2 can be expressed as

$$s = c' + \sigma'_{ff} \tan \phi' \quad (3.13)$$

where c' is the effective stress cohesion, σ'_{ff} the effective stress on the failure plane at failure, and ϕ' the effective stress angle of internal friction.

Sources of Shear Strength

If a shear load T is applied to the test specimen shown in Figure 3.1, the top of the shear box will move to the left relative to the bottom of the box. If the shear load is large enough, the clay will fail by shearing on the horizontal plane, and the displacement would be very large. Failure would be accompanied by development of a rupture zone, or break through the soil, along the horizontal plane.

As the upper half of the specimen moved to the left with respect to the lower half and the strength of the soil was mobilized, the particles within the rupture zone would be displaced from their original positions relative to adjacent particles. Interparticle bonds would be broken, some individual particles would be broken, particles would rotate and be reoriented into new positions, and particles would slide across their contacts with neighboring particles. These movements of the particles would be resisted by the strength of interparticle bonds, by frictional resistance to sliding, and by forces from adjacent particles resisting displacement and reorientation. These types of resistance are the sources of shear strength in soils.

The two most important factors governing the strengths of soils are the magnitude of the interparticle contact forces and the density of the soil. Larger interparticle contact forces (larger values of effective stress) and higher densities result in higher strengths. As τ increases, the shear displacement (Δx) between the top and the bottom of the shear box would increase, as shown in Figure 3.3. This shear displacement results from shear strains in the rupture zone. The shear displacements in direct shear tests can be measured easily, but shear strains cannot be determined, because the thickness of the shear zone is not known. While the direct shear test can be used to measure the shear strengths of soils, it provides only qualitative information about stress–strain behavior. It is possible to determine whether soils are ductile (shear resistance

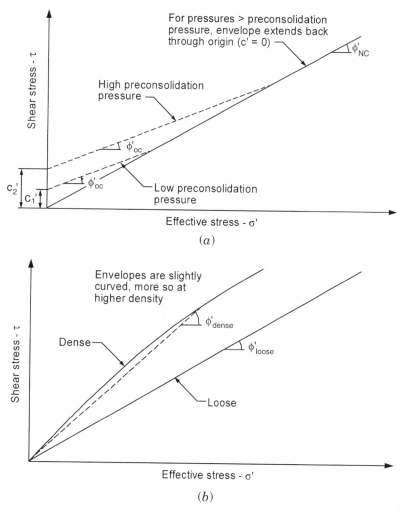

Figure 3.2 Effective stress shear strength envelopes: (*a*) for clay; (*b*) for sands, gravels, and rockfill.

remains high after failure) or brittle (shear resistance decreases after failure).

Drained Strength

Drained strength is the strength of the soil when it is loaded slowly enough so that no excess pore pressures are induced by applied loads. In the field, drained conditions result when loads are applied slowly to a mass of soil, or where they persist for a long enough time so that the soil can drain. In the laboratory, drained conditions are achieved by loading test specimens slowly so that excess pore pressures do not develop as the soil is loaded.

Imagine that the direct shear test specimen shown in Figure 3.1 reached a drained condition under the load of 200 N and was then loaded to failure by increasing

T slowly so that excess pore pressures did not develop. As shown by Eq. (3.11), the effective stress on the horizontal plane at equilibrium under the 200-N load would be 20.1 kPa, and it would remain constant as the clay was sheared slowly.

The strength of the specimen can be calculated using Eq. (3.13). If the clay is normally consolidated, c' would be zero. The value of ϕ' would probably be between 25 and 35° for normally consolidated sandy or silty clay. As an example, suppose that ϕ' is equal to 30°. The drained strength of the clay would be

$$s = c' + \sigma'_{ff} \tan \phi' = 0 + (20.1)(0.58) = 11.6 \text{ kPa}$$

$$(3.14)$$

where $c' = 0$ and $\tan \phi' = \tan 30° = 0.58$.

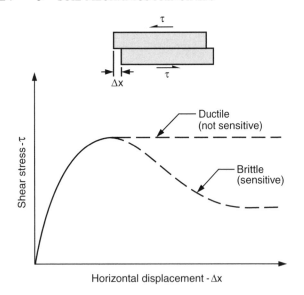

Figure 3.3   Shear stress–shear displacement curves for direct shear test.

Volume Changes During Drained Shear

Whether a soil tends to compress or dilate when sheared depends on its density and the effective stress that confines it. In dense soils the particles are packed tightly together, and tight packing results in a great deal of interference between particles when they move relative to one another. In very dense soils, particles cannot move relative to each other unless they ride up over each other, which causes dilation.

Higher effective stresses tend to prevent dilation, because work is required to cause the soil to expand against the effective confining pressure. If the effective confining pressure is high enough, the soil may not dilate. Instead, as shearing takes place, individual particles will be broken.

In soils with low densities the soil particles are farther apart on average, in a loose assemblage. As a loose soil is sheared, particles tend to fall into the gaps between adjacent particles, and the volume of the soil decreases.

The lower the density and the higher the effective stress, the more likely the soil is to compress when sheared. Conversely, the higher the density and the lower the confining pressure, the more likely the soil is to dilate. In clays, density is governed primarily by the highest effective stress to which the clay has been subjected.

A *normally consolidated soil* is one that has not been subjected to an effective stress higher than the present effective stress, and its density is the lowest possible for any given effective stress. As a result,

normally consolidated clays tend to compress when sheared.

An *overconsolidated clay* is one that was subjected previously to higher effective stress and thus has a higher density than that of a normally consolidated soil at the same effective stress. As a result, overconsolidated soils compress less when sheared than do normally consolidated soils, or if the previous maximum effective stress was much higher than the effective stress during shearing, the clay will dilate.

Pore Pressure Changes During Undrained Shear

The tendency of normally consolidated and lightly overconsolidated clays to compress when sheared results in increased pore pressures when shear stresses increase under undrained conditions. The tendency of heavily overconsolidated soils to dilate when sheared results in negative changes in pore pressures when shear stresses increase under undrained conditions. Thus, when clays are sheared under undrained conditions, the effective stress on the potential failure plane changes, becoming lower in normally consolidated soils and higher in heavily overconsolidated soils.

Undrained Strength

Undrained strength is the strength of the soil when loaded to failure under undrained conditions. In the field, conditions closely approximating undrained condtions result when loads are applied to a mass of soil faster than the soil can drain. In the laboratory, undrained conditions are achieved by loading test specimens so rapidly that they cannot drain, or by sealing them in impermeable membranes. (As noted previously, it is preferable to control drainage through the use of impermeable membranes rather than very high rates of loading, to avoid high strain rates that are not representative of field conditions.)

Imagine that the direct shear test specimen shown in Figure 3.1 reached a drained condition under the load of 200 N and was then loaded to failure by increasing T rapidly. As shown by Eq. (3.11), the effective stress on the horizontal plane at equilibrium under the 200-N load, before the shear load was increased, would be 20.1 kPa. The pore pressure before the shear load was increased would be 0.5 kPa, as shown by Eq. (3.4).

As the shear load T was applied without allowing time for drainage, the pore pressure would increase, because the clay is normally consolidated under the 20.1 kPa effective stress. As the shear load T is increased, the pore pressure within a specimen of a typical normally consolidated clay under these conditions would increase by about 12 kPa, and the effective normal stress on the failure plane at failure (σ'_{ff}) would decrease by the same amount. The effective stress on the failure plane at failure would thus be equal to 20.1 kPa − 12 kPa, or about 8 kPa. The undrained shear strength of the clay would thus be about 4.6 kPa:

$$s = c' + \sigma' \tan\phi' = 0 + (8.0)(0.58) = 4.6 \text{ kPa}$$

$$(3.15)$$

Figure 3.4 shows the stress paths and shear strengths for drained and undrained failure of the direct shear test specimen. The drained stress path is vertical, corresponding to an increase in shear stress and constant effective normal stress on the horizontal plane. The undrained stress path curves to the left, as the increase in shear stress is accompanied by a decrease in effective normal stress due to the increase in pore pressure.

As is typical for normally consolidated clays, the undrained strength is lower than the drained strength. This is due to the fact that the pore pressure increases and the effective stress decreases during undrained shear. For very heavily overconsolidated clays, the reverse is true: The undrained strength is greater than the drained strength, because pore pressure decreases and effective stress increases during undrained shear.

Strength Envelopes

Strength envelopes for soils are developed by performing strength tests on soils using a range of pressures and plotting the results on a Mohr stress diagram, as shown in Figure 3.5. Both effective stress and total stress strength envelopes can be developed. The strength envelopes shown in Figure 3.5 are representative of the results of tests on undisturbed specimens of clay, all trimmed from the same undisturbed sample and therefore all having the same preconsolidation

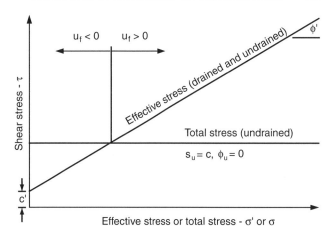

Figure 3.5 Drained and undrained strength envelopes for saturated clay.

pressure. The effective stress envelope represents the fundamental behavior of the clay, because the strength of the clay is controlled by effective stress and density. The total stress envelope reflects the pore pressures that develop during undrained shear as well as the fundamental behavior in terms of effective stresses.

Effective stress strength envelopes for clays consist of two parts. At high stresses the clay is normally consolidated, and the high-pressure part of the envelope extends back through the origin. At low stresses the clay is overconsolidated. The strength envelope in this range of pressures does not extend through the origin. The values of the effective stress shear strength parameters c' and ϕ' depend on whether the clay is normally consolidated or overconsolidated. If the clay is tested in a range of pressures where it is normally consolidated, c' is zero and ϕ' is constant. If the clay is tested in a range of pressures where it is overconsolidated, c' is greater than zero and ϕ' is smaller than the normally consolidated value. Because the values of c' and ϕ' that characterize the strength of the clay depend on the magnitude of the stresses in relation to the preconsolidation pressure, it is important that the range of stresses used in laboratory strength tests should correspond to the range of stresses involved in the problem being analyzed.

The total stress envelope is horizontal, representing shear strength that is constant and independent of the magnitude of the total stress used in the test. This behavior is characterized by these relationships:

$$c = s_u \qquad (3.16a)$$

$$\phi_u = 0 \qquad (3.16b)$$

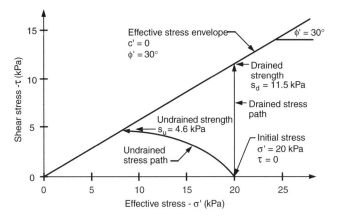

Figure 3.4 Drained and undrained stress paths and shear strengths.

where c is the total stress cohesion intercept, s_u the undrained shear strength, and ϕ_u the total stress friction angle.

The shear strength is the same for all values of total normal stress because the clay is saturated and undrained. Increasing or decreasing the total normal stress results only in a change in pore pressure that is equal in magnitude and opposite in sign to the change in normal stress. Thus, the effective stress is constant, and because the effective stress is constant the strength is constant because strength is controlled by effective stress. Although strength is controlled by effective stress, it is more convenient for some purposes to use the total stress envelope and the corresponding total stress parameters. Use of effective and total stress parameters in stability analyses is discussed later in the chapter.

If the clay were only partly saturated, the undrained strength envelope would not be horizontal. Instead, it would be inclined and shaped as shown in Figure 3.6. As the total normal stress increases, the strength also increases, because changes in total stress do not cause equal increases in pore pressure. As the total stress applied to a partly saturated specimen is increased, both the pore pressure and the effective stress increase. This occurs because, with both water and air in the voids of the clay, the pore fluid (the mixture of water and air) is not incompressible, and only part of the added total stress is carried by the pore fluid. The balance is carried by the soil skeleton, which results in an increase in effective stress.

How much of a change in total stress is borne by the change in pore pressure and how much by change in effective stress depends on the degree of saturation of the clay. At degrees of saturation in the range of 70% and lower, the change in pore pressure is negligible and virtually all of the change in total stress is reflected in change in effective stress. At degrees of saturation approaching 100%, the opposite is true: Virtually all of the change in total stress is reflected in change in pore pressure, and the change in effective stress is negligible. This behavior is responsible for the curvature of the total stress envelope in Figure 3.6: The degree of saturation increases as the total confining pressure increases. Therefore, at low values of total stress, where the degree of saturation is lower, the envelope is steeper because changes in effective stress are a larger portion of changes in total stress. At high values of total stress, where the degree of saturation is higher, the envelope is flatter because changes in effective stress are a smaller portion of changes in total stress.

Recapitulation

- *Shear strength* is defined as the maximum shear stress that the soil can withstand.
- The strength of soil is controlled by effective stresses, whether failure occurs under drained or undrained conditions.
- *Drained strength* is the strength corresponding to failure with no change in effective stress on the failure plane.
- *Undrained strength* is the strength corresponding to failure with no change in water content.
- Effective stress strength envelopes represent fundamental behavior, because strength is controlled by effective stress and density.
- Total stress strength envelopes reflect the pore pressures that develop during undrained shear, as well as fundamental behavior in terms of effective stress.
- Total stress strength envelopes for saturated clays are horizontal, corresponding to $c = s_u$, $\phi_u = 0$. Total stress envelopes for partly saturated clays are not horizontal, and ϕ_u is greater than zero.

BASIC REQUIREMENTS FOR SLOPE STABILITY ANALYSES

Whether slope stability analyses are performed for drained conditions or undrained conditions, the most basic requirement is that equilibrium must be satisfied in terms of total stresses. All body forces (weights), and all external loads, including those due to water pressures acting on external boundaries, must be included in the analysis. These analyses provide two useful results: (1) the total normal stress on the shear

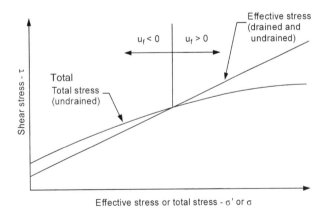

Figure 3.6 Drained and undrained strength envelopes for partly saturated clay.

surface and (2) the shear stress required for equilibrium.

The *factor of safety* for the shear surface is the ratio of the shear strength of the soil divided by the shear stress required for equilibrium. The normal stresses along the slip surface are needed to evaluate the shear strength: Except for soils with $\phi = 0$, the shear strength depends on the normal stress on the potential plane of failure.

In effective stress analyses, the pore pressures along the shear surface are subtracted from the total stresses to determine effective normal stresses, which are used to evaluate shear strengths. Therefore, to perform effective stress analyses, it is necessary to know (or to estimate) the pore pressures at every point along the shear surface. These pore pressures can be evaluated with relatively good accuracy for drained conditions, where their values are determined by hydrostatic or steady seepage boundary conditions. Pore pressures can seldom be evaluated accurately for undrained condtions, where their values are determined by the response of the soil to external loads.

In total stress analyses, pore pressures are not subtracted from the total stresses, because shear strengths are related to total stresses. Therefore, it is not necessary to evaluate and subtract pore pressures to perform total stress analyses. Total stress analyses are applicable only to undrained conditions. The basic premise of *total stress analysis* is this: The pore pressures due to undrained loading are determined by the behavior of the soil. For a given value of total stress on the potential failure plane, there is a unique value of pore pressure and therefore a unique value of effective stress. Thus, although it is true that shear strength is really controlled by effective stress, it is possible for the undrained condition to relate shear strength to total normal stress, because effective stress and total stress are uniquely related for the undrained condition. Clearly, this line of reasoning does not apply to drained conditions, where pore pressures are controlled by hydraulic boundary conditions rather than the response of the soil to external loads.

Analyses of Drained Conditions

Drained conditions are those where changes in load are slow enough, or where they have been in place long enough, so that all of the soils reach a state of equilibrium and no excess pore pressures are caused by the loads. In drained conditions pore pressures are controlled by hydraulic boundary conditions. The water within the soil may be static, or it may be seeping steadily, with no change in the seepage over time and no increase or decrease in the amount of water within the soil. If these conditions prevail in all the soils at a site, or if the conditions at a site can reasonably be approximated by these conditions, a drained analysis is appropriate. A *drained analysis* is performed using:

- Total unit weights
- Effective stress shear strength parameters
- Pore pressures determined from hydrostatic water levels or steady seepage analyses

Analyses of Undrained Conditions

Undrained conditions are those where changes in loads occur more rapidly than water can flow in or out of the soil. The pore pressures are controlled by the behavior of the soil in response to changes in external loads. If these conditions prevail in the soils at a site, or if the conditions at a site can reasonably be approximated by these conditions, an undrained analysis is appropriate. An *undrained analysis* is performed using:

- Total unit weights
- Total stress shear strength parameters

How Long Does Drainage Take?

As discussed earlier, the difference between undrained and drained conditions is time. The drainage characteristics of the soil mass, and its size, determine how long will be required for transition from an undrained to a drained condition. As shown by Eq. (3.1):

$$t_{99} = 4\,\frac{D^2}{c_v} \qquad (3.17)$$

where t_{99} is the time required to reach 99% of drainage equilibrium, D the length of the drainage path, and c_v the coefficient of consolidation.

Values of c_v for clays vary from about 1.0 cm²/h (10 ft²/yr) to about 100 times this value. Values of c_v for silts are on the order of 100 times the values for clays, and values of c_v for sands are on the order of 100 times the values for silts, and higher. These typical values can be used to develop some rough ideas of the lengths of time required to achieve drained conditions in soils in the field.

Drainage path lengths are related to layer thicknesses. They are half the layer thickness for layers that are bounded on both sides by more permeable soils, and they are equal to the layer thickness for layers that are drained only on one side. Lenses or layers of silt or sand within clay layers provide internal drainage, reducing the drainage path length to half of the thickness between internal drainage layers.

Values of t_{99} calculated using Eq. (3.17) are shown in Figure 3.7. For most practical conditions, many years or tens of years are required to reach drainage equilibrium in clay layers, and it is usually necessary to consider undrained conditions in clays. On the other hand, sand and gravels layers almost always reach drainage equilibrium quickly, and only drained conditions need be considered for these materials. Silts fall in between sands and clays, and it is often difficult to anticipate whether silt layers are better approximated as drained or undrained. When there is doubt whether a layer will be drained or undrained, the answer is to analyze both conditions, to cover the range of possibilities.

Short-Term Analyses

Short term refers to conditions during or following construction—the time immediately following the change in load. For example, if constructing a sand embankment on a clay foundation takes two months, the short-term condition for the embankment would be the end of construction, or two months. Within this period of time, it would be a reasonable approximation that no drainage would occur in the clay foundation, whereas the sand embankment would be fully drained.

For this condition it would be logical to perform a drained analysis of the embankment and an undrained

analysis of the clay foundation. There is no problem with performing a single analysis in which the embankment is considered to be drained and is treated in terms of effective stresses, and in which the foundation is considered to be undrained and is treated in terms of total stresses.

As discussed earlier, equilibrium in terms of total stresses must be satisfied for both total and effective stress analyses. The only differences between total and effective stress analyses relate to the strength parameters that are used and whether pore pressures are specified. In the case of short-term analysis of a sand embankment on a clay foundation, the strength of the sand would be characterized in terms of effective stresses (by a value of ϕ' for the sand), and the strength of the clay would be characterized in terms of total stresses (by values of $s_u = c$ varying with depth, with $\phi_u = 0$ for a saturated clay).

Pore pressures would be specified for the sand if the water table was above the top of the clay or if there was seepage through the embankment, but pore pressures would not be specified for the clay. There would, of course, be pore pressures in the clay. However, because the strength of the clay is related to total stress, it would be unneccesary to specify these nonzero values. Because most computer programs subtract pore pressures when they are specified, specifying pore

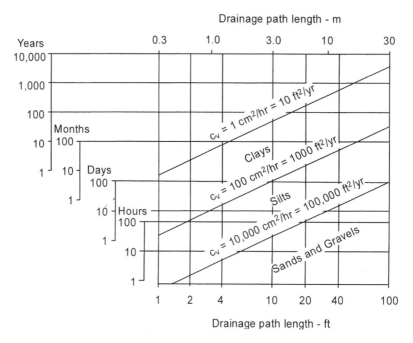

Figure 3.7 Time required for drainage of soil deposits (t_{99} based on Terzagh's theory of consolidation).

pressures for soils that are being treated as undrained can result in errors. Therefore, for soils that are treated in terms of total stresses, pore pressures should be set to zero, even though, in fact, they are not zero. (In the particular case of $\phi = 0$, no error will result if pore pressures are not specified as zero, because strengths are independent of normal stress, and misevaluating normal stress does not result in strengths that are wrongly evaluated.)

External water pressures acting on the surface of the foundation or the embankment would be specified for both materials, because external water pressures are a component of total stress, and they must be included to satisfy equilibrium in terms of total stress.

Long-Term Analyses

After a period of time, the clay foundation would reach a drained condition, and the analysis for this condition would be performed as discussed earlier under "Analyses of Drained Conditions," because *long term* and *drained conditions* carry exactly the same meaning. Both of these terms refer to the condition where drainage equilibrium has been reached and there are no excess pore pressures due to external loads.

For the long-term condition, both the sand embankment and clay foundation would be characterized in terms of effective stresses. Pore pressures, determined from hydrostatic water levels or steady seepage analyses, would be specified for both materials. External water pressures on the surface of the foundation or embankment would be specified for both materials, as always; these must be included to satisfy equilibrium in terms of total stress.

Progressive Failure

One of the fundamental assumptions of limit equilibrium analyses is that the strength of the soil can be mobilized over a wide range of strains, as shown by the curve labeled "ductile" in Figure 3.3. This implicit assumption arises from the fact that limit equilibrium analyses provide no information regarding deformations or strains.

Progressive failure is a strong possibility in the case of excavated slopes in overconsolidated clays and shales, particularly stiff-fissured clays and shales. These materials have brittle stress–strain characteristics, and they contain high horizontal stresses, often higher than the vertical stress. When an excavation is made in stiff fissured clay or shale, the excavated slope rebounds horizontally, as shown in Figure 3.8. Finite element studies by Duncan and Dunlop (1969), (1970) showed that shear stresses are very high at the toe of the slope, and there is a tendency for failure to begin at the toe and progress back beneath the crest, as shown in Figure 3.8.

Immediately after excavation of the slope (at time t_1), the stresses at point A might just have reached the peak of the stress–displacement curve, and the stresses at points B and C would be lower. With time, the slope would continue to rebound into the cut, due to a delayed response to the unloading from the excavation, and possibly also due to swelling of the clay as its water content increases following the reduction in stress. At a later time (t_2), therefore, the displacements at *A*, *B*, and *C* would all be larger, as shown in Figure 3.8. The shear stress at point *A* would decease as it moved beyond the peak, and the shear stresses at

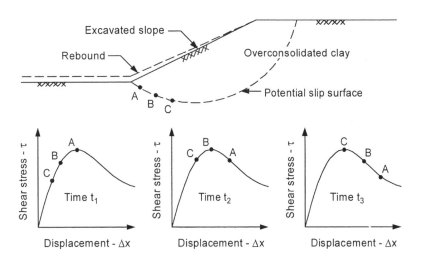

Figure 3.8 Mechanism of progressive failure of an excavated slope in overconsolidated clay.

points B and C would increase. At a later time (t_3), the displacement at point B would be large enough so that the shear stress there would fall below the peak. Through this process, progressively, failure would spread around the slip surface, without ever mobilizing the peak shear strength simultaneously at all points along the slip surface.

Because progressive failure can occur for soils with brittle stress–strain characteristics, peak strengths should not be used for these soils in limit equilibrium analyses; using peak strengths for brittle soils can lead to inaccurate and unconservative assessment of stability. As discussed in Chapter 5, experience with slopes in overconsolidated clays, particularly fissured clays, shows that *fully softened strengths* are appropriate for these materials in cases where slickensides have not developed, and *residual strengths* are appropriate in conditions where slickensides have developed.

Recapitulation

- Equilibrium must be satisfied in terms of total stress for all slope stability analyses.
- In effective stress analyses, pore pressures are subtracted from total stresses to evaluate the effective stresses on the shear surface.
- In total stress analyses, pore pressures are not subtracted. Shear strengths are related to total stresses.

- The basic premise of total stress analyses is that there is a unique relationship between total stress and effective stress. This is true only for undrained conditions.
- Total stress analyses are not applicable to drained conditions.
- The time required for drainage of soil layers varies from minutes for sands and gravels to tens or hundreds of years for clays.
- In short-term conditions, soils that drain slowly may best be characterized as undrained, while soils that drain more quickly are best characterized as drained. Analyses of such conditions can be performed by using effective stress strength parameters for the drained soils and total stress strength parameters for the undrained soils.
- When effective stress strength parameters are used, pore pressures determined from hydraulic boundary conditions are specified. When total stress strength parameters are used, no pore pressures are specified.
- An implicit assumption of limit equilibrium analyses is that the soils exhibit ductile stress–strain behavior. Peak strengths should not be used for materials such as stiff fissured clays and shales which have brittle stress–strain characteristics, because progressive failure can occur in these materials. Using peak strengths can result in inaccurate and unconservative evaluations of stability.

CHAPTER 4

Stability Conditions for Analyses

Variations of the loads acting on slopes, and variations of shear strengths with time, result in changes in the factors of safety of slopes. As a consequence, it is often necessary to perform stability analyses corresponding to several different conditions, reflecting different stages in the life of a slope.

When an embankment is constructed on a clay foundation, the embankment load causes the pore pressures in the foundation clay to increase. Over a period of time the excess pore pressures will dissipate, and eventually, the pore pressures will return to values governed by the groundwater conditions. As the excess pore pressures dissipate, the effective stresses in the foundation clay increase, the strength of the clay will increase, and the factor of safety of the embankment will also increase. Figure 4.1 illustrates these relationships. If, as shown, the embankment height stays constant and there is no external loading, the most critical condition occurs at the end of construction. In this case, therefore, it is only necessary to analyze the end-of-construction condition.

When a slope in clay is created by excavation, the pore pressures in the clay decrease in response to removal of the excavated material. Over time, the negative excess pore pressures dissipate and the pore pressures eventually return to values governed by the groundwater conditions. As the pore pressures increase, the effective stresses in the clay around the excavation decrease, and the factor of safety of the slope decreases with time. Figure 4.2 shows these relationships. If the depth of excavation is constant and there are no external loads, the factor of safety continually decreases, and its minimum value is reached when the pore pressures reach equilibrium with the groundwater seepage condition. In this case, therefore, the long-term condition is more critical than the end-of-construction condition.

In the case of a natural slope, not altered by either fill placement or excavation, there is no end-of-construction condition. The critical condition for a natural slope corresponds to whatever combination of seepage and external loading results in the lowest factor of safety. The higher the phreatic surface within the slope and the more severe the external loading condition, the lower is the factor of safety.

In the case of an embankment dam, several different factors affect stability. Positive pore pressures may develop during construction of clay embankments, particularly if the material is compacted on the wet side of optimum. The same is true of clay cores in zoned embankments. Over time, when water is impounded and seepage develops through the embankment, the pore pressures may increase or decrease as they come to equilibrium with steady seepage conditions. Reservoir levels may vary with time during operation of the dam. A rapid drop in reservoir level may create a critical loading condition on the upstream slope. A rise from normal pool level to maximum pool level may result in a new state of seepage through the embankment and a more severe loading condition on the downstream slope.

Earthquakes subject slopes to cyclic variations in load over a period of seconds or minutes that can cause instability or permanent deformations of the slope, depending on the severity of the shaking and its effect on the strength of the soil. As noted in Chapter 10, loose sands may liquefy and lose almost all shearing resistance as a result of cyclic loading. Other, more resistant soils may deform during shaking but remain stable.

END-OF-CONSTRUCTION STABILITY

Slope stability during and at the end of construction is analyzed using either drained or undrained strengths,

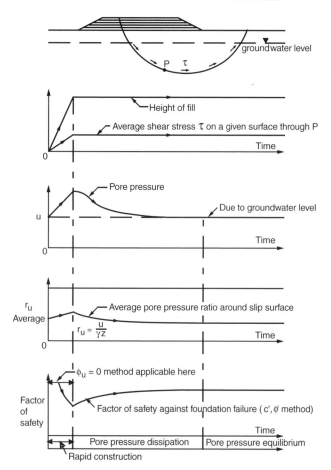

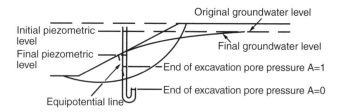

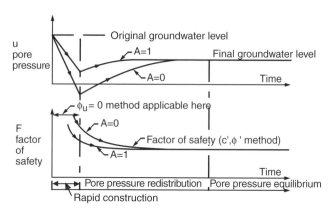

Figure 4.2 Variation with time of pore pressure and factor of safety during and after excavation of a slope in clay. (After Bishop and Bjerrum, 1960.)

Figure 4.1 Variations with time of shear stress, pore pressure, and factor of safety for an embankment on saturated clay. (After Bishop and Bjerrum, 1960.)

depending on the permeability of the soil. Many fine-grained soils are sufficiently impermeable that little drainage occurs during construction. This is particularly true for clays. For these fine-grained soils, undrained shear strengths are used, and the shear strength is characterized using total stresses. For soils that drain freely, drained strengths are used; shear strengths are expressed in terms of effective stresses, and pore water pressures are defined based on either water table information or an appropriate seepage analysis. Undrained strengths for some soils and drained strengths for others can be used in the same analysis.

For many embankment slopes the most critical condition is the end of construction. In some cases, however, there may be intermediate conditions during construction that might be more critical and should therefore be analyzed. In some fill placement operations, including some waste fills, the fill may be placed with a slope geometry such that the stability conditions during construction are more adverse than at the con-

clusion of construction. As discussed later, if an embankment is constructed in stages, and significant consolidation occurs between stages, each construction stage should be analyzed.

LONG-TERM STABILITY

Over time after construction the soil in slopes may either swell (with increase in water content) or consolidate (with decrease in water content). Long-term stability analyses are performed to reflect the conditions after these changes have occurred. Shear strengths are expressed in terms of effective stresses and the pore water pressures are estimated from the most adverse groundwater and seepage conditions anticipated during the life of the slope. Seepage analyses can be performed using either graphical techniques (flow nets) or numerical analyses (finite element, finite difference), depending on the complexity of the cross section.

RAPID (SUDDEN) DRAWDOWN

Rapid or *sudden drawdown* is caused by a lowering of the water level adjacent to a slope, at a rate so fast that the soil does not have sufficient time to drain significantly. Undrained shear strengths are assumed to apply for all but the coarsest free-draining materials ($k > 10^{-3}$ cm/s). If drawdown occurs during or immediately after construction, the undrained shear strength used in the drawdown analysis is the same as the undrained shear strength that applies to the end-of-construction condition. If drawdown occurs after steady seepage conditions have developed, the undrained strengths used in the drawdown analysis are different from those used in the end-of-construction analyses. For soils that expand when wetted, the undrained shear strength will be lower if drawdown occurs some time after construction than if it occurs immediately after construction. Rapid drawdown is discussed in Chapter 9.

EARTHQUAKE

Earthquakes affect the stability of slopes in two ways, as discussed in Chapter 10: (1) The acceleration produced by the seismic ground motion during an earthquake subjects the soil to cyclically varying forces, and (2) the cyclic strains induced by the earthquake loads may cause reduction in the shear strength of the soil.

If the strength of the soil is reduced less than 15% by cyclic loading, pseudostatic analyses of the earthquake loading can be used. In pseudostatic analyses, the effect of the earthquake is represented crudely by applying a static horizontal force to the potential sliding mass. This type of analysis, which is discussed in Chapter 10, provides a semi-empirical means of determining whether deformations due to an earthquake will be acceptably small.

If the strength of the soil is reduced more than 15% as a result of cyclic loading, dynamic analysis are needed to estimate the deformations that would result from earthquakes. Some engineers perform this type of analysis for all slopes, even if the strength reduction due to earthquake loading is less than 15%. These more complex analyses are highly specialized and are beyond the scope of this book.

In addition to analyses to estimating the potential for earthquake-induced deformation, analyses are also needed to evaluate post-earthquake stability. Strengths for these analyses are discussed in Chapter 5, and analysis procedures are discussed in Chapter 10.

PARTIAL CONSOLIDATION AND STAGED CONSTRUCTION

In cases where a clay foundation is so weak that it is unable to support the loads imposed by an embankment, the stability of the embankment can be improved by placing only a portion of the planned fill and allowing the foundation clay to consolidate and gain strength before additional fill is placed. In these cases, consolidation analyses are needed to estimate the increase in effective stresses due to consolidation of the foundation under the weight of the fill. The calculated values of effective stress are used to estimate the undrained shear strengths for use in total stress (undrained strength) analyses or are used directly in effective stress analyses. Procedures for analyses of staged constructions are discussed in Chapter 11.

Recapitulation

- End-of-construction stability is analyzed using drained or undrained strengths, depending on the permeability of the soil.
- Long-term stability analyses, which reflect conditions after swelling and consolidation are complete, are analyzed using drained strengths and pore water pressures corresponding to steady seepage conditions.
- Sudden drawdown removes the stabilizing effect of external water pressures and subjects the slope to increased shear stress. Either drained or undrained strengths are used, depending on the permeability of the soil.
- Earthquakes subject slopes to cyclically varying stresses and may cause reduction in the shear strength of the soil as a result of cyclic loading. Shear strengths measured in cyclic loading tests are appropriate for analyses of stability during earthquakes.
- Stability analyses for staged construction of embankments require consolidation analyses to estimate the increase in effective stresses that results from partial consolidation of the foundation.

OTHER LOADING CONDITIONS

The five loading conditions described above are those most frequently considered for earth slopes. There are, however, other loading conditions that may occur and should be considered. Two of these involve placement of surcharge loads at the top of a slope, and intermediate water levels producing partial submergence of a slope.

Surcharge Loading

Loads may be imposed on slopes as a result of either construction activities or operational conditions. The loads may be short term, such as passage of a heavy vehicle, or permanent, such as construction of a building. Depending on whether the load is temporary or permanent, and whether the soil drains quickly or slowly, undrained or drained strengths may be appropriate. If the surcharge loading occurs shortly after construction, the undrained strengths would be the same as those used for end-of-construction stability. However, if the load is imposed some time after construction, and the soil has had time to drain (consolidate or expand), the undrained strengths may be different and would be estimated using the same procedures as those used to estimate undrained strengths for rapid drawdown.

In many cases slopes will have a sufficiently high factor of safety that the effect of small surcharge loads is insignificant. Often, the loads imposed by even heavy vehicles and multistory buildings are negligible compared to the weight of the soil in the slope. For example, a typical one-story building will exert loads of about the same magnitude as an additional 1 ft of soil. If it is unclear whether a surcharge load will have a significant affect on stability, the condition should be analyzed.

Partial Submergence and Intermediate Water Levels

For the upstream slopes of dams and other slopes where the level of an adjacent body of water has an influence on stability, the lowest water level usually produces the most adverse conditions. In the case of slopes that contain zones of materials with different strength characteristics, the factor of safety of the upstream slope may be lower with a water level at some elevation between the top and the toe of the slope. The most critical water level for these conditions must be determined by repeated trials.

CHAPTER 5

Shear Strengths of Soil and Municipal Solid Waste

A key step in analyses of soil slope stability is measuring or estimating the strengths of the soils. Meaningful analyses can be performed only if the shear strengths used are appropriate for the soils and for the particular conditions analyzed. Much has been learned about the shear strength of soils within the past 60 years, often from surprising and unpleasant experience with the stability of slopes, and many useful research studies of soil strength have been performed. The amount of information that has been amassed on soil strengths is very large. The following discussion focuses on the principles that govern soil strength, the issues that are of the greatest general importance in evaluating strength, and strength correlations that have been found useful in practice. The purpose is to provide information that will establish a useful framework and a point of beginning for detailed studies of the shear strengths of soils at particular sites.

GRANULAR MATERIALS

The strength characteristics of all types of granular materials (sands, gravels, and rockfills) are similar in many respects. Because the permeabilities of these materials are high, they are usually fully drained in the field, as discussed in Chapter 3. They are cohesionless: The particles do not adhere to one another, and their effective stress shear strength envelopes pass through the origin of the Mohr stress diagram.

The shear strength of these materials can be characterized by the equation

$$s = \sigma' \tan \phi' \qquad (5.1)$$

where s is the shear strength, σ' the effective normal stress on the failure plane, and ϕ' the effective stress angle of internal friction. Measuring or estimating the drained strengths of these materials involves determining or estimating appropriate values of ϕ'.

The most important factors governing values of ϕ' for granular soils are density, confining pressure, grain-size distribution, strain boundary conditions, and the factors that control the amount of particle breakage during shear, such as the types of mineral and the sizes and shapes of particles.

Curvature of Strength Envelope

Mohr's circles of stress at failure for four triaxial tests on the Oroville Dam shell material are shown in Figure 5.1. Because this material is cohesionless, the Mohr–Coulomb strength envelope passes through the origin of stresses, and the relationship between strength and effective stress on the failure plane can be expressed by Eq. (5.1).

A *secant value* of ϕ' can be determined for each of the four triaxial tests. This value corresponds to a linear failure envelope going through the origin and passing tangent to the circle of stress at failure for the particular test, as shown in Figure 5.1. The dashed line in Figure 5.1 is the linear strength envelope for the test with the highest confining pressure. The secant value of ϕ' for an individual test is calculated as

$$\phi' = 2 \left[\left(\tan^{-1} \sqrt{\frac{\sigma'_{1f}}{\sigma'_{3f}}} \right) - 45° \right] \qquad (5.2)$$

where σ'_{1f} and σ'_{3f} are the major and minor principal stresses at failure. Secant values of ϕ' for the tests on Oroville Dam shell material shown in Figure 5.1 are given in Table 5.1, and the envelope for $\sigma'_3 = 4480$ kPa is shown in Figure 5.1.

35

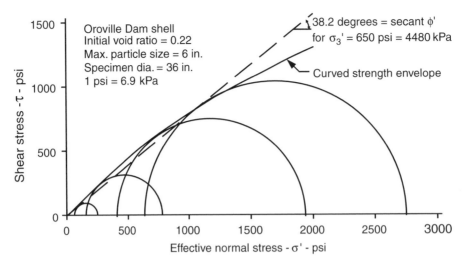

Figure 5.1 Mohr's circles of shear stress at failure and failure envelope for triaxial tests on Oroville Dam shell material. (Data from Marachi et al., 1969.)

Table 5.1   Stresses at Failure and Secant Values of ϕ' for Oroville Dam Shell Material

Test	σ_3' (kPa)	σ_1' (kPa)	ϕ' (deg)
1	210	1,330	46.8
2	970	5,200	43.4
3	2,900	13,200	39.8
4	4,480	19,100	38.2

The curvature of the envelope and the decrease in the secant value of ϕ' as the confining pressure increases are due to increased particle breakage as the confining pressure increases. At higher pressures the interparticle contact forces are larger. The greater these forces, the more likely it is that particles will be broken during shear rather than remaining intact and sliding or rolling over neighboring particles as the material is loaded. When particles break instead of rolling or sliding, it is because breaking requires less energy, and because the mechanism of deformation is changing as the pressures increase, the shearing resistance does not increase in exact proportion to the confining pressure. Even though the Oroville Dam shell material consists of hard amphibolite particles, there is significant particle breakage at higher pressures.

As a result of particle breakage effects, strength envelopes for all granular materials are curved. The envelope does pass through the origin, but the secant value of ϕ' decreases as confining pressure increases. Secant values of ϕ' for soils with curved envelopes can be characterized using two parameters, ϕ_0 and $\Delta\phi$:

$$\phi' = \phi_0 - \Delta\phi \log_{10} \frac{\sigma_3'}{p_a} \qquad (5.3)$$

where ϕ' is the secant effective stress angle of internal friction, ϕ_0 the value of ϕ' for $\sigma_3' = 1$ atm, $\Delta\phi$ the reduction in ϕ' for a 10-fold increase in confining pressure, σ_3' the confining pressure, and p_a = atmospheric pressure. This relationship between ϕ' and σ_3' is shown in Figure 5.2a. The variation of ϕ' with σ_3' for the Oroville Dam shell material is shown in Figure 5.2b.

Values of ϕ' should be selected considering the confining pressures involved in the conditions being analyzed. Some slope stability computer programs have provisions for using curved failure envelopes, which is an effective means of representing variations of ϕ' with confining pressure. Alternatively, different values of ϕ' can be used for the same material, with higher values of ϕ' in areas where pressures are low and lower values of ϕ' in areas where pressures are high. In many cases, sufficient accuracy can be achieved by using a single value of ϕ' based on the average confining pressure.

Effect of Density

Density has an important effect on the strengths of granular materials. Values of ϕ' increase with density. For some materials the value of ϕ_0 increases by 15° or more as the density varies from the loosest to the densest state. Values of $\Delta\phi$ also increase with density, varying from zero for very loose materials to 10° or more for the same materials in a very dense state. An example is shown in Figure 5.3 for Sacramento River sand, a uniform fine sand composed predominantly of

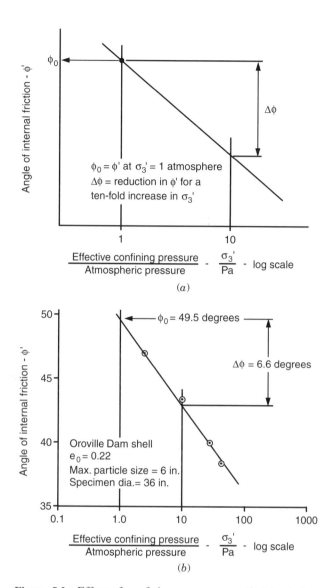

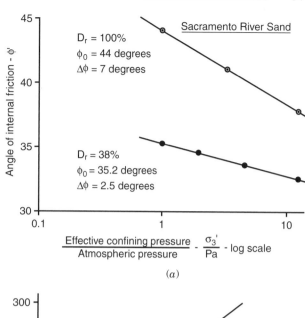

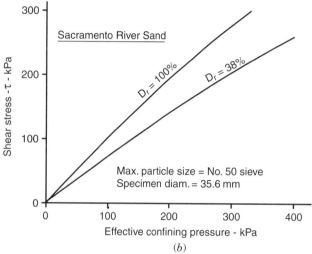

Figure 5.2 Effect of confining pressure on ϕ': (a) relationship between ϕ' and σ_3'; (b) variation of ϕ' and σ_3' for Oroville Dam shell.

Figure 5.3 Effect of density on strength of Sacramento River sand: (a) variations of ϕ' with confining pressure; (b) strength envelopes. (Data from Lee and Seed, 1967.)

feldspar and quartz particles. At a confining pressure of 1 atm, ϕ_0 increases from 35° for $D_r = 38\%$ to 44° for $D_r = 100\%$. The value of $\Delta\phi$ increases from 2.5° for $D_r = 38\%$ to 7° for $D_r = 100\%$

Effect of Gradation

All other things being equal, values of ϕ' are higher for well-graded granular soils such as the Oroville Dam shell material (Figures 5.1 and 5.2) than for uniformly graded soils such as Sacramento River sand (Figure 5.3). In well-graded soils, smaller particles fill gaps between larger particles, and as a result it is possible to form a denser packing that offers greater re-

sistance to shear. Well-graded materials are subject to segregation of particle sizes during fill placement and may form fills that are stratified, with alternating coarser and finer layers unless care is taken to ensure that segregation does not occur.

Plane Strain Effects

Most laboratory strength tests are performed using triaxial equipment, where a circular cylindrical test specimen is loaded axially and deforms with radial symmetry. In contrast, the deformations for many field conditions are close to plane strain. In plane strain, all displacements are parallel to one plane. In the field, this is usually the vertical plane. Strains and displacements perpendicular to this plane are zero. For example, in a long embankment, symmetry requires that all displacements are in vertical planes perpendicular to the longitudinal axis of the embankment.

The value of ϕ' for plane strain conditions (ϕ'_{ps}) is higher than the value for triaxial conditions (ϕ'_t). Becker et al. (1972) found that the value of ϕ'_{ps} was 1 to 6° larger than the value of ϕ'_t for the same material at the same density, tested at the same confining pressure. The difference was greatest for dense materials tested at low pressures. For confining pressures below 100 psi (690 kPa), they found that ϕ'_{ps} was 3 to 6° larger than ϕ'_t.

Although there may be a significant difference between values of ϕ' measured in triaxial tests and the values most appropriate for conditions close to plane strain, this difference is usually ignored. It is conservative to ignore the difference and use triaxial values of ϕ' for plane strain conditions. This conventional practice provides an intrinsic additional safety margin for situations where the strain boundary conditions are close to plane strain.

Strengths of Compacted Granular Materials

When cohesionless materials are used to construct fills, it is normal to specify the method of compaction or

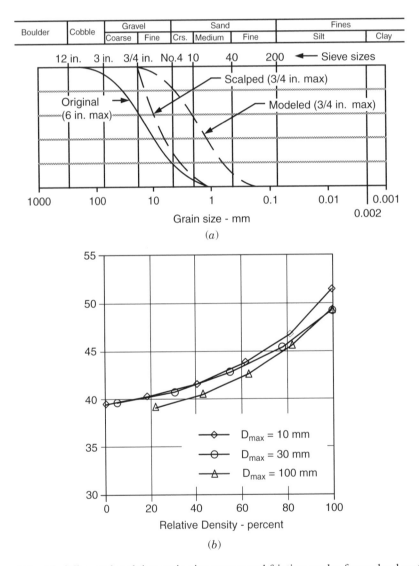

Figure 5.4 Modeling and scalping grain size curves and friction angles for scalped material: (*a*) grain-size curves for original, modeled, and scalped cobbely sandy gravel; (*b*) friction angles for scalped specimens of Goschenalp Dam rockfill. (After Zeller and Wullimann, 1957.)

the minimum acceptable density. Angles of internal friction for sands, gravels, and rockfills are strongly affected by density, and controlling the density of a fill is thus an effective way of ensuring that the fill will have the desired strength.

Minimum test specimen size. For the design of major structures such as dams, triaxial tests performed on specimens compacted to the anticipated field density are frequently used to determine values of ϕ'. The diameter of the triaxial test specimens should be at least six times the size of the largest soil particle, which can present problems for testing materials that contain large particles. The largest triaxial test equipment available in most soil mechanics laboratories is 100 to 300 mm (4 to 12 in.) in diameter. The largest particle sizes that can be tested with this equipment are thus about 16 to 50 mm (0.67 to 2 in.).

Modeling grain size curves and scalping. When soils with particles larger than one-sixth the triaxial specimen diameter are tested, particles that are too large must be removed. Becker et al. (1972) prepared test specimens with *modeled* grain-size curves. The curves for the modeled materials were parallel to the curve for the original material, as shown in Figure 5.4a. It was found that the strengths of the model materials were essentially the same as the strengths of the original materials, provided that the test specimens were prepared at the same relative density, D_r:

$$D_r = \frac{e_{max} - e}{e_{max} - e_{min}} \times 100\% \qquad (5.4)$$

where D_r is the relative density, e_{max} the maximum void ratio, e the void ratio, and e_{min} the minimum void ratio.

Becker et al. (1972) found that removing large particles changed the maximum and minimum void ratios of the material, and as a result, *the same relative density was not the same void ratio for the original and model materials.* The grain-size modeling technique used by Becker et al. (1972) can be difficult to use for practical purposes. When a significant quantity of coarse material has to be removed, there may not be enough fine material available to develop the model grain-size curve. An easier technique is *scalping,* where the large sizes are not replaced with smaller sizes. A scalped gradation is shown in Figure 5.4a.

The data in Figure 5.4b show that the value of ϕ' for scalped test specimens is essentially the same as for the original material, *provided that all specimens are prepared at the same relative density.* Again, *the*

same relative density will not be the same void ratio for the original and scalped materials.

Controlling field densities. Using relative density to control the densities of laboratory test specimens does not imply that it is necessary to use relative density for control of density in the field during construction. Controlling the density of granular fills in the field using relative density has been found to be difficult, especially when the fill material contains large particles. Specifications based on method of compaction, or on relative compaction, can be used for field control, even though relative density may be used in connection with laboratory tests.

Strengths of Natural Deposits of Granular Materials

It is not possible to obtain undisturbed samples of granular materials, except by exotic procedures such as freezing and coring the ground. In most cases friction angles for natural deposits of granular materials are estimated using the results of in situ tests such as the standard penetration test (SPT) or the cone penetration test (CPT). Correlations that can be used to interpret values of ϕ' from in situ tests are discussed below.

Strength correlations. Many useful correlations have been developed that can be used to estimate the strengths of sands and gravels based on correlations with relative densities or the results of in situ tests. The earliest correlations were developed before the influence of confining pressure on ϕ' was well understood. More recent correlations take confining pressure into account by correlating both ϕ_0 and $\Delta\phi$ with relative density or by including overburden pressure in correlations between ϕ' and the results of in situ tests.

Table 5.2 relates values of ϕ_0 and $\Delta\phi$ to relative density values for well-graded sands and gravels, poorly graded sands and gravels, and silty sands. Figures 5.5 and 5.6 can be used to estimate in situ relative density based on SPT blow count or CPT cone resistance. Values of relative density estimated using Figure 5.5 or 5.6 can be used together with Table 5.2 to estimate values of ϕ_0 and $\Delta\phi$ for natural deposits.

Figures 5.7 and 5.8 relate values of ϕ' to overburden pressure and SPT blow count or CPT cone resistance. Figure 5.9 relates ϕ' to relative density for sands. The values of ϕ' in Figure 5.9 correspond to confining pressures of about 1 atm, and are close to the values of ϕ_0 listed in Table 5.2. Tables 5.3 and 5.4 relate values of ϕ' to SPT blow count and CPT cone resistance. The correlations are easy to use, but they do not take the effect of confining pressure into account.

Recapitulation

- The drained shear strengths of sands, gravels, and rockfill materials can be expressed as $s = \sigma' \tan \phi'$.
- Values of ϕ' for these materials are controlled by density, gradation, and confining pressure.
- The variation of ϕ' with confining pressure can be represented by

$$\phi' = \phi_0 - \Delta\phi \log_{10} \frac{\sigma'_3}{p_a}$$

  where σ'_3 is the confining pressure and p_a is atmospheric pressure.
- When large particles are removed to prepare specimens for laboratory tests, the test specimens should be prepared at the same relative density as the original material, not the same void ratio.
- Values of ϕ' for granular materials can be estimated based on the Unified Soil Classification, relative density, and confining pressure.
- Values of ϕ' for granular materials can also be estimated based on results of standard penetration tests or cone penetration tests.

Table 5.2   Values of ϕ_0 and $\Delta\phi$ for Sands and Gravels

Unified classification	Standard Proctor RC[a] (%)	Relative density, D_r[b] (%)	ϕ_0[c] (deg)	$\Delta\phi$ (deg)
GW, SW	105	100	46	10
	100	75	43	8
	95	50	40	6
	90	25	37	4
GP, SP	105	100	42	9
	100	75	39	7
	95	50	36	5
	90	25	33	3
SM	100	—	36	8
	95	—	34	6
	90	—	32	4
	85	—	30	2

Source: Wong and Duncan (1974).
[a]RC = relative compaction = $\gamma_d / \gamma_{d\,max} \times 100\%$.
[b]$D_r = (e_{max} - e)/e_{max} - e_{min}) \times 100\%$.
[c]$\phi' = \phi_0 - \Delta\phi \log_{10} \sigma'_3/p_a$ where p_a is atmospheric pressure.

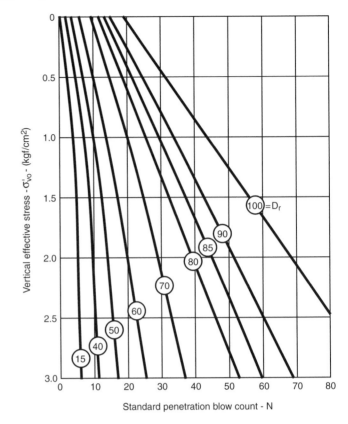

Figure 5.5 Relationship among SPT blow count, overburden pressure, and relative density for sands. (After Gibbs and Holtz, 1957, and U.S. Dept. Interior, 1974.)

SILTS

The shear strength of silts in terms of effective stress can be expressed by the Mohr–Coulomb strength criterion as

$$s = c' + \sigma' \tan \phi' \tag{5.5}$$

where s is the shear strength, c' the effective stress cohesion intercept, and ϕ' the effective stress angle of internal friction.

The behavior of silts has not been studied as extensively and is not as well understood as the behavior of granular materials or clays. Although the strengths of silts are governed by the same principles as the strengths of other soils, the range of their behavior is wide, and sufficient data are not available to anticipate or estimate their properties with the same degree of reliability as is possible in the case of granular soils or clays.

Silts encompass a broad range of behavior, from behavior that is very similar to the behavior of fine sands

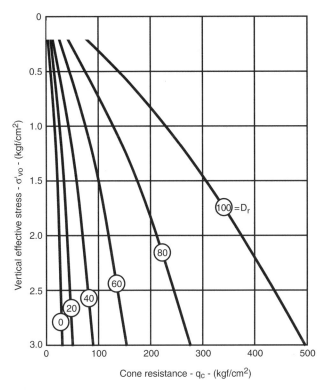

Figure 5.6 Relationship among CPT cone resistance, overburden pressure, and relative density of sands. (After Schmertmann, 1975.)

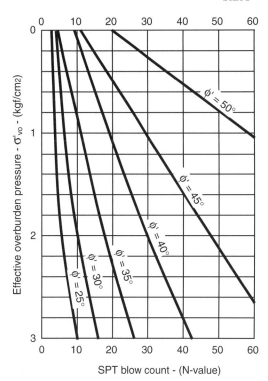

Figure 5.7 Relationship among SPT blow count, overburden pressure, and ϕ' for sands. (After DeMello, 1971, and Schmertmann, 1975.)

at one extreme to behavior that is essentially the same as the behavior of clays at the other extreme. It is useful to consider silts in two distinct categories: nonplastic silts, which behave more like fine sands, and plastic silts, which behave more like clays.

Nonplastic silts, like the silt of which Otter Brook Dam was constructed, behave similarly to fine sands. Nonplastic silts, however, have some unique characteristics, such as lower permeability, that influence their behavior and deserve special consideration.

An example of highly plastic silt is San Francisco Bay mud, which has a liquid limit near 90, a plasiticity index near 45, and classifies as MH (a silt of high plasticity) by the Unified Soil Classification System. San Francisco Bay mud behaves like a normally consolidated clay. The strength characteristics of clays discussed later in this chapter are applicable to materials such as San Francisco Bay mud.

Sample Disturbance

Disturbance during sampling is a serious problem in nonplastic silts. Although they are not highly sensitive by the conventional measure of sensitivity (sensitivity

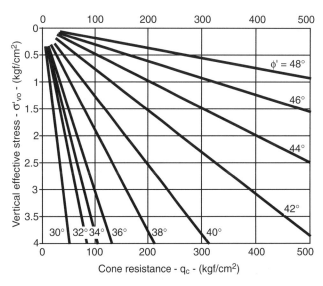

Figure 5.8 Relationship between CPT cone resistance, overburden pressure, and ϕ' for sands. (After Robertson and Campanella, 1983.)

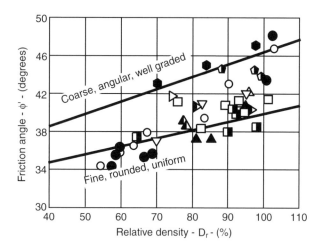

Figure 5.9 Correlation between friction angle and relative density for sands. (Data from Schmertmann, 1975, and Lunne and Kleven, 1982.)

Table 5.4   Correlation Among Relative Density, CPT Cone Resistance, and Angle of Internal Friction for Clean Sands

State of packing	Relative density, D_r (%)	q_c (tons/ft^2 or kgf/cm^2)	ϕ' (deg)
Very loose	< 20	< 20	< 32
Loose	20–40	20–50	32–35
Medium	40–60	50–150	35–38
Dense	60–80	150–250	38–41
Very dense	> 80	250–400	41–45

Source: Meyerhof (1976).

Table 5.3   Relationship Among Relative Density, SPT Blow Count, and Angle of Internal Friction for Clean Sands

State of packing	Relative density, D_r (%)	SPT blow count, N^a (blows/ft)	Angle of internal friction ϕ'^b (deg)
Very loose	< 20	< 4	< 30
Loose	20–40	4–10	30–35
Compact	40–60	10–30	35–40
Dense	60–80	30–50	40–45
Very dense	> 80	> 50	> 45

Source: Meyerhof (1956).

[a]$N = 15 + (N' - 15)/2$ for $N' > 15$ in saturated very fine or silty sand, where N is the blow count corrected for dynamic pore pressure effects during the SPT, and N' is the measured blow count.

[b]Reduce ϕ' by 5° for clayey sand; increase ϕ' by 5° for gravelly sand.

= undistured strength/remoulded strength), they are very easily disturbed. In a study of a silt from the Alaskan arctic (Fleming and Duncan, 1990), it was found that disturbance reduced the undrained strengths measured in unconsolidated–undrained tests by as much as 40%, and increased the undrained strengths measured in consolidated–undrained tests by as much as 40%. Although silts can usually be sampled using the same

techniques as those used for clays, the quality of samples should not be expected to be as good.

Cavitation

Unlike clays, nonplastic silts almost always tend to dilate when sheared, even if they are normally consolidated. In undrained tests, pore pressures decrease as a result of this tendency to dilate, and pore pressures can become negative. When pore pressures are negative, dissolved air or gas may come out of solution, forming bubbles within test specimens that greatly affect their behavior.

Figure 5.10 shows stress–strain and pore pressure–strain curves for consolidated–undrained triaxial tests on nonplastic silt from the Yazoo River valley. As the specimens were loaded, they tended to dilate, and the pore pressures decreased. As the pore pressures decreased, the effective confining pressures increased. The effective stresses stopped increasing when cavitation occurred, because from that point on the volume of the specimens increased as the cavitation bubbles expanded. The value of the maximum deviator stress for each sample was determined by the initial pore pressure (the back pressure), which determined how much negative change in pore pressure took place before cavitation occurred. The higher the back pressure, the greater was the undrained strength. These effects can be noted in Figure 5.10.

Drained or Undrained Strength?

Values of c_v for nonplastic silts are often in the range 100 to 10,000 cm^2/h (1000 to 100,000 ft^2/yr). It is often difficult to determine whether silts will be drained or undrained under field loading conditions,

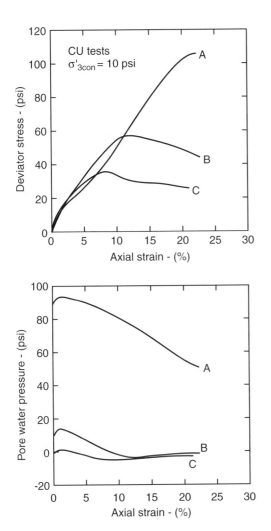

Figure 5.10 Effect of cavitation on undrained strength of reconstituted Yazoo silt. (From Rose, 1994.)

and in many cases it is prudent to consider both possibilities.

Strengths of Compacted Silts

Laboratory test programs for silts to be used as fills can be conducted following the principles that have been established for testing clays. Silts are moisture-sensitive and compaction characteristics are similar to those for clays. Densities can be controlled effectively using relative compaction. Undrained strengths of both plastic and nonplastic silts at the as-compacted condition are strongly influenced by water content.

Nonplastic silts have been used successfully as cores for dams and for other fills. Their behavior during compaction is sensitive to water content, and they become rubbery when compacted close to saturation. In

this condition they deform elastically under wheel loads, without failure and without further increase in density. Highly plastic silts, such as San Francisco Bay mud, have also been used as fills, but adjusting the moisture contents of highly plastic materials to achieve the water content and the degree of compaction needed for a high-quality fill is difficult.

Evaluating Strengths of Natural Deposits of Silt

Plastic and nonplastic silts can be sampled using techniques that have been developed for clays, although the quality of the samples is not as good. Disturbance during sampling is a problem for all silts, and care to minimize disturbance effects is important, especially for samples used to measure undrained strengths. Sample disturbance has a much smaller effect on measured values of the effective stress friction angle (ϕ') than it has on undrained strength.

Effective stress failure envelopes for silts can be determined readily using consolidated–undrained triaxial tests with pore pressure measurements, using test specimens trimmed from "undisturbed" samples. Drained direct shear tests can also be used. Drainage may occur so slowly in triaxial tests that performing drained triaxial tests may be impractical as a means of measuring drained strengths.

Correlations are not available for making reliable estimates of the undrained strengths of silts, because values of s_u/σ'_{1c} measured for different silts vary widely. A few examples are shown in Table 5.5.

Table 5.5   Values of s_u/σ'_{1c} for Normally Consolidated Alaskan Silts

Test[a]	k_c[b]	s_u/σ'_{1c}	Reference
UU	NA	0.25–0.30	Fleming and Duncan (1990)
UU	NA	0.18	Jamiolkowski et al. (1985)
IC-U	1.0	0.25	Jamiolkowski et al. (1985)
IC-U	1.0	0.30	Jamiolkowski et al. (1985)
IC-U	1.0	0.85–1.0	Fleming and Duncan (1990)
IC-U	1.0	0.30–0.65	Wang and Vivatrat (1982)
AC-U	0.84	0.32	Jamiolkowski et al. (1985)
AC-U	0.59	0.39	Jamiolkowski et al. (1985)
AC-U	0.59	0.26	Jamiolkowski et al. (1985)
AC-U	0.50	0.75	Fleming and Duncan (1990)

[a]UU, unconsolidated undrained triaxial; IC-U, isotropically consolidated undrained triaxial; AC-U, anisotropically consolidated undrained triaxial.

[b]$k_c = \sigma'_{3c}/\sigma'_{1c}$ during consolidation.

Additional studies will be needed to develop more refined methods of classifying silts and correlations that can be used to make reliable estimates of undrained strengths. Until more information is available, properties of silts should be based on conservative lower-bound estimates, or laboratory tests on the specific material.

Recapitulation

- The behavior of silts has not been studied as extensively, and is not as well understood, as the behavior of granular materials and clays.
- It is often difficult to determine whether silts will be drained or undrained under field loading conditions. In many cases it is prudent to consider both possibilities.
- Silts encompass a broad range of behavior, from fine sands to clays. It is useful to consider silts in two categories: nonplastic silts, which behave more like fine sands, and plastic silts, which behave like clays.
- Disturbance during sampling is a serious problem in nonplastic silts.
- Cavitation may occur during tests on nonplastic silts, forming bubbles within test specimens that greatly affect their behavior.
- Correlations are not available for making reliable estimates of the undrained strengths of silts.
- Laboratory test programs for silts to be used as fills can be conducted following the principles that have been established for testing clays.

CLAYS

Through their complex interactions with water, clays are responsible for a large percentage of problems with slope stability. The strength properties of clays are complex and subject to changes over time through consolidation, swelling, weathering, development of slickensides, and creep. Undrained strengths of clays are important for short-term loading conditions, and drained strengths are important for long-term conditions.

The shear strength of clays in terms of effective stress can be expressed by the Mohr-Coulomb strength criterion as

$$s = c' + \sigma' \tan \phi' \qquad (5.6)$$

where s is the shear strength, c' the effective stress cohesion intercept, and ϕ' the effective stress angle of internal friction.

The shear strength of clays in terms of total stress can be expressed as

$$s = c + \sigma \tan \phi \qquad (5.7)$$

where c and ϕ are the total stress cohesion intercept and the total stress friction angle.

For saturated clays, ϕ is equal to zero, and the undrained strength can be expressed as

$$s = s_u = c \qquad (5.8a)$$
$$\phi = \phi_u = 0 \qquad (5.8b)$$

where s_u is the undrained shear strength, independent of total normal stress, and ϕ_u is the total stress friction angle.

Factors Affecting Clay Strength

Low undrained strengths of normally consolidated and moderately overconsolidated clays cause frequent problems with stability of embankments constructed on them. Accurate evaluation of undrained strength, a critical factor in evaluating stability, is difficult because so many factors influence the results of laboratory and in situ tests for clays.

Disturbance. Sample disturbance reduces strengths measured in unconsolidated–undrained (UU) tests in the laboratory. Strengths measured using UU tests may be considerably lower than the undrained strength in situ unless the samples are of high quality. Two procedures have been developed to mitigate disturbance effects (Jamiolkowski et al., 1985):

1. The *recompression technique,* described by Bjerrum (1973), involves consolidating specimens in the laboratory at the same pressures to which they were consolidated in the field. This replaces the field effective stresses with the same effective stresses in the laboratory and squeezes out extra water that the sample may have absorbed as it was sampled, trimmed, and set up in the triaxial cell. This method is used extensively in Norway to evaluate undrained strengths of the sensitive marine clays found there.

2. The *SHANSEP technique,* described by Ladd and Foott (1974) and Ladd et al. (1977), involves consolidating samples to effective stresses that are higher than the in situ stresses, and interpreting the measured strengths in terms of the undrained strength ratio, s_u/σ'_v. Variations of s_u/σ'_v with OCR for six clays, determined from this type of testing, are shown in Figure 5.11. Data of the type shown in Figure 5.11, together with knowledge of the variations of σ'_v and OCR with

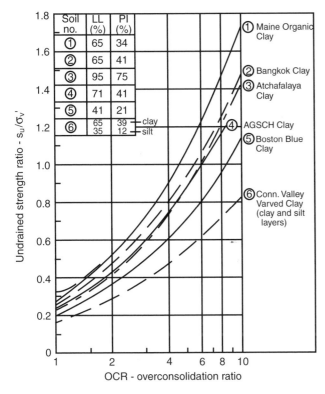

Figure 5.11 Variation of s_u/σ'_v with OCR for clays, measured in ACU direct simple shear tests. (After Ladd et al., 1977.)

inherent anisotropy and stress system-induced anisotropy.

Inherent anisotropy in intact clays results from the fact that plate-shaped clay particles tend to become oriented perpendicular to the major prinicpal strain direction during consolidation, which results in direction-dependent stiffness and strength. Inherent anisotropy in stiff-fissured clays also results from the fact that fissures are planes of weakness.

Stress system-induced anisotropy is due to the fact that the magnitudes of the stresses during consolidation vary depending on the orientation of the planes on which they act, and the magnitudes of the pore pressures induced by undrained loading vary with the orientation of the changes in stress.

The combined result of inherent and stress-induced anisotropy is that the undrained strengths of clays varies with the orientation of the principal stress at failure and with the orientation of the failure plane. Figure 5.12a shows orientations of principal stresses and failure planes around a shear surface. Near the top of the shear surface, sometimes called the *active zone*, the

depth, can be used to estimate undrained strengths for deposits of normally consolidated and moderately overconsolidated clays.

As indicated by Jamiolkowski et al. (1985), both the recompression and the SHANSEP techniques have limitations. The recompression technique is preferable whenever block samples (with very little disturbance) are available. It may lead to undrained strengths that are too low if the clay has a delicate structure that is subject to disturbance as a result of even very small strains (these are called *structured clays*), and it may lead to undrained strengths that are too high if the clay is less sensitive, because reconsolidation results in void ratios in the laboratory that are lower than those in the field. The SHANSEP technique is applicable only to clays without sensitive structure, for which undrained strength increases in direct proportion to the consolidation pressure. It requires detailed knowledge of past and present in situ stress conditions, because the undrained strength profile is constructed using data such as those shown in Figure 5.11, based on knowledge of σ'_v and OCR.

Anisotropy. The undrained strength of clays is *anisotopic;* that is, it varies with the orientation of the failure plane. Anisotropy in clays is due to two effects:

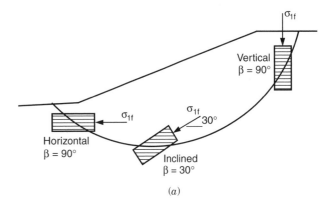

(a)

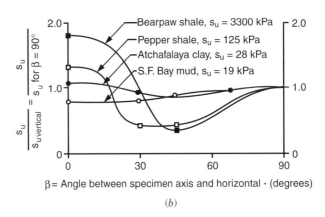

β = Angle between specimen axis and horizontal - (degrees)

(b)

Figure 5.12 Stress orientation at failure, and undrained strength anisotropy of clays and shales: (a) stress orientations at failure; (b) anisotropy of clays and shales—UU triaxial tests.

major principal stress at failure is vertical, and the shear surface is oriented about 60° from horizontal. In the middle part of the shear surface, where the shear surface is horizontal, the major principal stress at failure is oriented about 30° from horizontal. At the toe of the slope, sometimes called the *passive zone,* the major principal stress at failure is horizontal, and the shear surface is inclined about 30° past horizontal. As a result of these differences in orientation, the undrained strength ratio (s_u/σ_v') varies from point to point around the shear surface. Variations of undrained strengths with orientation of the applied stress in the laboratory are shown in Figure 5.12*b* for two normally consolidated clays and two heavily overconsolidated clay shales.

Ideally, laboratory tests to measure the undrained strength of clay would be performed on completely undisturbed plane strain test specimens, tested under unconsolidated–undrained conditions, or consolidated and sheared with stress orientations that simulate those in the field. However, equipment that can apply and reorient stresses to simulate these effects is highly complex and has been used only for research purposes. For practical applications, tests must be performed with equipment that is easier to use, even though it may not replicate all the various aspects of the field conditions.

Triaxial compression (TC) tests, often used to simulate conditions at the top of the slip surface, have been found to result in strengths that are 5 to 10% lower than vertical compression plane strain tests. Triaxial extension (TE) tests, often used to simulate conditions at the bottom of the slip surface, have been found to result in strengths that are significantly less (at least 20% less) than strengths measured in horizontal compression plane strain tests. Direct simple shear (DSS)

tests, often used to simulate the condition in the central portion of the shear surface, underestimate the undrained shear strength on the horizontal plane. As a result of these biases, the practice of using TC, TE, and DSS tests to measure the undrained strengths of normally consolidated clays results in strengths that are lower than the strengths that would be measured in ideally oriented plane strain tests.

Strain rate. Laboratory tests involve higher rates of strain than are typical for most field conditions. UU test specimens are loaded to failure in 10 to 20 minutes, and the duration of CU tests is usually 2 or 3 hours. Field vane shear tests are conducted in 15 minutes or less. Loading in the field, on the other hand, typically involves a period of weeks or months. The difference in these loading times is on the order of 1000. Slower loading results in lower undrained shear strengths of saturated clays. As shown in Figure 5.13, the strength of San Francisco Bay mud decreases by about 30% as the time to failure increases from 10 minutes to 1 week. It appears that there is no further decrease in undrained strength for longer times to failure.

In conventional practice, laboratory tests are not corrected for strain rate effects or disturbance effects. Because high strain rates increase strengths measured in UU tests and disturbance reduces them, these effects tend to cancel each other when UU laboratory tests are used to evaluate undrained strengths of natural deposits of clay.

Methods of Evaluating Undrained Strengths of Intact Clays

Alternatives for measuring or estimating undrained strengths of normally consolidated and moderately ov-

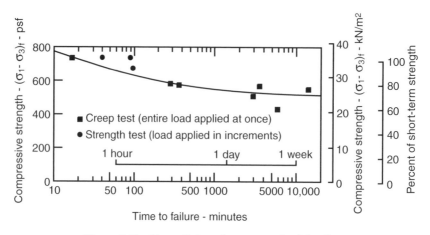

Figure 5.13 Strength loss due to sustained loading.

erconsolidated clays are summarized in Table 5.6. Samples used to measure strengths of natural deposits of clay should be as nearly undisturbed as possible. Hvorslev (1949) has detailed the requirements for good sampling, which include (1) use of thin-walled tube samplers (wall area no more than about 10% of sample area), (2) a piston inside the tube to minimize strains in the clay as the sample tube is inserted, (3) sealing samples after retrieval to prevent change in water content, and (4) transportation and storage procedures that protect the samples from shock, vibration, and excessive temperature changes. Block samples, carefully trimmed and sealed in moistureproof material, are the best possible types of sample. The consequence of poor sampling is scattered and possibly misleading data. One test on a good sample is better than 10 tests on poor samples.

Field vane shear tests. When the results of field vane shear tests are corrected for strain rate and anisotropy effects, they provide an effective method of measuring the undrained strength of soft and medium clays in situ. Bjerrum (1972) developed correction factors for vane shear tests by comparing field vane (FV) strengths with strengths back-calculated from slope failures. The value of the correction factor, μ, varies with the plasticity index, as shown in Figure 5.14. The data that form the basis for these corrections are rather widely scattered, and vane strengths should not be viewed as precise, even after correction. Nevertheless, the vane shear test avoids many of the problems involved in sampling and laboratory testing and has been found to be a useful tool for measuring the undrained strengths of normally consolidated and moderately overconsolidated clays.

Table 5.6   Methods of Measuring or Estimating the Undrained Strengths of Clays

Procedure	Comments
UU tests on vertical, inclined, and horizontal specimens to determine variation of undrained strength with direction of compression	Relies on counterbalancing effects of disturbance and creep. Empirical method of accounting for anisotropy gives results in agreement with vertical and horizontal plane strain compression tests for San Francisco Bay mud and with field behavior of Pepper shale.
AC-U triaxial compression, triaxial extension, and direct simple shear tests, using the recompression or SHANSEP technique	All three tests give lower undrained strengths than the ideal oriented plane strain tests they approximate. Creep strength loss tends to counterbalance these low strengths.
Field vane shear tests, corrected using empirical correction factors (see Figure 5.14)	Correction accounts for anisotropy and creep strength loss. The data on which the correction factor is based contain considerable scatter.
Cone penetration tests, with an empirical cone factor to evaluate undrained strength (see Figure 5.15)	Empirical cone factors can be determined by comparison with corrected vane strengths or estimated based on published data. Strengths based on CPT results involve at least as much uncertainty as strengths based on vane shear tests.
Standard Penetration Tests, with an empirical factor to evaluate undrained strength (see Figure 5.16)	The Standard Penetration Test is not a sensitive measure of undrained strengths in clays. Strengths based on SPT results involve a great deal of uncertainty.
Use $s_u = [0.23(OCR)^{0.8}]\sigma_v'$	This empirical formula, suggested by Jamiolkowski et al. (1985), reflects the influence of σ_v' (effective overburden pressure) and OCR (overconsolidation ratio), but merely approximates the average of the undrained strengths shown in Figure 5.11. The strengths of particular clays may be higher or lower.
Use $s_u = 0.22\sigma_p'$	This empirical formula, suggested by Mesri (1989) combines the influence of σ_v' and OCR in σ_p' (preconsolidation pressure), resulting in a simpler expression. The degree of approximation is essentially the same as for the formula suggested by Jamiolkowski et al. (1985).

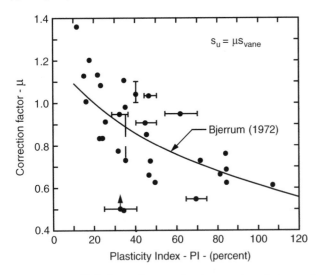

Figure 5.14 Variation of vane shear correction factor and plasticity index. (After Ladd et al., 1977.)

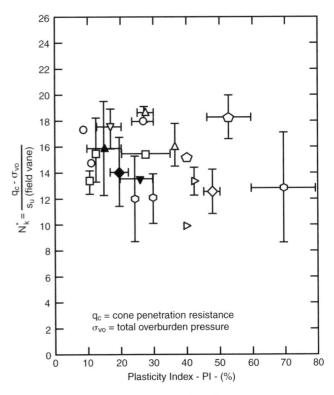

Figure 5.15 Variation of the ratio of net cone resistance $(q_c - \sigma_{vo})$ divided by vane shear strength (s_u vane) with plasticity index for clays. (After Lunne and Klevan, 1982.)

Cone penetration tests. Cone penetration tests (CPTs) are attractive as a means of evaluating undrained strengths of clays in situ because they can be performed quickly and at lower cost than field vane shear tests. The relationship between undrained strength and cone tip resistance is

$$s_u = \frac{q_c - \sigma_{vo}}{N_k^*} \qquad (5.9)$$

where s_u is the undrained shear strength, q_c the CPT tip resistance, σ_v the total overburden pressure at the test depth, and N_k^* the cone factor. The units for s_u, q_c, and σ_v in Eq. (5.9) must be the same.

Values of the cone factor N_k^* for a number of different clays are shown in Figure 5.15. These values were developed by comparing *corrected* vane strengths with cone penetration resistance. Therefore Eq. (5.9) provides values of s_u comparable to values determined from field vane shear tests *after correction*. It can be seen that there is little systematic variation of N_k^* with the plasticity index. A value of $N_k^* = 14 \pm 5$ is applicable to clays with any PI value.

A combination of field vane shear and CPT tests can often be used to good advantage to evaluate undrained strengths at soft clay sites. A few vane shear tests are performed close to CPT test locations, and a site-specific value of N_k^* is determined by comparing the results. The cone test is then used for production testing.

Standard Penetration Tests. Undrained strengths can be estimated very crudely based on the results of Standard Penetration Tests. Figure 5.16, which shows

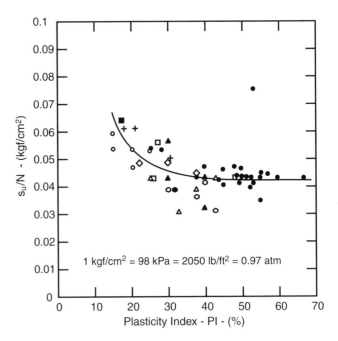

Figure 5.16 Variation of the ratio of undrained shear strength (s_u) divided by SPT blow count (N), with plasticity index for clay. (After Terzaghi et al., 1996.)

the variation of s_u/N with Plasticity Index, can be used to estimate undrained strength based on SPT blow count. In Figure 5.16 the value of s_u is expressed in kgf/cm² (1.0 kgf/cm² is equal to 98 kPa, or 1.0 ton per square foot). The Standard Penetration Test is not a sensitive indicator of the undrained strength of clays, and it is not surprising that there is considerable scatter in the correlation shown in Figure 5.16.

Typical Peak Friction Angles for Intact Clays

Tests to measure peak drained strengths of clays include drained direct shear tests and triaxial tests with pore pressure measurements to determine c' and ϕ'. The tests should be performed on undisturbed test specimens. Typical values of ϕ' for normally consolidated clays are given in Table 5.7. Strength envelopes for normally consolidated clays go through the origin of stresses, and $c' = 0$ for these materials.

Stiff-Fissured Clays

Heavily overconsolidated clays are usually stiff, and they usually contain fissures. The term *stiff-fissured clays* is often used to describe them. Terzaghi (1936) pointed out what has since been confirmed by many others—the strengths that can be mobilized in stiff-fissured clays in the field are less than the strength of the same material measured in the laboratory.

Skempton (1964, 1970, 1977, 1985), Bjerrum (1967), and others have shown that this discrepancy is due to swelling and softening that occurs in the field over long periods of time but does not occur in the laboratory within the period of time used to perform laboratory strength tests. A related factor is that fissures, which have an important effect on the strength of the clay in the field, are not properly represented in laboratory samples unless the test specimens are large

Table 5.7 Typical Values of Peak Friction Angle (ϕ') for Normally Consolidated Clays[a]

Plasticity index	ϕ' (deg)
10	33 ± 5
20	31 ± 5
30	29 ± 5
40	27 ± 5
60	24 ± 5
80	22 ± 5

Source: Data from Bjerrum and Simons (1960).
[a]$c' = 0$ for these materials.

enough to include a significant number of fissures. Unless the specimen size is several times the average fissure spacing, both drained and undrained strengths measured in laboratory tests will be too high.

Peak, fully softened, and residual strengths of stiff-fissured clays. Skempton (1964, 1970, 1977, 1985) investigated a number of slope failures in the stiff-fissured London clay and developed procedures for evaluating the drained strengths of stiff-fissured clays that have been widely accepted. Figure 5.17 shows stress–displacement curves and strength envelopes for drained direct shear tests on stiff-fissured clays. The undisturbed peak strength is the strength of undisturbed test specimens from the field. The magnitude of the cohesion intercept (c') depends on the size of the test specimens. Generally, the larger the test specimens, the smaller the value of c'. As displacement continues beyond the peak, reached at $\Delta x = 0.1$ to 0.25 in. (3 to 6 mm), the shearing resistance decreases. At displacements of 10 in. (250 mm) or so, the shearing resistance decreases to a residual value. In clays without coarse particles, the decline to residual strength is accompanied by formation of a slickensided surface along the shear plane.

If the same clay is remolded, mixed with enough water to raise its water content to the liquid limit, consolidated in the shear box, and then tested, its peak strength will be lower than the undisturbed peak. The strength after remolding and reconsolidating is shown by the NC (normally consolidated) stress–displacement curve and shear strength envelope. The peak is less pronounced, and the NC strength envelope passes through the origin, with c' equal to zero. As shearing displacement increases, the shearing resistance decreases to the same residual value as in the test on the undisturbed test specimen. The displacement required to reach the residual shearing resistance is again about 10 in. (250 mm).

Studies by Terzaghi (1936), Henkel (1957), Skempton (1964), Bjerrum (1967), and others have shown that factors of safety calculated using undisturbed peak strengths for slopes in stiff-fissured clays are larger than unity for slopes that have failed. It is clear, therefore, that laboratory tests on undisturbed test specimens do not result in strengths that can be used to evaluate the stability of slopes in the field.

Skempton (1970) suggested that this discrepancy is due to the fact that more swelling and softening occurs in the field than in the laboratory. He showed that the NC peak strength, also called the *fully softened strength*, corresponds to strengths back-calculated from *first-time slides,* slides that occur where there is no preexisting slickensided failure surface. Skempton also

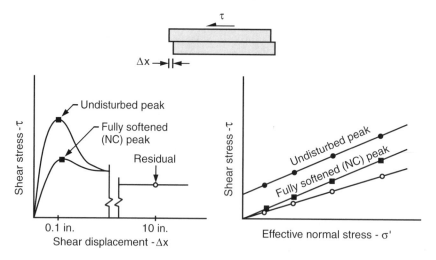

Figure 5.17 Drained shear strength of stiff fissured clay.

showed that once a failure has occurred and a continuous slickensided failure surface has developed, only the residual shear strength is available to resist sliding. Tests to measure fully softened and residual drained strengths of stiff clays can be performed using any representative sample, disturbed or undisturbed, because they are performed on remolded test specimens.

Direct shear tests have been used to measure fully softened and residual strengths. They are more suitable for measuring fully softened strengths because the displacement required to mobilize the fully softened peak strength is small, usually about 0.1 to 0.25 in. (2.5 to 6 mm). Direct shear tests are not so suitable for measuring residual strengths because it is necessary to displace the top of the shear box back and forth to accumulate sufficient displacement to develop a slickensided surface on the shear plane and reduce the shear strength to its residual value. Ring shear tests (Stark and Eid, 1993) are preferable for measuring residual shear strengths because unlimited shear displacement is possible through continuous rotation.

Figures 5.18 and 5.19 show correlations of fully softened friction angle and residual friction angle with liquid limit, clay-size fraction, and effective normal

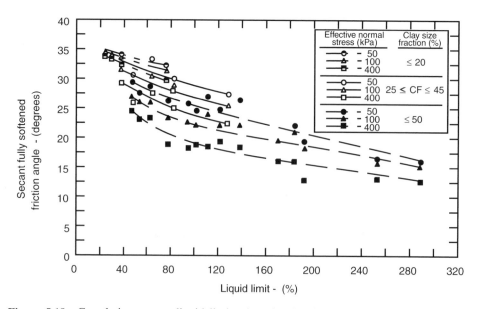

Figure 5.18 Correlation among liquid limit, clay size fraction, and fully softened friction angle. (From Stark and Eid, 1997.)

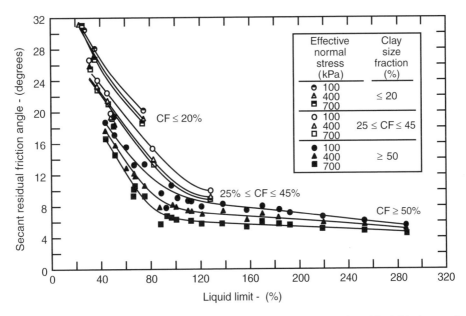

Figure 5.19 Correlation among liquid limit, clay size fraction, and residual friction angle. (From Stark and Eid, 1994.)

stress that were developed by Stark and Eid (1994, 1997). Both fully softened and residual friction angle are fundamental soil properties, and the correlations shown in Figures 5.18 and 5.19 have little scatter. Effective normal stress is a factor because the fully softened and residual strength envelopes are curved, as are the strength envelopes for granular materials. It is thus necessary to represent these strengths using nonlinear relationships between shear strength and normal stress, or to select values of ϕ' that are appropriate for the range of effective stresses in the conditions analyzed.

Undrained strengths of stiff-fissured clays. The undrained strength of stiff-fissured clays is also affected by fissures. Peterson et al. (1957) and Wright and Duncan (1972) showed that the undrained strengths of stiff-fissured clays and shales decreased as test specimen size increased. Small specimens are likely to be intact, with few or no fissures, and therefore stronger than a representative mass of the fissured clay. Heavily overconsolidated stiff fissured clays and shales are also highly anisotropic. As shown in Figure 5.12, inclined specimens of Pepper shale and Bearpaw shale, where

Table 5.8   Typical Peak Drained Strengths for Compacted Cohesive Soils

Unified classification	Relative compaction, RC[a] (%)	Effective stress cohesion, c' (kPa)	Effective stress friction angle, ϕ' (deg)
SM-SC	100	15	33
SC	100	12	31
ML	100	9	32
CL-ML	100	23	32
CL	100	14	28
MH	100	21	25
CH	100	12	19

Source: After U.S. Dept. Interior (1973).
[a]RC, relative compaction by USBR standard method, same energy as the Standard Proctor compaction test.

failure occurs on horizontal planes, are only 30 to 40% of the strengths of vertical specimens.

Compacted Clays

Compacted clays are used often to construct embankment dams, highway embankments, and fills to support buildings. When compacted well, at suitable water content, clay fills have high strength. Clays are more difficult to compact than are cohesionless fills. It is necessary to maintain their moisture contents during compaction within a narrow range to achieve good

---

Recapitulation

- The shear strength of clays in terms of effective stress can be expressed by the Mohr–Coulomb strength criterion as $s = c' + \sigma' \tan \phi'$.
- The shear strength of clays in terms of total stress can be expressed as $s = c + \sigma \tan \phi$.
- For saturated clays, ϕ is zero, and the undrained strength can be expressed as $s = s_u = c$, $\phi = \phi_u = 0$.
- Samples used to measure undrained strengths of normally consolidated and moderately overconsolidated clay should be as nearly undisturbed as possible.
- The strengths that can be mobilized in stiff fissured clays in the field are less than the strength of the same material measured in the laboratory using undisturbed test specimens.
- The normally consolidated peak strength, also called the fully softened strength, corresponds to strengths back calculated from first-time slides.
- Once a failure has occurred and a continuous slickensided failure surface has developed, only the residual shear strength is available to resist sliding.
- Tests to measure fully softened and residual drained strengths of stiff clays can be performed on remolded test specimens.
- Ring shear tests are preferable for measuring residual shear strengths, because unlimited shear displacement is possible through continuous rotation.
- Values of c' and ϕ' for compacted clays can be measured using consolidated–undrained triaxial tests with pore pressure measurements or drained direct shear tests.
- Undrained strengths of compacted clays vary with compaction water content and density and can be measured using UU triaxial tests performed on specimens at their as-compacted water contents and densities.

---

compaction, and more equipment passes are needed to produce high-quality fills. High pore pressures can develop in fills that are compacted wet of optimum, and stability during construction can be a problem in wet fills. Long-term stability can also be a problem, particularly with highly plastic clays, which are subject to swell and strength loss over time. It is necessary to consider both short- and long-term stability of compacted fill slopes in clay.

Drained strengths of compacted clays. Values of c' and ϕ' for compacted clays can be measured using consolidated–undrained triaxial tests with pore pressure measurements or drained direct shear tests. The values determined from either type of test are the same for practical purposes. The effective stress strength parameters for compacted clays, measured using samples that have been saturated before testing, are not strongly affected by compaction water content.

Table 5.8 lists typical values of c' and ϕ' for cohesive soils compacted to RC = 100% of the Standard Proctor maximum dry density. As the value of RC decreases below 100%, values of ϕ' remain about the same, and the value of c' decreases. For RC = 90%, values of c' are about half the values shown in Table 5.8.

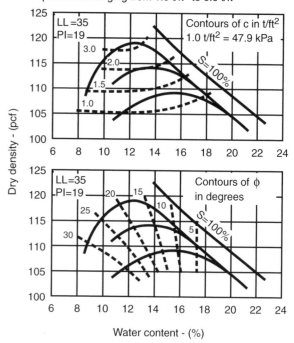

Figure 5.20 Strength parameters for compacted Pittsburgh sandy clay tested under UU test conditions. (From Kulhawy et al., 1969.)

Undrained strengths of compacted clays. Values of c and ϕ (total stress shear strength parameters) for the as-compacted condition can be determined by performing UU triaxial tests on specimens at their compaction water contents. Undrained strength envelopes for compacted, partially saturated clays tested are curved, as discussed in Chapter 3. Over a given range of stresses, however, a curved strength envelope can be approximated by a straight line and can be characterized in terms of c and ϕ. When this is done, it is especially important that the range of pressures used in the tests correspond to the range of pressures in the field conditions being evaluated. Alternatively, if the computer program used accommodates nonlinear strength envelopes, the strength test data can be represented directly.

Values of total stress c and ϕ for compacted clays vary with compaction water content and density. An example is shown in Figure 5.20 for compacted Pittsburgh sandy clay. The range of confining pressures used in these tests was 1.0 to 6.0 tons/ft^2. The value of c, the total stress cohesion intercept from UU tests, increases with dry density but is not much affected by compaction water content. The value of ϕ, the total stress friction angle, decreases as compaction water content increases, but is not so strongly affected by dry density.

If compacted clays are allowed to age prior to testing, they become stronger, apparently due to thixotropic effects. Therefore, undrained strengths measured using freshly compacted laboratory test specimens provide a conservative estimate of the strength of the fill a few weeks or months after compaction.

MUNICIPAL SOLID WASTE

Waste materials have strengths comparable to the strengths of soils. Strengths of waste materials vary depending on the amounts of soil and sludge in the waste, as compared to the amounts of plastic and other materials that tend to interlock and provide tensile strength (Eid et al., 2000). Larger amounts of materials that interlock increase the strength of the waste. Although solid waste tends to decompose or degrade with time, Kavazanjian (2001) indicates that the strength after degradation is similar to the strength before degradation.

Kavazanjian et al. (1995) used laboratory test data and back analysis of stable slopes to develop the lower-bound strength envelope for municipal solid waste shown in Figure 5.21. The envelope is horizontal with a constant strength $c = 24$ kPa, $\phi = 0$ at normal pres-

sures less than 37 kPa. At pressures greater than 37 kPa, the envelope is inclined at $\phi = 33°$ with $c = 0$.

Eid et al. (2000) used results of large-scale direct shear tests (300 to 1500 mm shear boxes) and back analysis of failed slopes in waste to develop the range of strength envelopes show in Figure 5.22. All three envelopes (lower bound, average, and upper bound) are inclined at $\phi = 35°$. The average envelope shown in Figure 5.22 corresponds to $c = 25$ kPa, and the lowest of the envelopes corresponds to $c = 0$.

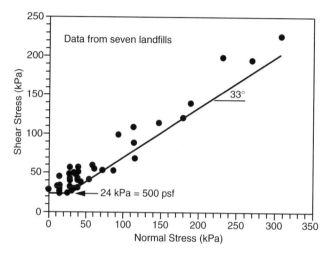

Figure 5.21 Shear strength envelope for municipal solid waste based on large-scale direct shear tests and back analysis of stable slopes. (After Kavazanjian et al., 1995.)

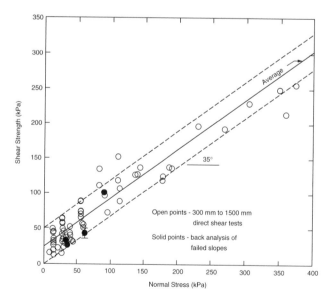

Figure 5.22 Range of shear strength envelopes for municipal solid waste based on large-scale direct shear tests and back analysis of failed slopes. (After Eid et al., 2000.)

CHAPTER 6

Mechanics of Limit Equilibrium Procedures

Once appropriate shear strength properties, pore water pressures, slope geometry and other soil and slope properties are established, slope stability calculations need to be performed to ensure that the resisting forces are sufficiently greater than the forces tending to cause a slope to fail. Calculations usually consist of computing a factor of safety using one of several limit equilibrium procedures of analysis. All of these procedures of analysis employ the same definition of the factor of safety and compute the factor of safety using the equations of static equilibrium.

DEFINITION OF THE FACTOR OF SAFETY

The *factor of safety, F,* is defined with respect to the shear strength of the soil as

$$F = \frac{s}{\tau} \qquad (6.1)$$

where s is the available shear strength and τ is the equilibrium shear stress. The equilibrium shear stress is the shear stress required to maintain a just-stable slope and from Eq. (6.1) may be expressed as

$$\tau = \frac{s}{F} \qquad (6.2)$$

The equilibrium shear stress is equal to the available shear strength divided (*factored*) by the factor of safety. The factor of safety represents the factor by which the shear strength must be reduced so that the reduced strength is just in equilibrium with the shear stress (τ) (i.e., the slope is in a state of just-stable *limiting equilibrium*. The procedures used to perform

such computations are known as *limit equilibrium procedures.*

The shear strength can be expressed by the Mohr–Coulomb equation. If the shear strength is expressed in terms of total stresses, Eq. (6.2) is written as

$$\tau = \frac{c + \sigma \tan \phi}{F} \qquad (6.3)$$

or

$$\tau = \frac{c}{F} + \frac{\sigma \tan \phi}{F} \qquad (6.4)$$

where c and ϕ are the cohesion and friction angle for the soil, respectively, and σ is the total normal stress on the shear plane. The same values for the factor of safety are applied to cohesion and friction in this equation. Equation (6.4) can also be written as

$$\tau = c_d + \sigma \tan \phi_d \qquad (6.5)$$

where

$$c_d = \frac{c}{F} \qquad (6.6)$$

$$\tan \phi_d = \frac{\tan \phi}{F} \qquad (6.7)$$

The quantities c_d and ϕ_d represent the developed (or *mobilized*) cohesion and friction angle, respectively.

If the shear strength is expressed in terms of effective stresses (e.g., drained shear strengths are being used), the only change from the above is that Eq. (6.3) is written in terms of effective stresses as

$$\tau = \frac{c' + (\sigma - u)\tan\phi'}{F} \qquad (6.8)$$

where c' and ϕ' represent the shear strength parameters in terms of effective stresses, and u is the pore water pressure.

To calculate the factor of safety, a slip surface is assumed and one or more equations of static equilibrium are used to calculate the stresses and factor of safety for each surface assumed. The term *slip surface* is used here to refer to an assumed surface along which sliding or rupture might occur. However, it is the intent of slope stability calculations that sliding and rupture not occur along such surfaces if the slope is designed adequately.

The factor of safety is assumed to be the same at all points along the slip surface. Thus, the value represents an average or overall value for the assumed slip surface. If failure were to occur, the shear stress would be equal to the shear strength at all points along the failure surface and the assumption that the factor of safety is constant would be valid. If, instead, the slope is stable, the factor of safety probably varies along the slip surface (e.g., Wright et al., 1973). However, this should not be of significant consequence as long as the overall factor of safety is suitably greater than 1 and the assumed shear strengths can be fully mobilized along the entire slip surface.

A number of slip surfaces must be assumed to find the slip surface that produces a minimum factor of safety. The surface with the minimum factor of safety is termed the *critical slip surface*. Such a critical surface and the corresponding minimum factor of safety represent the most likely sliding surface, presuming that all of the shear strengths have been determined in a comparable way and with comparable degrees of certainty. Although the slip surface with the minimum factor of safety may not represent a failure mechanism with a significant consequence, the minimum factor of safety is unique for a given problem and should be calculated as part of any analysis of stability. Other slip surfaces with higher factors of safety than the minimum may also be of interest and are discussed in Chapter 13.

Recapitulation

- The factor of safety is defined with respect to shear strength.
- The same factor of safety is applied to both cohesion (c, c') and friction ($\tan\phi$, $\tan\phi'$).
- The factor of safety is computed for an assumed slip surface.

- The factor of safety is assumed to be constant along the slip surface.
- A number of different slip surfaces must be assumed and the factor of safety computed for each to determine a critical slip surface with a minimum factor of safety.

EQUILIBRIUM CONDITIONS

Two different approaches are used to satisfy static equilibrium in the limit equilibrium analysis procedures. Some procedures consider equilibrium for the entire mass of soil bounded beneath by an assumed slip surface and above by the surface of the slope. In these procedures, equilibrium equations are written and solved for a single free body. The infinite slope procedure and the Swedish slip circle method are examples of such single-free-body procedures. In other procedures the soil mass is divided into a number of vertical slices and equilibrium equations are written and solved for each slice. These procedures, termed *procedures of slices,* include such methods as the Ordinary Method of Slices, the Simplified Bishop procedure, and Spencer's Procedure.

Three static equilibrium conditions are to be satisfied: (1) equilibrium of forces in the vertical direction, (2) equilibrium of forces in the horizontal direction, and (3) equilibrium of moments about any point. The limit equilibrium procedures all use at least some static equilibrium equations to compute the factor of safety. Some procedures use and satisfy all of the equilibrium equations, others use and satisfy only some. The Ordinary Method of Slices and Simplified Bishop procedure satisfy only some of the equilibrium requirements. In contrast, Spencer's procedure and the Morgenstern and Price procedure satisfy all the requirements for static equilibrium.

Regardless of whether equilibrium is considered for a single free body or a series of individual vertical slices making up the total free body, there are more unknowns (forces, locations of forces, factor of safety, etc.) than the number of equilibrium equations; the problem of computing a factor of safety is statically indeterminate. Therefore, assumptions must be made to achieve a balance of equations and unknowns. Different procedures make different assumptions to satisfy static equilibrium. Two procedures may even satisfy the same equilibrium conditions but make different assumptions and therefore produce different values for the factor of safety.

A number of limit equilibrium procedures are described in more detail in the following sections. Each procedure differs from the others in one or more of the ways described above. The different procedures may or may not divide the soil mass into slices, may satisfy different equilibrium conditions, and/or may make different assumptions to obtain a statically determinate solution. The procedures discussed in this chapter were selected because each has a particular advantage or usefulness depending on the slope geometry, the soil strength, and the purpose of the analysis.

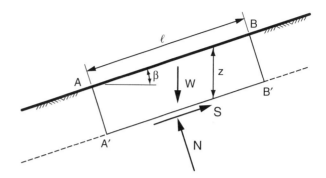

Figure 6.1 Infinite slope and plane slip surface.

Recapitulation

- Equilibrium may be considered either for a single free body or for individual vertical slices.
- Depending on the analysis procedure, complete static equilibrium may or may not be satisfied.
- Some assumptions must be made to obtain a statically determinate solution for the factor of safety.
- Different procedures make different assumptions, even when they may satisfy the same equilibrium equations.

SINGLE FREE-BODY PROCEDURES

The infinite slope, logarithmic spiral, and Swedish slip circle methods all consider equilibrium for a single free-body. These procedures are relatively simple to use and useful within their range of applicability.

Infinite Slope Procedure

As implied by its name, in the *infinite slope procedure* the slope is assumed to extend infinitely in all directions and sliding is assumed to occur along a plane parallel to the face of the slope (Taylor, 1948). Because the slope is infinite, the stresses will be the same on any two planes that are perpendicular to the slope, such as the planes A–A′ and B–B′ in Figure 6.1. Equilibrium equations are derived by considering a rectangular block like the one shown in Figire 6.1. For an infinite slope, the forces on the two ends of the block will be identical in magnitude, opposite in direction, and collinear. Thus, the forces on the ends of the block exactly balance each other and can be ignored in the equilibrium equations. Summing forces in directions perpendicular and parallel to the slip plane gives the following expressions for the shear force, S, and normal force, N, on the plane:

$$S = W \sin \beta \qquad (6.9)$$

$$N = W \cos \beta \qquad (6.10)$$

where β is the angle of inclination of the slope and slip plane, measured from the horizontal, and W is the weight of the block. For a block of unit thickness in the direction perpendicular to the plane of the cross section in Fig. 6.1, the weight is expressed as

$$W = \gamma l z \cos \beta \qquad (6.11)$$

where γ is the total unit weight of the soil, l the distance between the two ends of the block, measured parallel to the slope, and z the vertical depth to the shear plane. Substituting Eq. (6.11) into Eqs. (6.9) and (6.10) gives

$$S = \gamma l z \cos \beta \sin \beta \qquad (6.12)$$

$$N = \gamma l z \cos^2 \beta \qquad (6.13)$$

The shear and normal stresses on the shear plane are constant for an infinite slope and are obtained by dividing Eqs. (6.12) and (6.13) by the area of the plane ($l \cdot 1$), to give

$$\tau = \gamma z \cos \beta \sin \beta \qquad (6.14)$$

$$\sigma = \gamma z \cos^2 \beta \qquad (6.15)$$

Substituting these expressions for the stresses into Eq. (6.3) for the factor of safety for total stresses gives

$$F = \frac{c + \gamma z \cos^2 \beta \tan \phi}{\gamma z \cos \beta \sin \beta} \qquad (6.16)$$

For effective stresses, the equation for the factor of safety becomes

$$F = \frac{c' + (\gamma z \cos^2 \beta - u)\tan \phi'}{\gamma z \cos \beta \sin \beta} \qquad (6.17)$$

Equations for computing the factor of safety for an infinite slope are summarized in Figure 6.2 for both total stress and effective stress analyses and a variety of water and seepage conditions.

For a cohesionless (c, $c' = 0$) soil, the factor of safety calculated by an infinite slope analysis is independent of the depth, z, of the slip surface. For total stresses (or effective stresses with zero pore water pressure) the equation for the factor of safety becomes

$$F = \frac{\tan \phi}{\tan \beta} \qquad (6.18)$$

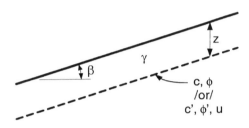

Total Stresses: $s = c + \sigma \tan \phi$

Subaerial (not-submerged) slopes:

$$F = \frac{c}{\gamma z} \frac{2}{\sin(2\beta)} + (\cot \beta) \tan \phi$$

Submerged slopes ($\phi = 0$ only):

$$F = \frac{c}{(\gamma - \gamma_w) z} \frac{2}{\sin(2\beta)}$$

Total Stresses: $s = c' + \sigma' \tan \phi'$

General case (subaerial slope):

$$F = \frac{c'}{\gamma z} \frac{2}{\sin(2\beta)} + \left[\cot \beta - \frac{u}{\gamma z} (\cot \beta + \tan \beta) \right] \tan \phi'$$

Submerged slopes – no flow:

$$F = \frac{c'}{(\gamma - \gamma_w) z} \frac{2}{\sin(2\beta)} + [\cot \beta] \tan \phi'$$

Subaerial slope – seepage parallel to slope face:

$$F = \frac{c'}{\gamma z} \frac{2}{\sin(2\beta)} + \left[\cot \beta - \frac{\gamma_w}{\gamma} (\cot \beta) \right] \tan \phi'$$

Subaerial slope – horizontal seepage:

$$F = \frac{c'}{\gamma z} \frac{2}{\sin(2\beta)} + \left[\cot \beta - \frac{\gamma_w}{\gamma} (\cot \beta + \tan \beta) \right] \tan \phi'$$

Subaerial slope – pore water pressures defined by $r_u = \frac{u}{\gamma z}$:

$$F = \frac{c'}{\gamma z} \frac{2}{\sin(2\beta)} + [\cot \beta - r_u (\cot \beta + \tan \beta)] \tan \phi'$$

Figure 6.2 Summary of equations for computing the factor of safety for an infinite slope using both total stresses and effective stresses.

Similarly for effective stresses, if the pore water pressures are proportional to the depth of slide, the factor of safety is expressed by

$$F = [\cot \beta - r_u(\cot \beta + \tan \beta)] \tan \phi' \qquad (6.19)$$

where r_u is the pore water pressure coefficient suggested by Bishop and Morgenstern (1960). The pore water pressure coefficient is defined as

$$r_u = \frac{u}{\gamma z} \qquad (6.20)$$

Because the factor of safety for a cohesionless slope is independent of the depth of the slip surface, it is possible for a slip surface that is only infinitesimally deep to have the same factor of safety as that for deeper surfaces. Regardless of the lateral extent of the slope, a slip surface can develop that is shallow with respect to the lateral dimensions of the slope. Any slope will constitute an infinite slope as long as the soil is cohesionless. Therefore, the infinite slope analysis procedure is the appropriate procedure to use for any slope in cohesionless soil.[1]

The infinite slope analysis is also applicable to slopes in cohesive soils provided that a firmer stratum parallel to the face of the slope limits the depth of the failure surface. If such a stratum exists at a depth that is small compared to the lateral extent of the slope, an infinite slope analysis provides a suitable approximation for stability calculations.

The infinite slope equations were derived by considering equilibrium of forces in two mutually perpendicular directions and thus satisfy all force equilibrium requirements. Moment equilibrium was not considered explicitly; however, the forces on the two ends of the block are collinear and the normal force acts at the center of the block. Thus, moment equilibrium is satisfied, and the infinite slope procedure can be considered to fully satisfy all the requirements for static equilibrium.

---

[1] An exception to this may occur for soils with curved Mohr failure envelopes that pass through the origin. Although there is no strength at zero normal stress, and thus the soil might be termed *cohesionless,* the factor of safety depends on the depth of slide and the infinite slope analysis may not be appropriate. Also see the example of the Oroville Dam presented in Chapter 7.

Recapitulation

- For a cohesionless slope the factor of safety is independent of the depth of the slip surface, and thus an infinite slope analysis is appropriate (exceptions may occur for nonhomogeneous slopes and/or curved Mohr failure envelopes).
- For cohesive soils the infinite slope analysis procedure may provide a suitable approximation provided that the slip surface is parallel to the slope and limited to a depth that is small compared to the lateral dimensions of the slope.
- The infinite slope analysis procedure fully satisfies static equilibrium.

Logarithmic Spiral Procedure

In the Logarithmic Spiral procedure, the slip surface is assumed to be a logarithmic spiral, as shown in Figure 6.3 (Frohlich, 1953). A center point and an initial radius, r_0, define the spiral. The radius of the spiral varies with the angle of rotation, θ, about the center of the spiral according to the expression

$$r = r_0 e^{\theta \tan \phi_d} \qquad (6.21)$$

where ϕ_d is the developed friction angle; ϕ_d depends on the friction angle of the soil and the factor of safety. The stresses along the slip surface consist of the normal stress (σ) and the shear stress (τ). The shear stress can be expressed in the case of total stresses by the normal stress, the shear strength parameters (c and ϕ), and the factor of safety. From Eq. (6.4),

$$\tau = \frac{c}{F} + \sigma \frac{\tan \phi}{F} \qquad (6.22)$$

or in terms of developed shear strengths,

$$\tau = c_d + \sigma \tan \phi_d \qquad (6.23)$$

A log spiral has the properties that the radius extended from the center of the spiral to a point on the slip surface intersects the slip surface at an angle, ϕ_d, to the normal (Figure 6.3). Because of this property, the resultant forces produced by the normal stress (σ) and the frictional portion of the shear stress ($\sigma \tan \phi_d$) act along a line through the center of the spiral and produce no net moment about the center of the spiral. The only forces on the slip surface that produce a moment about the center of the spiral are those due to the developed cohesion. An equilibrium equation may be

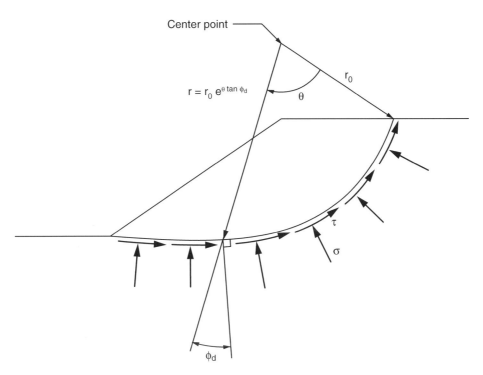

Figure 6.3 Slope and logarithmic spiral slip surface. (After Frohlich, 1993.)

written by summing moments about the center of the spiral, which involves only the factor of safety as the unknown. This equation may be used to compute the factor of safety.

In the logarithmic spiral procedure a statically determinant solution is achieved by assuming a particular shape (logarithmic spiral) for the slip surface. By assuming a logarithmic spiral, no additional assumptions are required. Force equilibrium is not considered explicitly in the logarithmic spiral procedure. However, there are an infinite number of combinations of normal and shear stresses along the slip surface that will satisfy force equilibrium. All of these combinations of shear and normal stress will yield the same value for the factor of safety that satisfies moment equilibrium. Thus, the logarithmic spiral implicitly satisfies complete static equilibrium. The logarithmic spiral and infinite slope procedures are the only two limit equilibrium procedures that satisfy complete equilibrium by assuming a specific slope geometry and shape for the slip surface.

Because the logarithmic spiral procedure fully satisfies static equilibrium, it is relatively accurate. Also, for homogeneous slopes, a logarithmic spiral appears to approximate the shape of the most critical potential sliding surface reasonably well. The logarithmic spiral procedure is theoretically the best limit equilibrium procedure for analyses of homogeneous slopes.

For cohesionless (c, $c' = 0$) slopes the critical logarithmic spiral that produces the minimum factor of safety has an infinite radius and the spiral coincides with the face of the slope (Figure 6.4). In this case the logarithmic spiral and infinite slope procedures produce identical values for the minimum factor of safety.

The logarithmic spiral equations are relatively complex and awkward for hand calculations, because of the assumed shape of the slip surface. However, the logarithmic spiral procedure is computationally efficient and well suited for implementation in computer calculations. The procedure is useful for performing the computations required to produce slope stability charts, and once such charts have been developed there is little need for using the detailed logarithmic spiral equations (Wright, 1969; Leshchinsky and Volk, 1985; Leschinsky and San, 1994). The logarithmic spiral procedure has also received recent interest and attention for use in software to analyze reinforced slopes, particularly for design software that must perform many repetitive calculations to find a suitable arrangement for reinforcement (Leshchinsky, 1997).

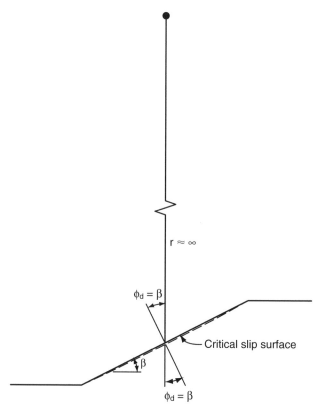

Figure 6.4   Critical logarithmic spiral slip surface for a cohesionless slope.

Recapitulation

- The logarithmic spiral procedure achieves a statically determinate solution by assuming a specific logarithmic spiral shape for the slip surface [Eq. (6.21)].
- The logarithmic spiral procedure explicitly satisfies moment equilibrium and implicitly satisfies complete force equilibrium. Because complete equilibrium is satisfied, the procedure is relatively accurate.
- The logarithmic spiral procedure is theoretically the best procedure for analysis for homogeneous slopes. Much of the effort required can be reduced by use of dimensionless slope stability charts (Leshchinsky, 1985, 1994).
- The logarithmic spiral procedure is used in several computer programs for design of reinforced slopes using geogrids, soil nails, and so on.

Swedish Circle/$\phi = 0$ Method

In the *Swedish Circle method* the slip surface is assumed to be a circular arc and moments are summed about the center of the circle to calculate a factor of safety. Some form of the method was apparently first used by Petterson in about 1916 (Petterson, 1955), but the method seems to have first been formalized for $\phi = 0$ by Fellenius in 1922 (Fellenius, 1922; Skempton, 1948). The friction angle is assumed to be zero and thus the shear strength is assumed to be due to "cohesion" only. For this reason, the Swedish Circle method is also called the $\phi = 0$ *method*.

The Swedish Circle or "$\phi = 0$" method is actually a special case of the logarithmic spiral procedure: When $\phi = 0$, a logarithmic spiral becomes a circle. However, the equilibrium equations for a circle are much simpler than those for a more general logarithmic spiral, and thus the Swedish Circle method is generally considered to be a separate method. The Swedish Circle method also seems to have preceded the development of the logarithmic spiral method for slope stability analysis.

Referring to the slope and circular slip surface shown in Figure 6.5, the driving (overturning) moment tending to produce rotation of the soil about the center of the circle is given by

$$M_d = Wa \qquad (6.24)$$

where W is the weight of the soil mass above the circular slip surface and a is the horizontal distance between the center of the circle and the center of gravity of the soil mass; a is the moment arm. The resisting moment is provided by the shear stresses (τ) acting along the circular arc. For a unit thickness of the cross section shown in Figure 6.5, the resisting moment is given by

$$M_r = \tau l r \qquad (6.25)$$

where l is the length of the circular arc and r is the radius. For equilibrium, the resisting and overturning moments must balance. Thus,

$$Wa = \tau l r \qquad (6.26)$$

The shear stress in this equation can be expressed in terms of the shear strength and factor of safety using Eq. (6.1). Introducing Eq. (6.1) and replacing the shear strength by the cohesion c yields

$$Wa = \frac{clr}{F} \qquad (6.27)$$

or, after rearranging,

$$F = \frac{clr}{Wa} \qquad (6.28)$$

Equation (6.28) is the equation used to compute the factor of safety by the Swedish Circle method.

The term clr in the numerator of Eq. (6.28) represents the available resisting moment; the term Wa in the denominator represents the driving moment. Therefore, the factor of safety in this case is equal to the

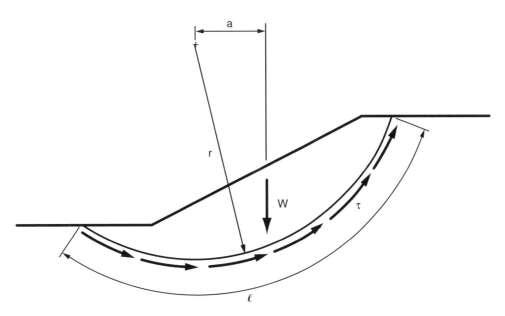

Figure 6.5 Slope and slip surface for the Swedish circle ($\phi = 0$) procedure.

available resisting moment, M_r, divided by the actual driving moment, M_d:

$$F = \frac{\text{available resisting moment}}{\text{actual driving moment}} \qquad (6.29)$$

The Swedish Circle method can be derived by starting with Eq. (6.29) as the definition of the factor of safety, and this approach is sometimes used. There are also other definitions that have been suggested for the factor of safety, and these are discussed later under the head "Alternative Definitions for the Factor of Safety." However, the authors prefer to begin with the definition of the factor of safety in terms of shear strength given by Eq. (6.1) rather than Eq. (6.29).

Because the Swedish Circle method is a special case of the logarithmic spiral method, it also satisfies complete static equilibrium. Both the logarithmic spiral and $\phi = 0$ methods use only the equilibrium equation for summation of moments about the center point of the slip surface, but all conditions of static equilibrium are implicitly satisfied. The $\phi = 0$ method achieves a statically determinate solution by assuming that $\phi = 0$ and a circular slip surface. No direct assumptions are made about the unknown forces that contribute to equilibrium.

Equation (6.28) was derived for a constant value of cohesion, but the equation is easily extended to cases where the cohesion varies. If c varies, the circular slip surface is subdivided into an appropriate number of segments of length, Δl_i, each with a corresponding average strength, c_i (Figure 6.6). The expression for the resisting moment becomes

$$M_r = \frac{\sum (c_i \, \Delta l_i r)}{F} \qquad (6.30)$$

where the summation is performed for the segments along the slip surface. The equation for the factor of safety is then

$$F = \frac{r \sum (c_i \, \Delta l_i)}{Wa} \qquad (6.31)$$

The term Wa in Eqs. (6.28) and (6.31) represents the driving moment due to the weight of the soil. To compute the moment arm, a, it is necessary to compute the center of gravity of the soil mass above the slip surface, which is not easy considering the usually odd shape of the soil mass. Instead, in practice, calculations are usually performed using either slope stability charts like those described in the Appendix or one of the procedures of slices that are described in the next section. The procedures of slices provide a more convenient way to calculate the driving moment, and for a circular slip surface the procedures of slices produce the same value for the factor of safety as the Swedish Circle method.

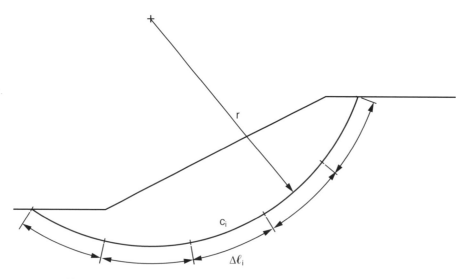

Figure 6.6 Circular slip surface subdivided into segments when the undrained shear strength varies.

<table>
<tr><td>

Recapitulation

- The Swedish Circle (or $\phi = 0$) method explicitly satisfies moment equilibrium and implicitly satisfies force equilibrium completely.
- The Swedish Circle (or $\phi = 0$) method is an accurate method of slope stability analysis for both homogeneous and inhomogeneous slopes in $\phi = 0$ soils, provided that the slip surface can be approximated by a circle.

</td></tr>
</table>

PROCEDURES OF SLICES: GENERAL

In the remaining procedures covered in this chapter, the soil mass above the slip surface is subdivided into a number of vertical slices. The actual number of slices used depends on the slope geometry and soil profile and is discussed in more detail in Chapter 14.

Some procedures of slices assume a circular slip surface while others assume an arbitrary (noncircular) slip surface. Procedures that assume a circular slip surface consider equilibrium of moments about the center of the circle for the entire free body composed of all slices. In contrast, the procedures that assume an arbitrary shape for the slip surface usually consider equilibrium in terms of the individual slices. It is appropriate to consider the procedures for circular and noncircular slip surfaces separately.

PROCEDURES OF SLICES: CIRCULAR SLIP SURFACES

Procedures based on a circular slip surface consider equilibrium of moments about the center of the circle. Referring to the slope and circular slip surface shown in Figure 6.7, the overturning moment can be expressed as

$$M_d = \sum W_i a_i \qquad (6.32)$$

where W_i is the weight of the ith slice and a_i is the horizontal distance between the center of the circle and the center of the slice. Distances toward the crest of the slope, to the right of the center shown in Figure 6.7, are positive; distances toward the toe of the slope, to the left of the center, are negative. Although theoretically the moment arm, a_i, is measured from the center of the circle to the center of gravity of the slice, a sufficient number of slices is generally used that the differences between the center (midwidth) and center of gravity of the slice are negligible. In most cases a_i is measured from the center of the circle to the center (midwidth) of the slice.

The moment arm, a_i, in Eq. (6.32) can be expressed in terms of the radius of the circle and the inclination of the bottom of the respective slice. Although the base of the slice is curved, the base can be assumed to be a straight line, as suggested in Figure 6.7, with negligible loss in accuracy. The inclination of the base of

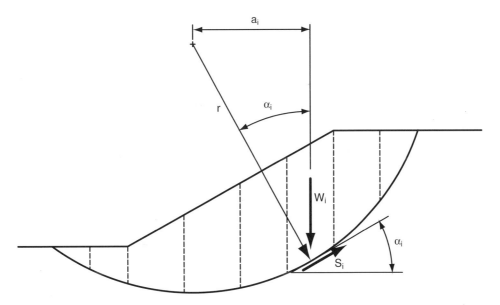

Figure 6.7 Circular slip surface with overlying soil mass subdivided into vertical slices.

the slice is represented by the angle, α_i, measured between the base of the slice and the horizontal. Positive and negative values are indicated in Figure 6.7. The angle between a line extended from the center of the circle to the center of the base of the slice and a vertical line is also equal to the angle, α_i (Figure 6.7). Thus, the moment arm (a_i) is expressed by

$$a_i = r \sin \alpha_i \qquad (6.33)$$

and the driving moment becomes

$$M_d = r \sum W_i \sin \alpha_i \qquad (6.34)$$

The radius in Eq. (6.34) has been moved outside the summation because the radius is constant for a circle.

The resisting moment is provided by the shear stresses (τ) on the base of each slice; normal stresses (σ) on the base of each slice act through the center of the circle and thus produce no moment. The total resisting moment for all slices is

$$M_r = \sum r S_i = r \sum S_i \qquad (6.35)$$

where S_i is the shear force on the base of the ith slice and the summation is performed for all slices. The shear force is the product of the shear stress, τ_i, and the area of the base of the slice, which for a slice of unit thickness is $\Delta l_i \cdot 1$. Thus,

$$M_r = r \sum \tau_i \, \Delta l_i \qquad (6.36)$$

The shear stress can be expressed in terms of the shear strength and the factor of safety by Eq. (6.1) to give

$$M_r = r \sum \frac{s_i \, \Delta l_i}{F} \qquad (6.37)$$

Equating the resisting moment [Eq. (6.37)] and the driving moment [Eq. (6.34)] and rearranging, the following equation can be written for the factor of safety:

$$F = \frac{\sum s_i \, \Delta l_i}{\sum W_i \sin \alpha_i} \qquad (6.38)$$

The radius has been canceled from both the numerator and denominator of this equation. However, the equation is still valid only for a circular slip surface.

At this point the subscript i will be dropped from use with the understanding that the quantities inside the summation are the values for an individual slice

and that the summations are performed for all slices. Thus, Eq. (6.38) is written as

$$F = \frac{\sum s \Delta l}{\sum W \sin \alpha} \qquad (6.39)$$

For total stresses the shear strength is expressed by

$$s = c + \sigma \tan \phi \qquad (6.40)$$

Substituting this into Eq. (6.39), gives

$$F = \frac{\sum (c + \sigma \tan \phi) \, \Delta l}{\sum W \sin \alpha} \qquad (6.41)$$

Equation (6.41) represents the static equilibrium equation for moments about the center of a circle. If ϕ is equal to zero, Eq. (6.41) becomes

$$F = \frac{\sum c \, \Delta l}{\sum W \sin \alpha} \qquad (6.42)$$

which can be solved for a factor of safety. Equations (6.42) and (6.31), derived earlier for the Swedish Circle ($\phi = 0$) method, both satisfy moment equilibrium about the center of a circle and make no assumptions other than that $\phi = 0$ and the slip surface is a circle. Therefore, both equations produce the same value for the factor of safety. The only difference is that Eq. (6.31) considers the entire free body as a single mass; Eq. (6.42) subdivides the mass into slices. Equation (6.42), based on slices, is more convenient to use than Eq. (6.31) because it avoids the need to locate the center of gravity of what may be an odd-shaped soil mass above the slip surface.

If the friction angle is not equal to zero, the equation presented above for the factor of safety [Eq. (6.41)] requires that the normal stress on the base of each slice be known. The problem of determining the normal stress is statically indeterminate and requires that additional assumptions be made in order to compute the factor of safety. The Ordinary Method of Slices and the Simplified Bishop procedures described in the next two sections make two different sets of assumptions to obtain the normal stress on the base of the slices and, subsequently, the factor of safety.

Ordinary Method of Slices

The Ordinary Method of Slices is a procedure of slices that neglects the forces on the sides of the slice. The Ordinary Method of Slices has also been referred to as the "Swedish Method of Slices" and the "Fellenius Method." This method should not, however, be confused with the U.S. Army Corps of Engineers' Modified Swedish method, which is described later. Similarly, the method should not be confused with other methods of slices that Fellenius developed, including a method of slices that fully satisfies static equilibrium (Fellenius, 1936).

Referring to the slice shown in Figure 6.8 and resolving forces perpendicular to the base of the slice, the normal force for the Ordinary Method of Slices can be expressed as

$$N = W \cos \alpha \qquad (6.43)$$

The normal force expressed by Eq. (6.43) is the same as the normal force that would exist if the resultant force due to the forces on the sides of the slice acted in a direction parallel to the base of the slice (Bishop, 1955). However, it is impossible for this to occur and for the forces on the slice to be in equilibrium unless the interslice forces are zero.

The normal stress on the base of a slice is obtained by dividing the normal force by the area of the base of the slice $(1 \cdot \Delta l)$, to give

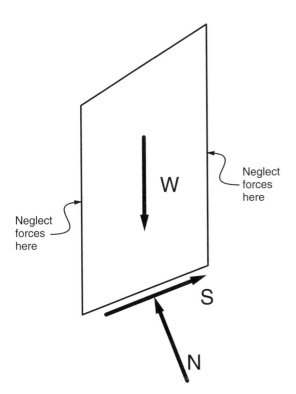

Figure 6.8 Slice with forces considered in the Ordinary Method of Slices.

$$\sigma = \frac{W \cos \alpha}{\Delta l} \qquad (6.44)$$

Substituting this expression for the normal force into Eq. (6.41), derived above for the factor of safety from moment equilibrium, gives the following equation for the factor of safety:

$$F = \frac{\sum (c \, \Delta l + W \cos \alpha \tan \phi)}{\sum W \sin \alpha} \qquad (6.45)$$

Equation (6.45) is the equation for the factor of safety by the Ordinary Method of Slices when the shear strength is expressed in terms of total stresses.

When the shear strength is expressed in terms of effective stresses the equation for the factor of safety from moment equilibrium is

$$F = \frac{\sum (c' + \sigma' \tan \phi') \Delta l}{\sum W \sin \alpha} \qquad (6.46)$$

where σ' is the effective normal stress, $\sigma - u$. From Eq. (6.44) for the total normal stress, the effective normal stress can be expressed as

$$\sigma' = \frac{W \cos \alpha}{\Delta l} - u \qquad (6.47)$$

where u is the pore water pressure on the slip surface. Substituting this expression for the effective normal stress [Eq. (6.47)] into the equation for the factor of safety (6.46) and rearranging gives

$$F = \frac{\sum [c' \, \Delta l + (W \cos \alpha - u \, \Delta l) \tan \phi']}{\sum W \sin \alpha} \qquad (6.48)$$

Equation (6.48) represents an expression for the factor of safety by the Ordinary Method of Slices for effective stresses. However, the assumption involved in this equation $(\sigma = W \cos \alpha / \Delta l)$ can lead to unrealistically low, even negative values for the effective stresses on the slip surface. This can be demonstrated as follows: Let the weight of the slice be expressed as

$$W = \gamma h b \qquad (6.49)$$

where h is the height of the slice at the centerline and b is the width of the slice (Figure 6.9). The width of the slice is related to the length of the base of the slice, Δl, as

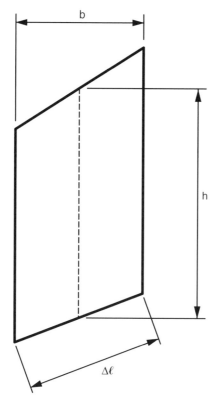

Figure 6.9 Dimensions for an individual slice.

$$b = \Delta l \cos \alpha \qquad (6.50)$$

Thus, Eq. (6.49) can be written as

$$W = \gamma h \, \Delta l \cos \alpha \qquad (6.51)$$

Substituting this expression for the weight of the slice into Eq. (6.48) and rearranging gives

$$F = \frac{\sum [c' \Delta l + (\gamma h \cos^2 \alpha - u) \Delta l \tan \phi']}{\sum W \sin \alpha} \qquad (6.52)$$

The expression in parentheses, $\gamma h \cos^2 \alpha - u$, represents the effective normal stress, σ', on the base of the slice. Therefore, we can also write

$$\frac{\sigma'}{\gamma h} = \cos^2 \alpha - \frac{u}{\gamma h} \qquad (6.53)$$

where the ratio $\sigma'/\gamma h$ is the ratio of effective normal stress to total overburden pressure and $u/\gamma h$ is the ratio of pore water pressure to total vertical overburden pressure. Let's now suppose that the pore water pres-

sure is equal to one-third the overburden pressure (i.e., $u/\gamma h = \frac{1}{3}$). Further suppose that the slip surface is inclined upward at an angle, α, of 60° from the horizontal. Then, from Eq. (6.53),

$$\frac{\sigma'}{\gamma h} = \cos^2 (60°) - \frac{1}{3} = -0.08 \qquad (6.54)$$

which indicates that the effective normal stress is negative! Negative values will exist for the effective stress in Eq. (6.52) as the pore water pressures become larger and the slip surface becomes steeper (α becomes large). The negative values occur because the forces on the sides of the slice are ignored in the Ordinary Method of Slices and there is nothing to counteract the pore water pressure.

By first expressing the weight of the slice in terms of an effective weight and then resolving forces perpendicular to the base of the slice; a better expression for the factor of safety can be obtained for the Ordinary Method of Slices (Turnbull and Hvorslev, 1967). The effective slice weight, W', is given by

$$W' = W - ub \qquad (6.55)$$

The term ub represents the vertical *uplift force* due to the pore water pressure on the bottom of the slice. The uplift force acts to counterbalance the weight of the slice. Resolving forces due to the effective stresses in a direction perpendicular to the base of the slice gives the effective normal force, N',

$$N' = W' \cos \alpha \qquad (6.56)$$

or from Eqs. (6.50) and (6.55),

$$N' = W \cos \alpha - u \, \Delta l \cos^2 \alpha \qquad (6.57)$$

The effective normal stress, σ', is obtained by dividing this force by the area of the base of the slice, which gives

$$\sigma' = \frac{W \cos \alpha}{\Delta l} - u \cos^2 \alpha \qquad (6.58)$$

Finally, introducing Eq. (6.58) for the effective normal stress into Eq. (6.46) for the factor of safety derived from moment equilibrium gives

$$F = \frac{\sum [c' \, \Delta l + (W \cos \alpha - u \, \Delta l \cos^2\alpha)\tan \phi']}{\sum W \sin \alpha}$$

$$(6.59)$$

This alternative expression for the factor of safety by the Ordinary Method of Slices does not result in negative effective stresses on the slip surface as long as the pore water pressures are less than the total vertical overburden pressure, a condition that must clearly exist for any reasonably stable slope.

Recapitulation

- The Ordinary Method of Slices assumes a circular slip surface and sums moments about the center of the circle; the method only satisfies moment equilibrium.
- For $\phi = 0$ the Ordinary Method of Slices gives exactly the same value for the factor of safety as does the Swedish Circle method.
- The Ordinary Method of Slices permits the factor of safety to be calculated directly. All of the other procedures of slices described subsequently require an iterative, trial-and-error solution for the factor of safety. Thus, the method is convenient for hand calculations.
- The Ordinary Method of Slices is less accurate than are other procedures of slices. The accuracy is less for effective stress analyses and decreases as the pore water pressures become larger.
- Accuracy of the Ordinary Method of Slices can be improved by using Eq. (6.59) rather than Eq. (6.48) for effective stress analyses.

Simplified Bishop Procedure

In the Simplified Bishop procedure the forces on the sides of the slice are assumed to be horizontal (i.e., there are no shear stresses between slices). Forces are summed in the vertical direction to satisfy equilibrium in this direction and to obtain an expression for the normal stress on the base of each slice. Referring to the slice shown in Figure 6.10 and resolving forces in the vertical direction, the following equilibrium equation can be written for forces in the vertical direction:

$$N \cos \alpha + S \sin \alpha - W = 0 \qquad (6.60)$$

Forces are considered positive when they act upward. The shear force in Eq. (6.60) is related to the shear stress by

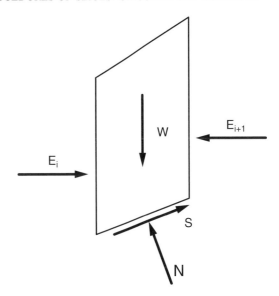

Figure 6.10 Slice with forces for the Simplified Bishop procedure.

$$S = \tau \, \Delta l \qquad (6.61)$$

or in terms of the shear strength and factor of safety [Eq. (6.2)], we can write

$$S = \frac{s \, \Delta l}{F} \qquad (6.62)$$

For shear strengths expressed in terms of effective stresses with the Mohr–Coulomb strength equation, we can write

$$S = \frac{1}{F}[c' \, \Delta l + (N - u \, \Delta l)\tan \phi'] \qquad (6.63)$$

Combining Eqs. (6.60) and (6.63) and solving for the normal force, N, we obtain

$$N = \frac{W - (1/F)(c' \, \Delta l - u \, \Delta l \tan\phi')\sin \alpha}{\cos \alpha + (\sin \alpha \tan \phi')/F} \qquad (6.64)$$

the effective normal stress on the base of the slice is given by

$$\sigma' = \frac{N}{\Delta l} - u \qquad (6.65)$$

Combining Eqs. (6.64) and (6.65) and introducing them into the equation for equilibrium of moments about the center of a circle for effective stresses [Eq. (6.46)], we can write, after rearranging terms,

$$F = \frac{\sum\left[\dfrac{c' \, \Delta l \cos \alpha + (W - u \, \Delta l \cos \alpha)\tan \phi'}{\cos \alpha + (\sin \alpha \tan \phi')/F}\right]}{\sum W \sin \alpha}$$

(6.66)

Equation (6.66) represents the equation for the factor of safety for the Simplified Bishop procedure.

Equation (6.66) was derived with the shear strength expressed in terms of effective stresses. The only distinction between total and effective stresses that is made in deriving any equation for the factor of safety is in whether the shear strength is expressed in terms of total stresses or effective stresses [e.g., Eq. (6.3) vs. Eq. (6.8)]. An equation for the factor of safety based on total stresses can be obtained from the equation for effective stresses by replacing the effective stress shear strength parameters (c' and ϕ') by their total stress equivalents (c and ϕ) and setting the pore water pressure term (u) to zero. Thus, the equation for the factor of safety in terms of total stresses for the Simplified Bishop procedure is

$$F = \frac{\sum\left[\dfrac{c \, \Delta l \cos \alpha + W \tan \phi}{\cos \alpha + (\sin \alpha \tan \phi)/F}\right]}{\sum W \sin \alpha}$$

(6.67)

In many problems, the shear strength will be expressed in terms of total stresses (e.g., UU strengths) for some materials and in terms of effective stresses (e.g., CD strengths) for other materials. Thus, the terms being summed in the numerator of Eq. (6.66) or (6.67) will contain a mixture of effective stresses and total stresses, depending on the applicable drainage conditions along the slip surface (base of each slice).

For saturated soils and undrained loading, the shear strength may be characterized using total stresses with $\phi = 0$. In this case Eq. (6.67) reduces further to

$$F = \frac{\sum c\Delta l}{\sum W \sin \alpha}$$

(6.68)

Equation (6.68) is identical to Eq. (6.42) derived for the Ordinary Method of Slices. In this case ($\phi = 0$) the logarithmic spiral, Swedish Circle, Ordinary Method of Slices, and Simplified Bishop procedures all give the same value for the factor of safety. In fact, any procedure that satisfies equilibrium of moments about the center of a circular slip surface will give the

same value for the factor of safety for $\phi = 0$ conditions.

Although the Simplified Bishop procedure does not satisfy complete static equilibrium, the procedure gives relatively accurate values for the factor of safety. Bishop (1955) showed that the procedure gives improved results over the Ordinary Method of Slices, especially when analyses are being performed using effective stresses and the pore water pressure are relatively high. Also, good agreement has been shown between the factors of safety calculated by the Simplified Bishop procedure and limit equilibrium procedures that fully satisfy static equilibrium (Bishop, 1955; Fredlund and Krahn, 1977; Duncan and Wright, 1980). Also Wright et al. (1973) have shown that the factor of safety calculated by the Simplified Bishop procedure agrees favorably (within about 5%) with the factor of safety calculated using stresses computed independently using finite element procedures. The primary practical limitation of the Simplified Bishop procedure is that it is restricted to circular slip surfaces.

Recapitulation

- The Simplified Bishop procedure assumes a circular slip surface and horizontal forces between slices. Moment equilibrium about the center of the circle and force equilibrium in the vertical direction for each slice are satisfied.
- For $\phi = 0$ the Simplified Bishop procedure gives the same, identical value for the factor of safety as the Swedish Circle and Ordinary Method of Slices procedures because all these procedures satisfy moment equilibrium about the center of a circle and that produces a unique value for the factor of safety.
- The Simplified Bishop procedure is more accurate than the Ordinary Method of Slices, especially for effective stress analyses with high pore water pressures.

Inclusion of Additional Known Forces

The equations presented above for the factor of safety by both the Ordinary Method of Slices and Simplified Bishop procedures are based on the assumption that the only driving forces are due to the weight of the soil mass, and the only resisting forces are those due to the shear strength of the soil. Frequently, there are additional known driving and resisting forces: Slopes that have water adjacent to them or support additional surcharge loads due to traffic or stockpiled materials are subjected to additional loads. Also, the pseudostatic analyses for seismic loading, which are discussed in Chapter 10, involve an additional horizontal body force on the slices to represent earthquake loading. Finally, stability computations for reinforced slopes include additional forces to represent the reinforcement. All these forces are considered to be known forces; that is, they are prescribed as part of the definition of the problem and must be included in the equilibrium equations to compute the factor of safety. However, because the additional forces are known, they can be included in the equilibrium equations without requiring any additional assumptions to achieve a statically determinate solution. The inclusion of additional forces is shown below using the Simplified Bishop procedure for illustration.

Consider first the equation of overall moment equilibrium about the center of a circle. With only forces due to the weight and shear strength of the soil, equilibrium is expressed by

$$r \sum \frac{s_i \, \Delta l_i}{F} - r \sum W_i \sin \alpha_i = 0 \qquad (6.69)$$

where counterclockwise (resisting) moments are considered positive and clockwise (overturning) moments are considered negative. Instead, if there are also seismic forces, kW_i, and forces due to soil reinforcement, T_i (Figure 6.11), the equilibrium equation might be written as

$$r \sum \frac{s_i \, \Delta l_i}{F} - r \sum W_i \sin \alpha_i$$
$$- \sum kW_i d_i + \sum T_i h_i = 0 \quad (6.70)$$

where k is the seismic coefficient, d_i the vertical distance between the center of the circle and the center of gravity of the slice, T_i represents the force in the

reinforcement where the reinforcement crosses the slip surface, and h_i is the moment arm of the reinforcement force about the center of the circle. The summation, $\Sigma \, kW_i d_i$, is performed for all slices, while the summation, $\Sigma \, T_i h_i$, applies only to slices where the reinforcement intersects the slip surface. The reinforcement shown in Figure 6.11 is horizontal, and thus the moment arm is simply the vertical distance between the reinforcement and the center of the circle. This, however, may not always be the case. For example, in Figure 6.12 a slice is shown where the reinforcement force is inclined at an angle, ψ, from the horizontal.

Because the forces represented by the last two summations in Eq. (6.70) involve only known quantities, it is convenient to replace these summations by a single term, M_n, that represents the net moment due to the known forces. The known forces may include seismic forces, reinforcement forces, and in the case of the slice shown in Figure 6.12, an additional moment due to a force, P, on the top of the slice. Equation (6.70) is then written as

$$r \sum \frac{s_i \, \Delta l_i}{F} - r \sum W_i \sin \alpha_i + M_n = 0 \quad (6.71)$$

Positive values for M_n represent a net counterclockwise moment; negative values represent a net clockwise mo-

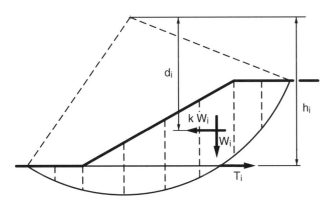

Figure 6.11 Slope with additional known seismic and reinforcement forces.

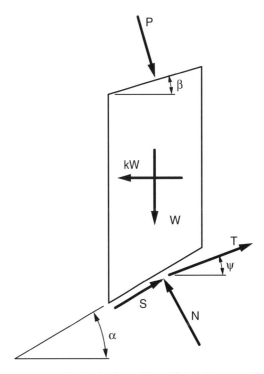

Figure 6.12 Individual slice with additional known forces.

ment. The equation for the factor of safety that satisfies moment equilibrium then becomes

$$F = \frac{\sum s_i \, \Delta l_i}{\sum W_i \sin \alpha_i - M_n/r} \qquad (6.72)$$

If the shear strength, s, is expressed in terms of effective stresses, and the subscripts i are now dropped with the understanding that the terms inside each summation apply to an individual slice, Eq. (6.72) can be written as

$$F = \frac{\sum [c' + (\sigma - u) \tan\phi'] \, \Delta l}{\sum W \sin \alpha - M_n/r} \qquad (6.73)$$

To determine the normal stress, $\sigma \, (= N/\Delta l)$ in Eq. (6.73), the equation for equilibrium of forces in the vertical direction is used again. Suppose that the slice contains the known forces shown in Figure 6.12. The known forces in this instance consist of a seismic force, kW, a force, P, due to water loads on the surface of the slope, and force, T, due to reinforcement intersecting the base of the slice. The force P acts perpendicular to the top of the slice, and the reinforcement force is inclined at an angle, ψ, from the horizontal. Summation of forces in the vertical direction gives

$$N \cos \alpha + S \sin \alpha - W - P \cos \beta + T \sin \psi = 0 \qquad (6.74)$$

where β is the inclination of the top of the slice and ψ represents the inclination of the reinforcement from the horizontal. Equation (6.74) is based on the Simplified Bishop assumption that there are no shear forces on the sides of the slice (i.e., the interslice forces are horizontal). Note that because the seismic force is assumed to be horizontal, it does not contribute to equilibrium in the vertical direction; however, if there were a seismic force component in the vertical direction, the vertical component would appear in Eq. (6.74). It is again convenient to combine the contribution of the known forces into a single quantity, represented in this case by a vertical force, F_v, which includes the vertical components of all of the known forces except the slice weight,[2] that is,

$$F_v = -P \cos \beta + T \sin \psi \qquad (6.75)$$

Positive forces are assumed to act upward; negative forces act downward. The summation of forces in the vertical direction then becomes

$$N \cos \alpha + S \sin \alpha - W + F_v = 0 \qquad (6.76)$$

Introducing the Mohr–Coulomb strength equation, which includes the definition of the factor of safety [Eq. (6.63)], into Eq. (6.76) and then solving for the normal force, N, gives

$$N = \frac{W - F_v - (1/F)(c' \, \Delta l - u \, \Delta l \tan \phi')\sin \alpha}{\cos \alpha + (\sin \alpha \tan \phi')/F} \qquad (6.77)$$

Combining Eq. (6.77) for the normal force with Eq. (6.73) for the factor of safety then gives

$$F = \frac{\sum \left[\dfrac{c'\Delta l \cos \alpha + (W - F_v - u\Delta l \cos \alpha)\tan \phi'}{\cos \alpha + (\sin \alpha \tan \phi')/F} \right]}{\sum W \sin \alpha - M_n/r} \qquad (6.78)$$

The term M_n represents moments due to all known forces except the weight, including moments produced by the seismic forces (kW), external loads (P), and reinforcement (T) on the slice in Figure 6.12.

Equation (6.78) is the equation for the factor of safety by the Simplified Bishop procedure extended to include additional known forces like those due to seismic loads, reinforcement, and external water pressures. However, because only vertical and not horizontal force equilibrium is considered, the method largely neglects any contribution to the normal stresses on the slip surface from horizontal forces, such as a seismic force and horizontal reinforcement forces. Horizontal forces are included in Eq. (6.78) only indirectly through their contribution to the moment, M_n. Consequently, care should be exercised if the Simplified Bishop procedure is used where there are significant horizontal forces that contribute to stability. However, it has been the writers' experience that even when there are significant horizontal forces, the Simplified Bishop procedure produces results comparable to those obtained by procedures that satisfy all conditions of equilibrium.

The Simplified Bishop procedure is often used for analysis of reinforced slopes. If the reinforcement is horizontal, the reinforcement contributes in the equa-

---

[2] The weight, W, could also be included in the force, F_v, but for now the weight will be kept separate to make these equations more easily compared with those derived previously with no known forces except the slice weight.

tion of moment equilibrium but does not contribute in the equation for equilibrium of forces in the vertical direction. Thus, the effect of the reinforcement can be neglected in the equation of vertical force equilibrium [Eq. (6.74)]. However, if the reinforcement is inclined, the reinforcement contributes to both moment equilibrium and vertical force equilibrium. Some engineers have ignored the contribution of inclined reinforcement in the equation of vertical force equilibrium [Eq. (6.74)], while others have included its effect. Consequently, different results have been obtained depending on whether or not the contribution of vertical reinforcement forces is included in the equation of vertical force equilibrium (Wright and Duncan, 1991). It is recommended that the contribution always be included, as suggested by Eq. (6.74), and when reviewing the work of others it should be determined whether or not the force has been included.

Equations similar to those presented above for the Simplified Bishop procedure can be derived using the Ordinary Method of Slices. However, because of its relative inaccuracy, the Ordinary Method of Slices is generally not used for analyses of more complex conditions, such as those involving seismic loading or reinforcement. Therefore, the appropriate equations for additional known loads with the Ordinary Method of Slices are not presented here.

Complete Bishop Procedure

Bishop (1955) originally presented two different procedures for slope stability analysis. One procedure is the "Simplified" procedure described above; the other procedure considered all of the unknown forces acting on a slice and made sufficient assumptions to fully satisfy static equilibrium. The second procedure is often referred to as the *Complete Bishop procedure*. For the complete procedure, Bishop outlined what steps and assumptions would be necessary to fully satisfy static equilibrium; however, no specific assumptions or details were stated. In fact, Bishop's second procedure was similar to a procedure that Fellenius (1936) described much earlier. Neither of these "procedures" consists of a well-defined set of assumptions and steps like those of the other procedures described in this chapter. Because neither the Complete Bishop procedure nor Fellenius's rigorous method has been described completely, these methods are not considered further. Since the pioneering contributions of Bishop and Fellenius, several procedures have been developed that set forth a distinct set of assumptions and steps for satisfying static equilibrium. These newer procedures are discussed later in this chapter.

PROCEDURES OF SLICES: NONCIRCULAR SLIP SURFACES

Up to this point, all of the procedures of slices as well as the single free-body procedures that have been presented are based on relatively simple shapes for the slip surface: a plane, a logarithmic spiral, or a circle. Many times the slip surface is more complex, often following zones or layers of relatively weak soil or weak interfaces between soil and other materials, such as geosynthetics. In such cases it is necessary to compute stability using more complex shapes for the slip surface. Several procedures have been developed for analyses of more complex, noncircular slip surfaces. These procedures are all procedures of slices. Some of the procedures consider all of the conditions of static equilibrium; others consider only some of them. Several procedures are based on satisfying only the requirements of force equilibrium; these are known as *force equilibrium procedures*. Most of the other procedures for analysis with noncircular slip surfaces consider all of the requirements for static equilibrium and are referred to as *complete equilibrium procedures*. The force equilibrium and complete equilibrium procedures are discussed separately below.

Force Equilibrium (Only) Procedures

Force equilibrium procedures satisfy only the conditions of force equilibrium and ignore moment equilibrium. These procedures use the equations for equilibrium of forces in two mutually perpendicular directions to compute the factor of safety, the forces on the base of each slice, and the resultant interslice forces. Usually, forces are resolved either vertically and horizontally or parallel and perpendicular to the base of the slice. To obtain a solution that is statically determinate (equal number of equations and unknowns), the inclinations of the forces between each slice are assumed. Once the inclination of the force is assumed, the factor of safety can be calculated.

Interslice force assumptions.   Various authors have suggested different assumptions for the interslice force inclinations to be used in force equilibrium procedures. Three of the most recognized assumptions are summarized in Table 6.1 along with the names commonly associated with each assumption.

Of the three assumptions shown in Table 6.1, the one suggested by Lowe and Karafiath (1959), that the interslice forces act at the average of the inclination of the slope and slip surface, seems to produce the best results. Factors of safety calculated using Lowe and Karafiath's procedure are generally in closest agree-

Table 6.1 Interslice Force Assumptions Used in Force Equilibrium Procedures

Procedure/assumption	Description
Lowe and Karafiath (Lowe and Karafiath, 1959)	The interslice forces are assumed to be inclined at the average slope of the ground surface and slip surface. The inclination varies from slice to slice, depending on where the slice boundaries are located.
Simplified Janbu (Janbu et al., 1956; Janbu, 1973)	The side forces are horizontal; there is no shear stress between slices. Correction factors are used to adjust (increase) the factor of safety to more reasonable values.
U.S. Army Corps of Engineers' modified Swedish method (U.S. Army Corps of Engineers, 1970).	The side forces are parallel to the average embankment slope. Although not clearly stated in their 1970 manual, the Corps of Engineers has established that the interslice force inclination will be the same for all slices[a].

[a]During the development of the UTEXAS2 and UTEXAS3 slope stability software, the U.S. Army Corps of Engineers' CAGE Committee made the decision that in the Modified Swedish procedure, all side forces would be assumed to be parrallel.

ment with the factors of safety calculated using procedures that satisfy complete equilibrium.

The Simplified Janbu procedure is based on the assumption that the interslice forces are horizontal. This assumption alone almost always produces factors of safety that are smaller than those obtained by more rigorous procedures that satisfy complete equilibrium. To account for this, Janbu et al. (1956) proposed the correction factors shown in Figure 6.13. These correction factors are based on a number of slope stability computations using both the simplified procedure with horizontal interslice forces and the more rigorous GPS procedure described later. The correction factors are only approximate, being based on analyses of 30 to 40 cases; however, the correction factors seem to provide an improved value for the factor of safety for many slopes.[3] Caution should be used in evaluating analyses that are reported using the Simplified Janbu procedure: Some analyses and computer programs automatically apply the correction factor to the computed factor of safety; others do not. Whenever results are reported for the Simplified Janbu procedure, it should be determined whether the correction factor has been applied, as the correction can have a noticeable effect on the results.

[3]Janbu (1973) indicates that the correction factors are based on a comprehensive investigation of some 40 different soil profiles. In a more recent personal communication (2003) between the authors and Janbu, Janbu confirmed that the correction factors were established based on analyses of approximately 30 cases in the files of the Norwegian Geotechnical Institute.

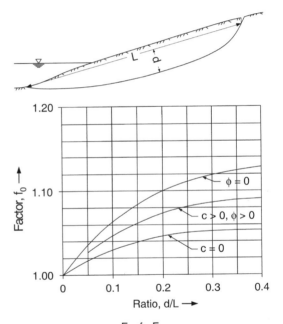

$$F = f_0 \cdot F_0$$

F_0 = Factor of safety from force equilibrium solution with horizontal interslice forces

Figure 6.13 Correction factors for Janbu's simplified procedure of slices.

The U.S. Army Corps of Engineers' Modified Swedish procedure is based on the assumption that the interslice forces act at the "average inclination of the embankment slope" (U.S. Army Corps of Engineers, 1970). This can be interpreted in at least three different ways, as illustrated in Figure 6.14. As shown in the

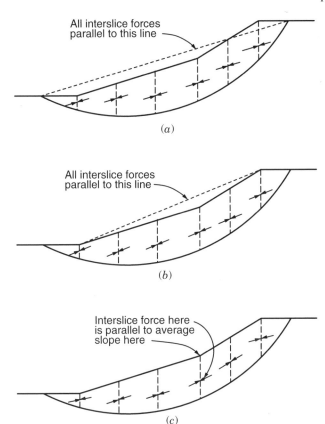

All interslice forces
parallel to this line

(a)

All interslice forces
parallel to this line

(b)

Interslice force here
is parallel to average
slope here

(c)

Figure 6.14 Candidate interpretations of the U.S. Army Corps of Engineers' assumption for the interslice force inclinations in the Modified Swedish procedure—average inclinations of the embankment slope: (a) interpretation 1; (b) interpretation 2; (c) interpretation 3.

figure, the interslice force inclinations may be interpreted either as being the same for every slice (Figure 6.14a and b) or differing from slice to slice (Figure 6.14c). To the writers' knowledge all three interpretations that are shown in Figure 6.14 have been used. However, currently the standard practice is to assume that the interslice forces all have the same inclination.[4] Regardless of the interpretation, any of the three assumptions illustrated in Figure 6.14 can lead to factors of safety that are larger than those obtained by more rigorous procedures of analysis that fully satisfy moment equilibrium, and thus the factors of safety are unconservative. Accordingly, some engineers elect to assume flatter angles for the interslice forces than those suggested by the U.S. Army Corps of Engineers

(1970).[5] If in the extreme the interslice forces are assumed to be horizontal, the Modified Swedish procedure becomes identical to the Simplified Janbu procedure (without the correction factor) and the procedure then probably underestimates the factor of safety. As with the Simplified Janbu procedure, it is recommended that the details of the interslice force assumptions be determined whenever reviewing the results of calculations performed by others using the Modified Swedish procedure.

One of the principal limitations of force equilibrium procedures is that the procedures are sensitive to what is assumed for the interslice force inclination. To illustrate this sensitivity, calculations were performed for the short-term stability of the homogeneous slope shown in Figure 6.15a. The slope is composed of saturated clay. The undrained shear strength is 400 psf at the elevation of the crest of the slope and increases linearly with depth below the crest at the rate of 7.5 psf per foot of depth. Stability calculations were performed using force equilibrium procedures with parallel interslice forces (i.e., all interslice forces had the same inclination). The interslice force inclination was varied from horizontal to 21.8°. An inclination of 21.8° represents interslice forces parallel to the slope face—the Corps of Engineers' Modified Swedish assumption. For each interslice force inclination assumed, the minimum factor of safety was computed using circular slip surfaces for simplicity. The factors of safety computed are plotted in Figure 6.15b versus the assumed interslice force inclination. The values range from approximately 1.38 to 1.74, a difference of approximately 25%. As discussed earlier, there is a unique value for the factor of safety calculated from moment equilibrium that satisfies complete static equilibrium for circular slip surfaces when $\phi = 0$. The factor of safety for moment equilibrium is 1.50. The factors of safety computed from the force equilibrium solutions where the interslice force inclinations were varied range from approximately 8% less (Simplified Janbu without correction factor) to 16% greater (Corps of Engineers Modified Swedish procedure) than the value satisfying complete static equilibrium. These differences are for the relatively simple slope and problem chosen for illustration; even larger differences should be anticipated for other slopes and soil properties.

The factor of safety computed by Spencer's procedure, which is described later in this chapter, is also

[4] During development of the UTEXAS2 and UTEXAS3 slope stability software, a decision was made by the Corps of Engineers' CAGE Committee that all interslice forces would be assumed to be parallel.

[5] In discussions with various Corps of Engineers' personnel, the writers have learned that flatter angles have been used at the engineers' discretion in recognition of the fact that steeper angles may lead to too high a value for the factor of safety.

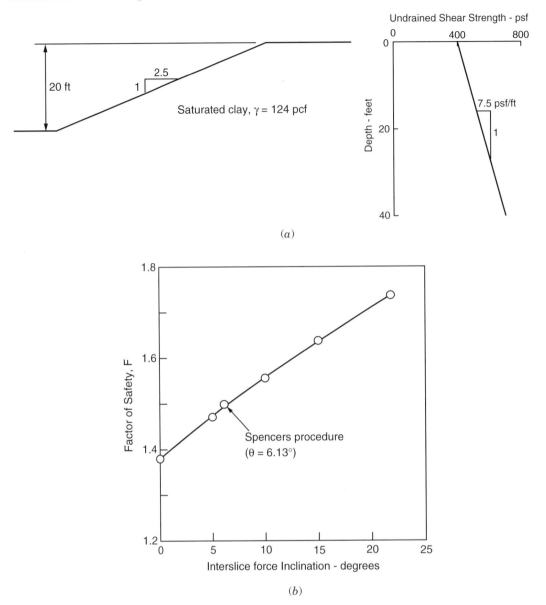

Figure 6.15 Influence of interslice force inclination on the computed factor of safety for force equilibrium with parallel interslice forces.

plotted in Figure 6.15*b*. Spencer's procedure assumes that the interslice forces are parallel but solves for the interslice force inclination that satisfies *both* force and moment equilibrium. For this problem the minimum factor of safety by Spencer's procedure is 1.50, as expected, and the corresponding interslice force inclination is 6.11°. The factor of safety is plotted in Figure 6.15*b* versus the interslice force inclination (6.13°). Somewhat surprisingly, the factor of safety by Spencer's procedure, which satisfies complete equilibrium, is larger than what would be expected for a force equilibrium solution for a interslice force inclination of

6.13°; the point corresponding to Spencer's procedure plots slightly above the line from the force equilibrium solutions. The difference occurs because the critical circle for Spencer's procedure, where the interslice force inclination may be different for each circle, is not the same as the critical circle for the force equilibrium solution where the interslice force inclination is the same for all circles.

Solution procedure: general.   All of the force equilibrium procedures require that trial-and-error methods be used to solve the equilibrium equations and calculate the factor of safety. Calculations using force equilibrium procedures have been performed for many years, and the procedures were first used long before electronic calculators and computers were readily available. Thus, originally the equilibrium equations were "solved" using trial-and-error graphical methods rather than numerical methods. The graphical methods consisted of constructing force equilibrium polygons (vector diagrams) repeatedly until the assumed factor of safety satisfied equilibrium (i.e., the force polygons "closed"). Such graphical procedures are now seldom used and, instead, have been replaced by hand calculators, spreadsheets, or more sophisticated software programs. However, the graphical methods provide a useful insight into the force equilibrium procedures and the results that are obtained. In some instance, construction of the force equilibrium polygons is helpful in understanding the forces acting to stabilize or destabilize a slope and can even provide insight into numerical problems that may develop. Graphical methods are also helpful in explaining the force equilibrium procedures.

Graphical solutions. A graphical solution for the factor of safety is begun by assuming a trial value for the factor of safety. Once a trial value is assumed, the equilibrium requirements for the first slice are used to determine the magnitudes of the unknown normal force on the base of the slice and the interslice force between the first and second slices. It is convenient to illustrate this procedure with the example shown in Figures 6.16, 6.17, and 6.18. The equilibrium force polygon for the first slice is shown in Figure 6.16. The forces on the first slice include:

1. The weight for the slice (W_1).
2. A force ($u \, \Delta l$) representing the effect of the pore water pressure on the base of the slice, which is considered separately from the effective normal force.
3. A force (R') representing the resultant due to the effective normal force (N') and the component of the shear force due to friction ($N' \tan \phi'_d$). The resulting force acts at an angle of ϕ'_d from a line perpendicular to the base of the slice. Because the factor of safety has been assumed, ϕ'_d is defined.
4. A force due to the mobilized cohesion ($c'_d \Delta l$).
5. A interslice force (Z_2) on the right side of the slice acting at an inclination, θ. The value of θ is assumed before starting a solution.

All of the information about the forces is known except for the magnitudes of the resulting forces R' on the slip surface and Z_2 on the vertical slice boundary. The directions of both of these forces are known. From the equilibrium force polygon the magnitudes of R' and Z_2 are then determined by the requirement that the force polygon must close. Once R' is found, the effective normal force, N', is simply calculated from $N' = R' \cos \phi'_d$.

For the second slice an undrained shear strength expressed in terms of total stresses is used. Because the soil is saturated, ϕ is zero. Thus, while effective stresses were used for the first slice, total stresses are used for the second slice. The forces on the second slice include:

1. The weight for the slice (W_2).
2. A force due to the mobilized cohesion ($c_d \, \Delta l$). Again, because a factor of safety has been assumed, c_d can be calculated.
3. A force, R, representing the resulting force due to the normal force, N, and the component of shear strength due to friction. However, because $\phi = 0$, the resultant force, R, is the same as the normal force, N.
4. The interslice forces (Z_2 and Z_3) on the left and right sides of the slice, respectively.

Again the only two unknown quantities are the magnitudes of the force, N, on the base of the slice and the force, Z_3, on the right side of the slice. Thus, the magnitude of N and Z_3 can be determined.

The steps above are repeated, constructing equilibrium force polygons slice-by-slice until the last slice is reached. For the last slice only the value of the resulting force, R, is unknown because there is no interslice force on the right of this slice. A value for this force (R) may or may not be found that closes the force polygon. As shown in Figure 6.18, the force polygon can be made to close only by introducing an additional force. The magnitude of the additional force required to close the force polygon will depend on the direction assumed. If all interslice forces are parallel, the additional force is generally assumed to have the same inclination as the interslice forces; if the inclination of the interslice forces varies from slice to slice, the imbalanced force is often assumed to be horizontal. In Figure 6.18 the additional force is assumed to be horizontal and is represented as $Z_{\text{imbalance}}$. The horizontal force indicates that the factor of safety that was assumed at the outset is not correct. The force $Z_{\text{imbalance}}$ provides a measure of the error in the assumed factor of safety. In the case of a slope facing to the left, like the one shown in Figure 6.18, if the imbalanced force acts to the left, it indicates that an additional force (one pushing the potential slide mass downslope) is needed

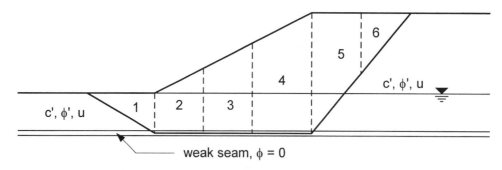

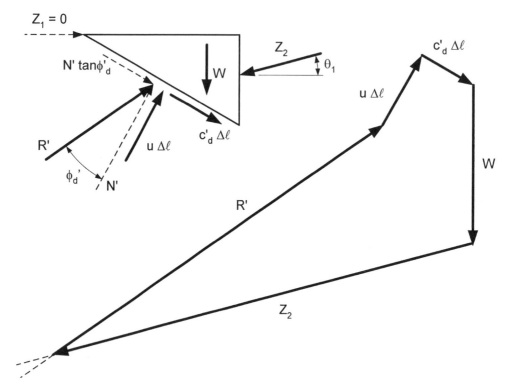

Figure 6.16   Force equilibrium polygon (vector diagram) of forces acting on the first slice for a force equilibrium solution by the graphical method.

to produce the factor of safety that was assumed. Because such a force does not actually exist, the value that was assumed for the factor of safety must have been too low, and a larger value of F should be assumed for the next trial.

In the graphical procedure, values for the factor of safety are assumed repeatedly and the equilibrium force polygons are drawn. This process is repeated until closure of the force polygons with negligible error (force imbalance) is achieved (i.e., until $Z_{imbalance} = 0$).

Although today, graphical procedures have largely been abandoned in favor of electronic means of calculation, the graphical procedures provide useful insight into the forces contributing to slope stability. The

relative magnitudes of forces, like those due to cohesion and friction, for example, provide insight into the relative importance of each. Even though force polygons may no longer be used as the means of solving for the factor of safety, they provide a useful means for examining a solution graphically.

Analytical solutions.   Today, most analyses performed using force equilibrium procedures are carried out by performing calculations using a spreadsheet or other computer program. In this case the equilibrium equations are written as algebraic equations. Consider the slice shown in Figure 6.19. Summation of forces in the vertical direction for an individual slice produces the following equilibrium equation:

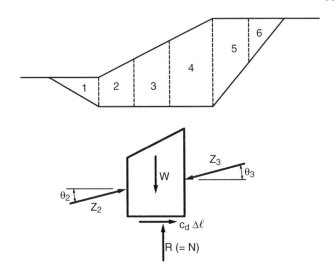

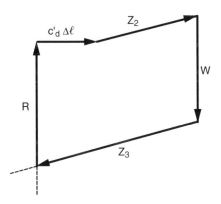

Figure 6.17 Force equilibrium polygon (vector diagram) of forces acting on the second slice for a force equilibrium solution by the graphical method.

The quantity F_h represents the net sum of all known forces acting on the slice in the horizontal direction; forces acting to the right are considered positive. If there are no seismic forces, external loads or reinforcement forces, the force, F_h, will be zero; for seismic loading alone, $F_h = -kW$.

Equations (6.79) and (6.80) can be combined with the Mohr–Coulomb equation for the shear force [Eq. (6.63)] to eliminate the shear and normal forces (S and N) and obtain the following equation for the interslice force, Z_{i+1}, on the right side of a slice:

$$Z_{i+1} = \frac{\begin{array}{c} F_v \sin \alpha + F_h \cos \alpha + Z_i \cos(\alpha - \theta) \\ -[F_v \cos \alpha - F_h \sin \alpha + u \, \Delta l \\ + Z_i \sin(\alpha - \theta)](\tan \phi'/F) + c' \, \Delta l/F \end{array}}{\cos(\alpha - \theta_{i+1}) + [\sin(\alpha - \theta_{i+1}) \tan \phi']/F}$$

(6.81)

By first assuming a trial value for the factor of safety, Eq. (6.81) is used to calculate the interslice force, Z_{i+1}, on the right of the first slice where $Z_i = 0$. Proceeding to the next slice, where Z_i is equal to the value of Z_{i+1} calculated for the previous slice, the interslice force on the right of the second slice is calculated. This process is repeated slice by slice for the rest of the slices from left to right until a force on the right of the last slice is calculated. If the force, Z_{i+1}, on the right of the last slice is essentially zero, the assumed factor of safety is correct because there is no "right side" on the last slice, which is triangular. If the force is not zero, a new trial value is assumed for the factor of safety and the process is repeated until the force on the right of the last slice is acceptably small.

Janbu's Generalized Procedure of Slices. At this point it is appropriate to return to the procedure known as Janbu's Generalized Procedure of Slices (GPS) (Janbu, 1954, 1973a). There has been some debate as to whether this procedure satisfies complete equilibrium or only force equilibrium. In the GPS procedure the vertical components of the interslice forces are assumed based on a numerical approximation of the following differential equation for equilibrium of moments for a slice of infinitesimal width[6]:

$$X = -E \tan \theta_t + h_t \frac{dE}{dX}$$

(6.82)

$$F_v + Z_i \sin \theta_i - Z_{i+1} \sin \theta_{i+1}$$

$$+ N \cos \alpha + S \sin \alpha = 0 \qquad (6.79)$$

where Z_i and θ_i represent the respective magnitudes and inclinations of the inteslice force at the left of the slice, Z_{i+1} and θ_{i+1} represent the corresponding values at the right of the slice, and F_v represents the sum of all known forces in the vertical direction, including the weight of the slice. In the absence of any surface loads and reinforcement forces, F_v is equal to $-W$. Forces are considered positive when they act upward. Summation of forces in the horizontal direction yields the following, second equation of force equilibrium:

$$F_h + Z_i \cos \theta_i - Z_{i+1} \cos \theta_{i+1}$$

$$- N \sin \alpha + S \cos \alpha = 0 \qquad (6.80)$$

---

[6]Additional terms appear in this equation when there are external forces and other known forces acting on the slice; these additional terms are omitted here for simplicity.

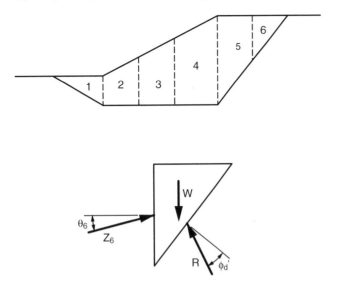

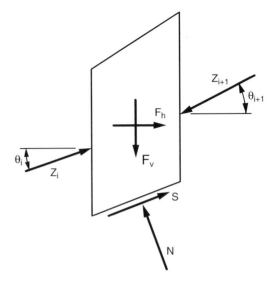

Figure 6.19 Slice with forces for force equilibrium procedures.

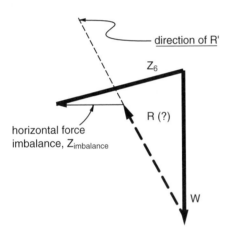

Figure 6.18 Force equilibrium polygon (vector diagram) of forces acting on the last slice for a force equilibrium solution by the graphical method.

The quantities X and E represent the vertical and horizontal components, respectively, of the interslice forces. The quantity h_t represents the height of the line of thrust above the slip surface. The line of thrust is the imaginary line drawn through the points where the interslice forces, E (or Z), act (Figure 6.20). The term θ_t is an angle, measured from the horizontal, that represents the slope of the line of thrust. In the GPS procedure the location of the line of thrust is assumed by the user. The derivative dE/dx in Eq. (6.82) is approximated numerically in the GPS procedure, and Eq. (6.82) is written in difference form as

$$X = -E \tan \theta_t + h_t \frac{E_{i+1} - E_{i-1}}{X_{i+1} - X_{i-1}} \qquad (6.83)$$

Equation (6.83) is based on considerations of moment equilibrium. However, Eq. (6.83) does not rigorously satisfy moment equilibrium in this discrete form; only Eq. (6.82) rigorously satisfies moment equilibrium. The other procedures of slices that are considered next satisfy complete equilibrium rigorously for a discrete set of slices. Thus, they are considered complete equilibrium procedures, whereas the GPS procedure is not.

The factor of safety is computed in the GPS procedure by performing successive force equilibrium solutions similar to those described in previous sections. Initially, the interslice forces are assumed to be horizontal and the unknown factor of safety and horizontal interslice forces, E, are calculated. Using this initial set of interslice forces, E, new interslice shear forces, X, are calculated from Eq. (6.83) and the force equilibrium solution is repeated. This process is repeated, each time making a revised estimate of the vertical component (X) of the interslice force and calculating the unknown factor of safety and horizontal interslice forces, until the solution converges (i.e., until there is not a significant change in the factor of safety). The GPS procedure frequently produces a factor of safety that is nearly identical to values calculated by procedures that rigorously satisfy complete static equilibrium. However, the procedure does not always produce a stable numerical solution that converges within an acceptably small error.

The GPS procedure satisfies moment equilibrium in only an approximate way [Eq. (6.83) rather than (6.82)]. It can be argued that once the approximate solution is obtained, a solution can be forced to satisfy moment equilibrium by summing moments for each

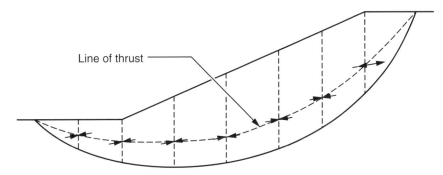

Figure 6.20 Line of thrust describing the locations of the interslice forces on slice boundaries.

slice individually and calculating a location for the normal force (N) on the base of the slice, which will then satisfy moment equilibrium rigorously. This, however, can be done with any of the force equilibrium procedures described in this chapter, but by summing moments only after a factor of safety is calculated, there is no influence of moment equilibrium on the computed factor of safety. Summing moments to compute the location of the normal force on the base of slices does not appear to be particularly useful.

Complete Equilibrium Procedures

Several different procedures of slices satisfy static equilibrium completely. Each of these procedures makes different assumptions to achieve a statically determinate solution. Several of these procedures are described in this section.

Spencer's procedure. Spencer's (1967) procedure is based on the assumption that the interslice forces are parallel (i.e., all interslice forces have the same inclination). The specific inclination of the interslice forces is unknown and is computed as one of the unknowns in the solution of the equilibrium equations. Spencer's procedure also assumes that the normal force (N) acts at the center of the base of each slice. This assumption has negligible influence on the computed values for the unknowns provided that a reasonably large number of slices is used; virtually all calculations with Spencer's procedure are performed by computer and a sufficiently large number of slices is easily attained.[7]

Spencer originally presented his procedure for circular slip surfaces, but the procedure is easily extended to noncircular slip surfaces. Noncircular slip surfaces are assumed here. In Spencer's procedure, two equilibrium equations are solved first. The equations represent overall force and moment equilibrium for the entire soil mass, consisting of all slices.[8] The two equilibrium equations are solved for the unknown factor of safety, F, and interslice force inclination, θ. Because the interslice forces are assumed to be parallel, there is only one unknown inclination for interslice forces to be solved for.

The equation for force equilibrium can be written as

$$\sum Q_i = 0 \qquad (6.84)$$

where Q_i is the resultant of the interslice forces, Z_i and Z_{i+1}, on the left and right, respectively, of the slice (Figure 6.21). That is,

$$Q_i = Z_i - Z_{i+1} \qquad (6.85)$$

Because the interslice forces are assumed to be parallel, Q_i, Z_i, and Z_{i+1} have the same direction and Q_i is simply the scalar difference between the interslice forces on the left and right of the slice.

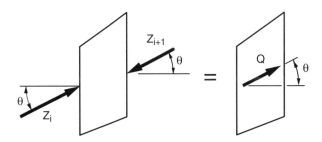

Figure 6.21 Interslice forces and resultant when interslice forces are parallel.

[8]The equations for overall force equilibrium in the horizontal and vertical directions reduce to a single equation when interslice forces are parallel. Thus, there is only one force equilibrium equation considered at this stage.

For moment equilibrium, moments can be summed about any arbitrary point. Taking moments about the origin ($x = 0$, $y = 0$) of a Cartesian coordinate system, the equation for moment equilibrium is expressed as

$$\sum Q(x_b \sin \theta - y_Q \cos \theta) = 0 \qquad (6.86)$$

where x_b is the x (horizontal) coordinate of the center of the base of the slice and y_Q is the y (vertical) co-ordinate of the point on the line of action of the force, Q, directly above the center of the base of the slice (Figure 6.22). The coordinate y_Q can be expressed in terms of the y coordinate of the point on the center of the base of the slice (y_b) by

$$y_Q = y_b + \frac{M_o}{Q \cos \theta} \qquad (6.87)$$

where M_o is the moment produced by any known forces about the center of the base of the slice. In the absence of forces due to seismic loads, loads on the surface of the slope, and any internal forces due to reinforcement, the moment M_o is zero and $y_Q = y_b$[9]. Each of the quantities in the summation shown for Eq. (6.86) represents the value for an individual slice. The subscript i has been omitted for simplicity and will be omitted in subsequent discussion with the understanding that the quantities Q, x_b, y_b, and so on, represent values for individual slices.

The expression for Q in the equilibrium equations [(6.84) and (6.86)] is obtained from the equations of force equilibrium for individual slices (Figure 6.23). Summing forces in directions perpendicular and parallel to the base of the slice gives the following two equilibrium equations:

$$N + F_v \cos \alpha - F_h \sin \alpha - Q \sin(\alpha - \theta) = 0 \qquad (6.88)$$

$$S + F_v \sin \alpha + F_h \cos \alpha + Q \cos(\alpha - \theta) = 0 \qquad (6.89)$$

The quantities F_h and F_v represent *all* known horizontal and vertical forces on the slice, including the weight of the slice, seismic loads, forces due to distributed and concentrated surface loads, and reinforcement forces. Combining these two force equilibrium equations [(6.88) and (6.89)] with the Mohr–Coulomb equation for the shear force, S [Eq. (6.63)] and solving for Q gives

$$Q = \frac{\begin{array}{c} -F_v \sin \alpha - F_h \cos \alpha - (c' \, \Delta l / F) \\ + (F_v \cos \alpha - F_h \sin \alpha + u \, \Delta l)(\tan \phi' / F) \end{array}}{\cos(\alpha - \theta) + [\sin(\alpha - \theta) \tan\phi' / F]} \qquad (6.90)$$

Equations (6.87) for y_Q and (6.90) for Q can be substituted into the equilibrium equations [(6.84) and (6.86)] to give two equations with two unknowns: the factor of safety, F, and the interslice force inclination,

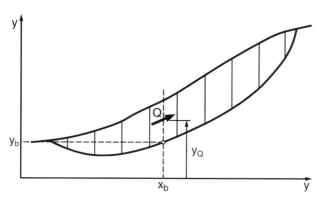

Figure 6.22 Coordinates for noncircular slip surface used in Spencer's procedure.

[9]The forces W, S, and N all act through a common point on the center of the base of the slice, and thus Q must also act through this point unless there are additional forces on the slice. In Spencer's (1967) original derivation M_o is zero, and thus $y_Q = y_b$.

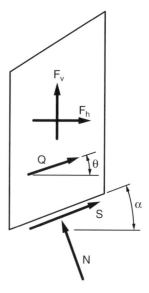

Figure 6.23 Slice with all known and unknown forces for Spencer's procedure.

θ. Trial-and-error procedures are used to solve Eqs. (6.84) and (6.86) for F and θ. Values of F and θ are assumed repeatedly until these two equations are satisfied within acceptable levels of error (force and moment imbalance). Once the factor of safety and interslice force inclination are computed, the equations of force and moment equilibrium for the individual slices are used to calculate the values of the normal force (N) on the base of the slice, the individual interslice force resultants (Z) between slices, and the location (y_t) of the interslice forces on the vertical boundary between the slices.

Morgenstern and price procedure. The Morgenstern and Price (1965) procedure assumes that the shear forces between slices are related to the normal forces as

$$X = \lambda f(x)E \qquad (6.91)$$

where X and E are the vertical and horizontal forces between slices, λ is an unknown scaling factor that is solved for as part of the unknowns, and $f(x)$ is an assumed function that has prescribed values at each slice boundary. In the Morgenstern and Price procedure the location of the normal force on the base of the slice is also explicitly or implicitly assumed. In the original formulation of the Morgenstern and Price procedure, stresses were integrated across each slice assuming that $f(x)$ varied linearly across the slice (Morgenstern and Price, 1967). This implicitly fixed the distribution of the normal stresses, including the location of the normal force on the base of the slice. In more recent implementations of the Morgenstern and Price procedure, discrete formulations have been used for slices and the location of the normal force has been assumed. Typically, the normal force is assumed to act at the midpoint of the base of the slice or at a point on the base of the slice that is directly below the center of gravity.

The unknowns that are solved for in the Morgenstern and Price procedure are the factor of safety (F), the scaling parameter (λ), the normal forces on the base of the slice (N), the horizontal interslice force (E), and the location of the interslice forces (line of thrust). The vertical component of the interslice force, X, is known [defined by Eq. (6.91)]; that is, once the unknowns are calculated using the equilibrium equations, the vertical component of the interslice forces is calculated from the independent equation (6.91).

Morgenstern and Price's procedure is similar to Spencer's procedure. The only difference in terms of unknowns is that Spencer's procedure involves a single interslice force inclination whereas Morgenstern and

Price's procedure involves a single "scaling" parameter, λ. If the function $f(x)$ is assumed to be constant in Morgenstern and Price's procedure it produces results essentially identical to those using Spencer's procedure.[10] The major difference between the two procedures is that Morgenstern and Price's procedure provides added flexibility in the assumptions for the interslice force inclinations. The added flexibility allows the assumption regarding the interslice forces to be changed. However, the assumptions generally appear to have little effect on the computed factor of safety when static equilibrium is satisfied, and thus there is little practical difference among Spencer's, Morgenstern and Price's, and all the other complete equilibrium procedures of slices.

Chen and Morgenstern procedure. The Chen and Morgenstern (1983) procedure represents a refinement of the Morgenstern and Price procedure that attempts to account better for the stresses at the ends of a slip surface. Chen and Morgenstern suggested that at the ends of the slip surface the interslice forces must become parallel to the slope. This leads to the following relationship between the shear (X) and horizontal (E) forces on the side of the slice:

$$X = [\lambda f(x) + f_0(x)]E \qquad (6.92)$$

where $f(x)$ and $f_0(x)$ are two separate functions that define the distribution of the interslice force inclinations. The function $f(x)$ is zero at each end of the slip surface, and the function $f_0(x)$ is equal to the tangent of the slope inclination at each end of the slip surface. The variations of both $f(x)$ and $f_0(x)$ between the two ends of the slip surface are assumed by the engineer. Chen and Morgenstern's procedure restricts the range of admissible interslice force inclinations and thus reduces the range of possible solutions.

Sarma's procedure. Sarma's (1973) procedure is different from all the other procedures discussed in this chapter because it considers the seismic coefficient (k) to be unknown, and the factor of safety is considered to be known. A value for the factor of safety is assumed and the seismic coefficient required to produce this factor of safety is solved for as an unknown. Usually, the factor of safety is assumed to be 1 and the seismic coefficient that is then calculated represents the seismic coefficient required to cause sliding, referred to in Chapter 10 as the *seismic yield coefficient*. In

---

[10]There may be subtle differences in the two procedures in this case because of slightly different locations assumed for the normal forces on the base of the slice. However, these differences are negligible for all practical purposes.

Sarma's procedure the shear force between slices is related to shear strength by the relationship

$$X = \lambda f(x) S_v \qquad (6.93)$$

where S_v is the available shear force on the slice boundary, λ is an unknown scaling parameter, and $f(x)$ is an assumed function with prescribed values at each vertical slice boundary. The shear force, S_v, depends on the shear strength parameters (c, c' and ϕ, ϕ') for the soil along the slice boundary and for frictional materials (ϕ, $\phi' > 0$) on the normal (horizontal) interslice force, E. For effective stress analyses the shear force also depends on the pore water pressure on the slice boundary.

Sarma's procedure was developed for evaluations of seismic stability and offers some advantage over other procedures for this purpose. In Sarma's procedure the seismic coefficient and other unknowns can be calculated directly; no iterative, trial-and-error procedure is required to calculate the unknowns. Sarma's procedure can also be used to calculate a factor of safety by repeatedly assuming different values for the factor of safety and calculating the seismic coefficient. The process is repeated until the value assumed for the seismic coefficient matches the value for which the factor of safety is desired. For slopes with no seismic loads the target seismic coefficient is zero. However, to compute a factor of safety, Sarma's procedure requires trial and error and thus offers no advantage over other complete equilibrium procedures.

The function, $f(x)$, and scaling parameter, λ, in Sarma's procedure are similar, but not identical, to the corresponding quantities in the Morgenstern and Price (1965) and the Chen and Morgenstern (1983) procedures. Depending on the assumption for $f(x)$ in these procedures, different inclinations will be found for the interslice forces. Depending on the range of assumed $f(x)$ patterns, there will probably be some overlap among solutions by the three (Morgenstern and Price, Chen and Morgenstern, and Sarma) procedures. Except for small differences in the values for $f(x)$ and λ, the three procedures should produce similar results for either the seismic coefficient required to produce a given factor of safety or the factor of safety corresponding to a given seismic coefficient. Sarma's procedure is easier to use to calculate a seismic coefficient for a prescribed factor of safety. On the other hand, the Morgenstern and Price procedure involves an assumption for the interslice forces that is much simpler and easier to use.

Sarma's procedure requires that the shear strength, S_v, along vertical slice boundaries be determined. For complex slopes with several materials and complex distributions of pore water pressure, Sarma's procedure becomes relatively complex. For frictional materials (ϕ, $\phi' > 0$), additional assumptions must then be made about what fractions of the total normal force (E) between slices is distributed to each different material along the slice boundary. If the shear strength is represented by effective stresses, the distribution of pore water pressures along the slice boundary must also be considered. This makes the procedure excessively complex for many practical problems and difficult to implement in computer software. The major utility of Sarma's procedure seems to be for hand calculations for slopes with relatively simple geometries.

Discussion. All of the complete equilibrium procedures of slices have been shown to give very similar values for the factor of safety (Fredlund and Krahn, 1977; Duncan and Wright, 1980). Thus, no complete equilibrium procedure is significantly more or less accurate than another. Spencer's procedure is the simplest of the complete equilibrium procedures for calculating the factor of safety, while Sarma's procedure may be simplest for calculating the seismic coefficient required to produce failure (i.e., the seismic yield coefficient).

Morgenstern and Price's and Chen and Morgenstern's procedures are the most rigorous and flexible of the complete equilibrium procedures and may be useful for cases where interslice forces might have a significant effect on stability. In most cases interslice force inclinations have little effect on the factor of safety computed, provided that complete equilibrium is satisfied. The writers are aware of few cases where the assumptions regarding interslice forces have a noticeable effect on either the computed factor of safety or the numerical stability of a solution in the complete equilibrium procedures. Two cases where the assumptions regarding interslice force inclinations can be important are:

1. When the slip surface is forced to change direction abruptly, due to the geometry and properties of the slope cross section (Figure 6.24a)
2. For slopes with significant forces due to reinforcement or external loads whose direction is very different from the usual direction of the interslice forces (Figure 6.24b and c).

In these two cases, procedures that will allow the interslice force assumptions to be varied are useful in

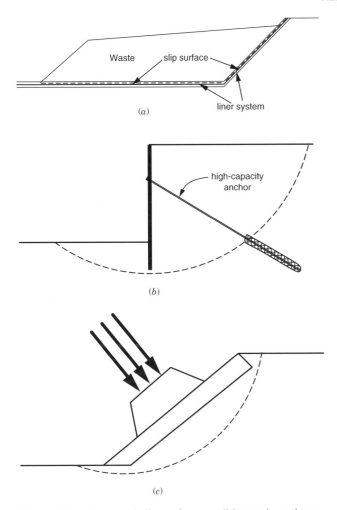

(a)

(b)

(c)

Figure 6.24 Slope and slip surface conditions where the assumptions pertaining to the interslice forces may have a significant effect on the results of slope stability computations by complete equilibrium procedures.

establishing the amount of uncertainty and probable ranges in the factor of safety.

ASSUMPTIONS, EQUILIBRIUM EQUATIONS, AND UNKNOWNS

As noted at the beginning of this chapter, all of the limit equilibrium procedures employ the equations of static equilibrium to compute a factor of safety. Assumptions are required to make the problem statically determinate and to obtain a balance between the number of equations and the number of unknowns that are solved for. Table 6.2 lists the various procedures discussed in this chapter along with the assumptions that are made, the equilibrium equations that are satisfied,

and the unknowns. In each case there is an equal number of equations and unknowns.

Thirteen different procedures of limit equilibrium analysis have been discussed in this chapter. In general, the procedures that satisfy complete static equilibrium are the most accurate and preferred when all other things are equal. However, there are numerous instances where simpler, though less accurate procedures are useful. All of the procedures examined in this chapter are summarized in Table 6.3 along with the range or conditions of practical usefulness for each procedure.

REPRESENTATION OF INTERSLICE FORCES (SIDE FORCES)

Up to this point the interslice forces have been assumed to represent all of the forces transmitted across a slice boundary, including the forces due to effective stresses in the soil, pressures in the pore water, and forces in any internal soil reinforcing. The interslice forces have been represented either in terms of their vertical and horizontal components (X and E), or the resultant force and its inclination (Z and θ). No distinction has been made between the various components that make up the total force on an interslice boundary. However, because some of the forces, such as the force due to water pressures, may be known, it is possible to separate the forces into various components, which are then treated independently.

Soil and Water Forces

For ordinary unreinforced slopes the interslice forces represent the forces due to effective stresses and pore water pressures. These forces can be represented separately as an effective force in the soil and a force in the water. Possible representations of the interslice forces are illustrated in Figure 6.25. When the forces are represented as total forces, as they have been in previous sections of the chapter, the shear and normal components, X and E, respectively, are treated as unknowns and their values are either assumed or calculated from the equilibrium equations (Figure 6.25b). If, instead, the forces are represented by the forces due to effective stresses (E' and X) plus the force due to water pressure (U), the water pressures are assumed to be known and the effective force components are treated as either unknowns that are calculated or they are assumed (Figure 6.25c). In either representation there are two forces that must be calculated or assumed on each slice boundary: E and X, or E' and X.

Table 6.2   Assumptions, Equilibrium Conditions, and Unknowns in Limit Equilibrium Procedures

Procedure	Assumptions	Equilibrium equations satisfied	Unknowns solved for
Infinite Slope	A slope of infinite extent; slip surface parallel to slope face.	1 Σ Forces perpendicular to slope 1 Σ Forces parallel to slope	1 Factor of safety (F) 1 Normal force on shear surface (N)
		2 Total equations (Moment equilibrium is implicitly satisfied)	2 Total unknowns
Logarithmic Spiral	The slip surface is a logarithmic spiral.	1 Σ Moments about center of spiral	1 Factor of safety (F)
		1 Total equations (Force equilibrium is implicitly satisfied)	1 Total unknown
Swedish Circle ($\phi = 0$)	The slip surface is circular; the friction angle is zero.	1 Σ Moments about center of circle	1 Factor of safety (F)
		1 Total equations (Force equilibrium is implicitly satisfied)	1 Total unknown
Ordinary Method of Slices (also known as Fellenius's Method; Swedish Method of Slices)	The slip surface is circular; the forces on the sides of the slices are neglected.	1 Σ Moments about center of circle	1 Factor of safety (F)
		1 Total equations	1 Total unknown
Simplified Bishop	The slip surface is circular; the forces on the sides of the slices are horizontal (i.e., there is no shear force between slices).	1 Σ Moments about center of circle n Σ Forces in the vertical direction.	1 Factor of safety (F) n Normal force on the base of slices (N)
		$n + 1$ Total equations	$n + 1$ total unknowns
Force Equilibrium (Lowe and Karafiath, Simplified Janbu, Corps of Engineer's Modified Swedish, Janbu's GPS procedure)	The inclinations of the interslice forces are assumed; assumptions vary with procedure.	n Σ Forces in the horizontal direction n Σ Forces in the vertical direction	1 Factor of safety (F) n Normal force on the base of slices (N) $n - 1$ Resultant interslice forces (Z)
		$2n$ Total equations	$2n$ Total unknowns
Spencer	Interslice forces are parallel, (i.e., all have the same inclination). The normal force (N) acts at the center of the base of the slice (typically).	n Σ Moments about any selected point n Σ Forces in the horizontal direction n Σ Forces the vertical direction	1 Factor of safety (F) 1 Interslice force inclination (θ) n Normal force on the base of slices (N) $n - 1$ Resultant interslice forces (Z) $n - 1$ Location of side forces (line of thrust)
		$3n$ Total equations	$3n$ Total unknowns

Table 6.2 (*Continued*)

Procedure	Assumptions	Equilibrium equations satisfied		Unknowns solved for	
Morgenstern and Price	Interslice shear force is related to interslice normal force by $X = \lambda f(x)E$; the normal force (N) acts at the center of the base of the slice (typically).	n	Σ Moments about any selected point	1	Factor of safety (F)
		n	Σ Forces in the horizontal direction	1	Interslice force inclination "scaling" factor (λ)
		n	Σ Forces in the vertical direction	n	Normal force on the base of slices (N)
		$3n$	Total equations	$n-1$	Horizontal interslice forces (E)
				$n-1$	Location of interslice forces (line of thrust)
				$3n$	Total unknowns
Chen and Morgenstern	Interslice shear force is related to interslice normal force by $X = [\lambda f(x) + f_o(x)]E$; the normal force ($N$) acts at the center of the base of the slice (typically).	n	Σ Moments about any selected point	1	Factor of safety (F)
		n	Σ Forces in the horizontal direction	1	Interslice force inclination "scaling" factor (λ)
		n	Σ Forces in the vertical direction	n	Normal force on the base of slices (N)
		$3n$	Total equations	$n-1$	Horizontal interslice forces (E)
				$n-1$	Location of interslice forces (line of thrust)
				$3n$	Total unknowns
Sarma	Interslice shear force is related to the interslice shear strength, S_v, by $X = \lambda f(x)S_v$; interslice shear strength depends on shear strength parameters, pore water pressures, and the horizontal component of interslice force; the normal force (N) acts at the center of the base of the slice (typically).	n	Σ Moments about any selected point	1	Seismic coefficient (k) [or factor of safety (F) if trial and error is used]
		n	Σ Forces in the horizontal direction	1	Interslice force scaling factor (λ)
		n	Σ Forces in the vertical direction	n	Normal force on the base of slices (N)
		$3n$	Total equations	$n-1$	Horizontal interslice forces (E)
				$n-1$	Location of side forces (line of thrust)
				$3n$	Total unknowns

Table 6.3   Summary of Procedures for Limit Equilibrium Slope Stability Analysis and Their Usefulness

Procedure	Use
Infinite Slope	Homogeneous cohesionless slopes and slopes where the stratigraphy restricts the slip surface to shallow depths and parallel to the slope face. Very accurate where applicable.
Logarithmic Spiral	Applicable to homogeneous slopes; accurate. Potentially useful for developing slope stability charts and used some in software for design of reinforced slopes.
Swedish Circle; $\phi = 0$ method	Applicable to slopes where $\phi = 0$ (i.e., undrained analyses of slopes in saturated clays). Relatively thick zones of weaker materials where the slip surface can be approximated by a circle.
Ordinary Method of Slices	Applicable to nonhomogeneous slopes and $c-\phi$ soils where slip surface can be approximated by a circle. Very convenient for hand calculations. Inaccurate for effective stress analyses with high pore water pressures.
Simplified Bishop procedure	Applicable to nonhomogeneous slopes and $c-\phi$ soils where slip surface can be approximated by a circle. More accurate than Ordinary Method of Slices, especially for analyses with high pore water pressures. Calculations feasible by hand or spreadsheet.
Force Equilibrium procedures (Lowe and Karafiath's side force assumption recommended)	Applicable to virtually all slope geometries and soil profiles. The only procedures suitable for hand calculations with noncircular slip surfaces. Less accurate than complete equilibrium procedures and results are sensitive to assumed inclinations for interslice forces.
Spencer's procedure	An accurate procedure applicable to virtually all slope geometries and soil profiles. The simplest complete equilibrium procedure for computing the factor of safety.
Morgenstern and Price's procedure	An accurate procedure applicable to virtually all slope geometries and soil profiles. Rigorous, well-established complete equilibrium procedure.
Chen and Morgenstern's procedure	Essentially an updated Morgenstern and Price procedure. A rigorous and accurate procedure applicable to any shape of slip surface and slope geometry, loads, etc.
Sarma's procedure	An accurate procedure applicable to virtually all slope geometries and soil profiles. A convenient complete equilibrium procedure for computing the seismic coefficient required to produce a given factor of safety. Side force assumptions are difficult to implement for any but simple slopes.

Fundamentally, it seems logical to represent the interslice forces by the known component due to water pressures and the unknown components due to effective stresses (Figure 6.25c). If the water pressures along an interslice boundary are simply hydrostatic, as suggested in Figure 6.26a, it is relatively easy to calculate the force due to water pressure and include it in stability computations. However, if the water pressures vary in a more complex manner, perhaps with distinctly different pressure regimes in different strata, as suggested in Figure 6.26b, it can be difficult to compute the force due to water pressures and its location. For complex groundwater conditions and subsurface soil profiles it is impractical to compute the forces due to water pressures, on each interslice boundary. Even though it may be possible to compute the force due to water pressures, the added effort complicates the logic and coding of computer programs that are used to per-

form such computations. For these practical reasons most slope stability formulations and computations are based on representing the interslice forces as total forces that include the force due to both the effective stress and the water pressures in the soil.

For procedures of slices that satisfy complete static equilibrium, it makes very little difference whether the unknown interslice forces include the water pressure force or the water pressure force is considered as a known force separately from the unknown force due to effective stress. The submerged slope shown in Figure 6.27 can be used to illustrate this. The factor of safety for this slope was calculated two different ways using Spencer's procedure of slices. In the first case the factor of safety was calculated with the interslice forces representing the total forces between slices. In the second case the water pressures on the side of the slice were computed and treated independent of the

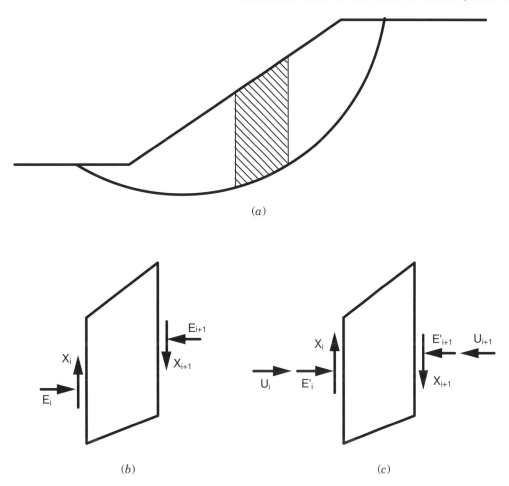

Figure 6.25 (*a*) Slope and slice. Representations of interslice forces as (*b*) total forces and (*c*) effective forces with water pressure forces.

unknown interslice forces due to the effective stresses. Computations were performed for two different heights of water above the top of the slope: 30 and 60 ft. Computations were performed using both Spencer's complete equilibrium procedure and the Corps of Engineers' Modified Swedish force equilibrium procedure. For the Modified Swedish procedure the interslice forces (total or effective) were assumed to be parallel to the slope face (i.e., they were inclined at an angle of 21.8° from the horizontal). The factors of safety calculated for each case by both procedures are summarized in Table 6.4. The inclinations, θ, of the interslice forces are also shown in Table 6.4. For Spencer's procedure it can be seen that the factors of safety calculated with total and effective interslice forces are almost identical. It can also be seen that the calculated inclination of the interslice forces is less when the interslice forces represent the total force (soil + water) rather than the effective force (soil only). The force due to water always acts horizontally, and it is there-

fore logical that the total force due to soil and water should be inclined at a flatter angle than the force due to effective stresses alone.

The factors of safety shown in Table 6.4 for the Modified Swedish procedure are very different depending on whether the interslice forces were represented using total or effective forces. Representation of the forces as effective forces with the water pressures treated separately produced factors of safety in close agreement with those calculated by Spencer's procedure, while representation of the interslice forces as total forces produced factors of safety that were from 47 to 60% higher! Representation of the interslice forces as total forces with the same inclination as the effective forces meant that the total forces were inclined more steeply than when the effective and water pressure forces were considered separately. As already shown, the steeper the inclination of the interslice forces, generally the higher the factor of safety (e.g., Figure 6.15). The results presented in Table 6.4 suggest

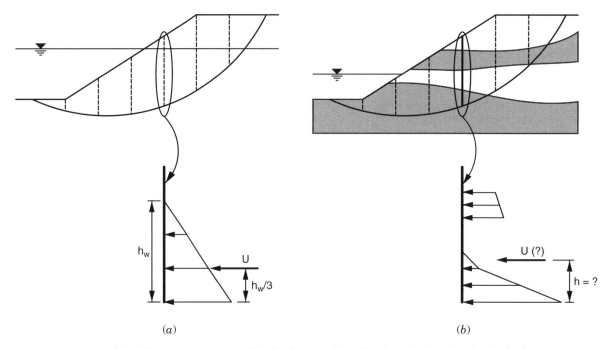

(a)   (b)

Figure 6.26   Pore water pressure distributions on interslice boundaries; (a) simple hydrostatic pressures; (b) complex groundwater and pore water pressure conditions.

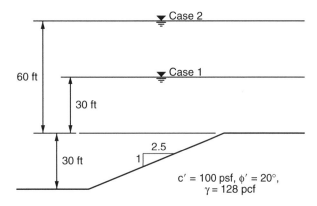

Figure 6.27   Submerged slope analyzed with total and effective stress representations of the interslice forces.

that it would be better to express the interslice forces in terms of effective forces and forces due to water pressures. However, as discussed earlier, this is difficult and impractical for complex slopes.

Soil-Water and Reinforcement Forces

For slopes like the one shown in Figure 6.28a, with reinforcement such as geogrids, geotextiles, piles, soil nails, and tieback anchors, the interslice forces include both the forces in the soil and water, as well as the forces transmitted across the interslice boundaries

through the reinforcing elements. This allows additional choices pertaining to the representation of the interslice forces. One possibility is to let the interslice forces represent the forces both in the soil and in the reinforcing (Figure 6.28b); another possibility is to separate the interslice forces due to the reinforcement from the forces due to the soil and water (Figure 6.28c). If the interslice forces represent all the forces between the slices, the total interslice forces (X and E, or Z and θ) are considered unknown. Once the interslice forces are calculated, the amount of force carried by just the soil and the water can be determined by subtracting the known forces due to the reinforcement from the total interslice forces. If, on the other hand, the reinforcement forces are considered separately from the forces due to the soil and the water, the reinforcement forces are determined first and treated as known interslice forces and the remaining (soil + water) interslice forces are treated as unknown forces. For hand calculations it is usually easiest (requires fewer calculations) to let the interslice forces represent all the force, soil + water + reinforcement, together as one set of forces. However, for calculations using a computer program, it is probably more appropriate to treat the forces in the reinforcement separately from the forces due to the soil and water.

To compute the stability of a reinforced slope in a computer program, it is necessary to define the forces

Table 6.4   Summary of Calculations for Slope Using Total Interslice Forces and Effective Interslice Forces with Water Pressures[a]

Procedure of analysis	30 ft of water above slope crest		60 ft of water above slope crest	
	Total forces	Effective forces and water forces	Total forces	Effective forces and water forces
Spencer's	$F = 1.60$	$F = 1.60$	$F = 1.60$	$F = 1.60$
	$\theta = 1.7°$	$\theta = 17.2°$	$\theta = 1.1°$	$\theta = 17.2°$
Corps of Engineers'	$F = 2.38$	$F = 1.62$	$F = 2.60$	$F = 1.62$
Modified Swedish	$\theta = 21.8°$	$\theta = 21.8°$	$\theta = 21.8°$	$\theta = 21.8°$

[a] θ, inclination of interslice forces.

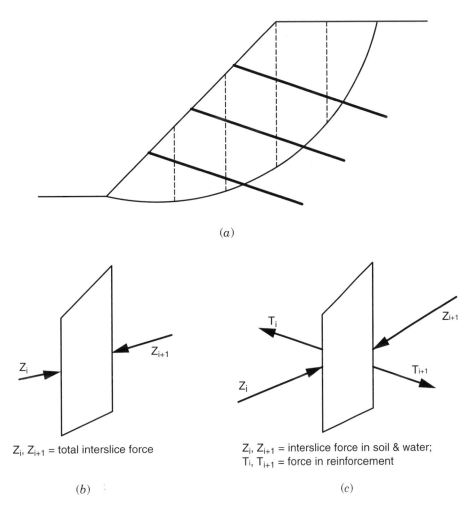

(a)

Z_i, Z_{i+1} = total interslice force

(b)

Z_i, Z_{i+1} = interslice force in soil & water;
T_i, T_{i+1} = force in reinforcement

(c)

Figure 6.28   Alternative representations of interslice forces for a slope with internal reinforcing elements; (a) slope with internal reinforcing elements; (b) reinforcement and soil + water forces combined; (c) reinforcement and soil + water forces considered separately.

for any point along the reinforcement and to be able
to compute the force wherever the reinforcement
crosses a boundary. There is very little difference between the scheme used to compute the force where the
reinforcement crosses the slip surface and the scheme
used to compute the force where the reinforcement
crosses a slice boundary. Therefore, once a suitable
computational scheme has been developed, it can be
used to compute the reinforcement force at points
where the reinforcement crosses both the interslice
boundaries and the slip surface. Unlike calculating the
water pressure force on an interslice boundary, it is
relatively easy to compute the reinforcement force on
an interslice boundary. Thus, it is also relatively easy
to consider the reinforcement force separately from the
remaining interslice forces. Also, the reinforcement
forces may act in directions that are very different from
the interslice forces in the soil and water. By treating
the reinforcement forces separately from the forces due
to the soil and water, more rational assumptions can
be made about the inclinations of the interslice forces.

Separation of the reinforcement force from the force
due to the soil and water at the interslice boundaries
also produces a more realistic set of internal forces and
usually a more stable numerical solution to the equilibrium equations. If the reinforcement forces are applied where the reinforcement intersects both the slip
surface and each interslice boundary, the forces that
are applied to each slice will be more realistic. For
example, for a slice like the one shown in Figure 6.29b,
the actual force exerted on the slice by the reinforcement will be equal to the difference between the force
where the reinforcing element enters the slice at the
left interslice boundary and exits the slice (slip surface)
at the bottom. The net reinforcement force, $T_{i+1} - T_i$,
on the slice in this case may be quite small. In contrast,
if only the reinforcement force, T_{i+1}, acting on the base
of the slice is applied to the slice, the applied reinforcement force could be quite large. For equilibrium
to be satisfied with the reinforcement force applied
only to the base of the slice, the unknown interslice
force on the left of the slice might need to be much
larger than the unknown interslice force on the right
of the slice, and the inclination of the interslice forces
on the left and right of the slice might be very different. Such abrupt changes in the magnitude and inclination of the interslice forces can lead to numerical
problems in the solution of the equilibrium equations.

The reason for separating the interslice forces due
to reinforcement from those in the soil can be further
seen by considering a slice such as the one shown in
Figure 6.29c where the reinforcing element passes entirely across the slice, intersecting both vertical bound-

aries. The actual force exerted on the slice by the
reinforcement will be the force that is transferred by
shear (load transfer) between the reinforcing element
and the soil as the reinforcing element passes through
the slice. This force is properly represented as the difference between the forces in the reinforcing element
at the two sides of the slice (i.e., $T_{i+1} - T_i$). If the
reinforcement forces on each side of the slice (T_i and
T_{i+1}) are computed and applied as known forces, separately from the unknown interslice forces in the soil,
the slice will be assured of receiving the proper contribution of the reinforcement.

COMPUTATIONS WITH ANISOTROPIC SHEAR STRENGTHS

Earlier, the shear strength has been expressed by appropriate values of cohesion (c, c') and friction angle
(ϕ, ϕ'). If the shear strengths are anisotropic, the values of cohesion and friction angle may depend on the
orientation of the failure plane. Typically, the shear
strength in such cases is the undrained shear strength
and ϕ will be equal to zero, so the anisotropic shear
strength is expressed by the variation in undrained
shear strength, s_u, with orientation of the failure plane.
To perform stability computations when the strengths
vary with the failure plane orientation, it is simply necessary to assign appropriate values of cohesion and
friction to each slice based on the inclination of the
bottom of the slice (slip surface). Once the strengths
are assigned, computations proceed in the normal
manner.

COMPUTATIONS WITH CURVED FAILURE ENVELOPES

All of the equations presented previously assume that
the shear strength will be represented by a linear
Mohr–Coulomb failure envelope defined by an intercept (cohesion) and a slope (friction angle). However,
failure envelopes are often curved. Curved failure envelopes require somewhat more effort in a stability
analysis. If the Mohr failure envelope is curved, the
shear strength varies with the normal stress (σ, σ'),
and the normal stress must be known before the shear
strength is known. Only the Ordinary Method of Slices
allows the normal stress to be computed without knowing the shear strength; all other procedures require that
the shear strength be known before the normal stress

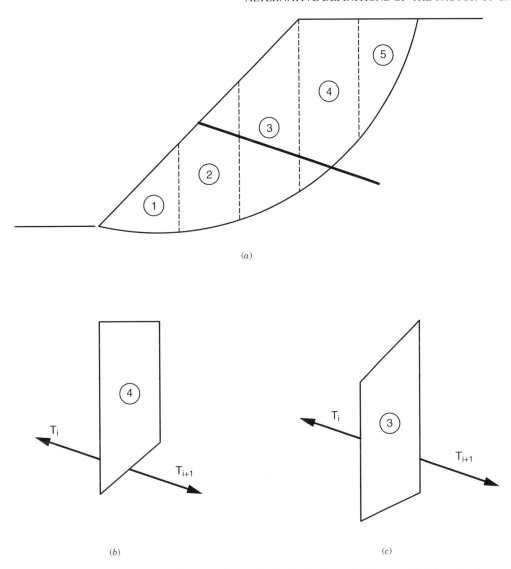

Figure 6.29 Reinforcement forces acting on individual slices: (*a*) slope with internal reinforcing elements; (*b*) reinforcement forces for slice 4; (*c*) reinforcement forces for slice 3.

can be computed (solved for). In these procedures it is necessary to estimate the shear strength first and then perform the necessary stability computations to calculate the normal stress (and factor of safety). Once the normal stress is found, new shear strengths can be estimated and the process repeated until convergence is reached. Shear strengths may be assigned as a pure cohesion ($\phi = 0$), based on the estimated normal stress (Figure 6.30*a*). Alternatively, the shear strength may be represented by a cohesion and a friction angle representing a failure envelope tangent to the curved envelope at the estimated normal stress (Figure 6.30*b*). The second approach of using an envelope tangent to the curved envelope is more complex but produces

faster convergence in the necessary trial-and-error procedures. However, both methods produce identical results once convergence is reached.

ALTERNATIVE DEFINITIONS OF THE FACTOR OF SAFETY

Up to this point the factor of safety has been defined with respect to the shear strength of the soil. This definition is the one generally used for slope stability analyses, and this definition [Eq. (6.1)] is recommended and used throughout the book unless otherwise noted. However, other definitions for the factor of

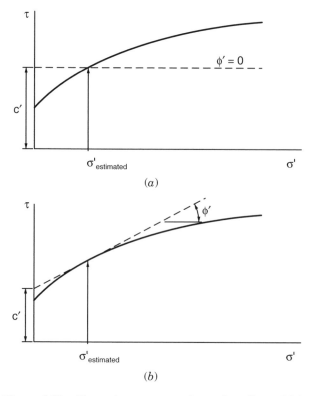

Figure 6.30 Alternative representations of nonlinear Mohr failure envelopes by equivalent values of cohesion and friction angle: (a) representation by equivalent strength c ($\phi = 0$); (b) representation by equivalent tangent values of cohesion and friction.

safety have sometimes been used. These usually lead to different values for the factor of safety. Two such definitions for the factor of safety are discussed below.

Factor of Safety for Load

One definition of the factor of safety is with respect to load. The case where the factor of safety is usually defined with respect to load is bearing capacity. The factor of safety for bearing capacity defined in terms of load is

$$F = \frac{\text{load required to cause failure}}{\text{actual applied load}} \quad (6.94)$$

To illustrate the difference between this definition of the factor of safety and the factor of safety defined with respect to shear strength, consider the footing shown in Figure 6.31. The footing is 8 ft wide, rests on cohesionless soil, and exerts a bearing pressure of 10,000 psf on the soil. The ultimate bearing pressure, q_{ult}, required to cause failure of the footing is expressed as

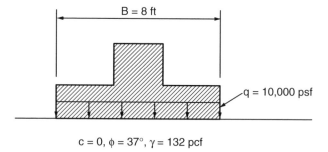

$$c = 0, \phi = 37°, \gamma = 132 \text{ pcf}$$

Figure 6.31 Footing problem used to illustrate differences between factors of safety applied to load and to soil shear strength.

$$q_{ult} = \tfrac{1}{2}\gamma B N_\gamma \quad (6.95)$$

where N_γ is a bearing capacity factor, which depends only on the angle of internal friction, ϕ. For a friction angle of 37° the value of N_γ is 53 and the ultimate bearing capacity is then

$$q_{ult} = \tfrac{1}{2}\gamma B N_\gamma = \tfrac{1}{2}(132)(8)(53) = 27{,}984 \approx 28{,}000 \text{ lb}$$
$$(6.96)$$

Thus, the factor of safety for load is

$$F = \frac{28{,}000}{10{,}000} = 2.80 \quad (6.97)$$

Equation (6.96) is actually a limit equilibrium equation that relates the shear strength (ϕ) to the bearing pressure that produces equilibrium when the shear strength of the soil is fully developed. If, instead, we consider some fraction of the shear strength being developed such that

$$\tan \phi_d = \frac{\tan \phi}{F} \quad (6.98)$$

where F is the factor of safety on shear strength and ϕ_d is the developed friction angle, we can write

$$q_{equil} = \tfrac{1}{2}\gamma B N_{\gamma\text{-developed}} \quad (6.99)$$

where q_{equil} is the equilibrium bearing pressure and $N_{\gamma\text{-developed}}$ is the value of N_γ based on the developed friction angle, ϕ_d. If we let the applied bearing pressure of 10,000 psf shown in Figure 6.31 be the equilibrium pressure, we can find the value of the factor of safety with respect to shear strength (and ϕ_d) that satisfies equilibrium [Eq. (6.99)]. This was done by trial and error, assuming a factor of safety and computing the corresponding equilibrium bearing pressure. The results are summarized in Table 6.5. Referring to Table 6.5, it can be seen that a factor of safety of 1.25 applied to the shear strength produces equilibrium of the footing under the applied load of 10,000 psf. This value (1.25) for the factor of safety applied to shear strength is considerably less than the corresponding value of 2.80 applied to load, which was calculated earlier [Eq. (6.97)]. Factors of safety that are applied to load for bearing capacity are thus not comparable to the factors of safety applied to shear strength, as used for slope stability analyses.

The factor of safety with respect to shear strength is closer to 1.0 than the factor of safety with respect to load; however, the magnitude of the difference varies significantly depending on the value of ϕ. Consider, for example, the slope shown in Figure 6.32. Factors of safety were calculated for the three different sets of shear strength parameters shown in this figure. The values of the shear strength parameters were selected so that the factor of safety with respect to shear strength [Eq. (6.1)] was approximately 1.5. For each slope the factors of safety both with respect to shear strength and with respect to load were calculated. The factor of

safety with respect to load was calculated by multiplying the unit weight of the soil by a factor of safety until the slope was in just-stable equilibrium with the shear strength fully developed. The factors of safety for shear strength and load are summarized for the three slopes in Table 6.6. For the first set of shear strengths ($\phi = 0$) the factors of safety for shear strength and load are identical. This will always be the case when $\phi = 0$, for slope stability as well as for bearing capacity problems such as the footing shown previously. For the second set of shear strength parameters, the factor of safety with respect to load was approximately 11, whereas the factor of safety for shear strength was approximately 1.5. This represents a difference of over 700%. Finally, for the third set of shear strength parameters, $\phi = 36.9°$ and $c = 0$, the factor of safety with respect to shear strength was 1.5, whereas the factor of safety with respect to load is infinite (i.e., no matter how large the weight of soil, the shear strength always remains greater than the shear stress). In summary, the factors of safety for shear strength and for load can vary from being the same to being very different for large values of ϕ, and the two values are not comparable. Because the soil shear strength is one of the largest unknowns in a slope stability analysis—certainly it presents greater uncertainty than the unit weight of soil in almost all instances—it seems logical to apply the factor of safety to shear strength.

Factor of Safety for Moments

Another definition that has been suggested for the factor of safety is one based on moments. In this case the factor of safety is defined as the ratio of the available resisting moment divided by the actual driving moment:

$$F = \frac{\text{available resisting moment}}{\text{actual driving moment}} = \frac{M_r}{M_d} \quad (6.100)$$

This can be rearranged and written as

Table 6.5 Summary of Computed Equilibrium Bearing Pressures for Various Factors of Safety on Shear Strength

Assumed F	ϕ_d (deg)	$N_{\gamma\text{-developed}}$	q_{equil} (psf)
1.00	37.0	53	28,000
1.25	31.1	19	10,000
1.50	26.7	9	4,800

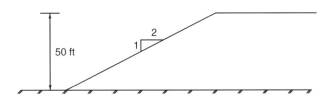

Property Set	c (psf)	φ (degrees)	γ (pcf)
1	1065	0	115
2	285	23.7	128
3	0	36.9	125

Figure 6.32 Slope and soil properties used to compute factors of safety applied to shear strength and applied to load.

Table 6.6 Sumary of Computed Factors of Safety Applied to Shear Strength and Applied to Load for Example Slope and Three Sets of Soil Properties

	Factor of safety applied to:	
Property set	Shear strength	Load
1 ($\phi = 0$)	1.50	1.50
2 ($c > 0$, $\phi > 0$)	1.50	11.00
3 ($c = 0$)	1.50	Infinite

$$M_d - \frac{M_r}{F} = 0 \qquad (6.101)$$

Equation (6.101) is an equilibrium equation that expresses a balance between the driving moment and a developed resisting moment that is equal to the total available resisting moment factored by the factor of safety. If the resisting moment is due entirely to the shear strength of the soil, the factor of safety applied to the resisting moment is the same as the factor of safety defined earlier with respect to shear strength. In fact, it was shown earlier for the Swedish Circle method [Eq. (6.28)] that the factor of safety defined with respect to shear strength was equal to the ratio of moments expressed by Eq. (6.100). In this case there is no difference between the factors of safety defined with respect to shear strength [Eq. (6.1)] and with respect to moments [Eq. (6.100)].

If instead of a simple slope, where all resistance is from the shear strength of the soil, there are additional forces due to reinforcement, the two definitions of factor of safety [Eqs. (6.1) and (6.100)] can be quite different. Also, the definition of the factor of safety as a ratio of moments can be ambiguous. To illustrate this, consider the slope and circular slip surface shown in Figure 6.33. This slope has a single layer of reinforcement. Let the moment taken about the center of the circle due to the reinforcement be designated as M_t. Let the corresponding moments due to the available

shear strength be designated as M_s, and the moment due to the weight of soil be designated as M_d. We can now define a factor of safety with respect to resisting and driving moments. If we choose to add the moment due to the reinforcement to the resisting moment due to the shear strength of the soil, we can write

$$F = \frac{M_s + M_t}{M_d} \qquad (6.102)$$

Alternatively, we can choose to subtract the restoring moment due to the reinforcement from the driving moment due to the soil weight and write

$$F = \frac{M_s}{M_d - M_t} \qquad (6.103)$$

Equations (6.102) and (6.103) both represent legitimate definitions for the factor of safety defined as a ratio of moments; however, the two definitions give different values for the factor of safety. Equations (6.102) and (6.103) are more easily interpreted if we rewrite them as follows: For Eq. (6.102) we can write

$$M_d = \frac{M_s}{F} + \frac{M_t}{F} \qquad (6.104)$$

and for Eq. (6.103) we can write

$$M_d = \frac{M_s}{F} + M_t \qquad (6.105)$$

Both of these equations can be interpreted as equilibrium equations. The first equation, where the contribution of the reinforcement was added to the resisting moment, states that the driving moment is balanced by moments due to the developed shear strength and the developed reinforcement forces, where developed values are the available values reduced by a factor of safety, F. Thus, the factor of safety in Eq. (6.104) is applied equally to the reinforcement forces and the shear strength. The second of the two equilibrium equations [Eq. (6.105)], where the reinforcement contribution was used to reduce the driving moments, states that the driving moment is in equilibrium with the full reinforcement force, plus the factored resistance due to the shear strength of the soil. In this case the factor of safety is applied only to the soil shear strength.

To illustrate the differences in values computed for the factors of safety of reinforced slopes depending on how the factor of safety is defined, consider the slope

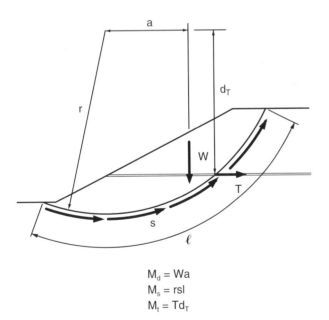

$M_d = Wa$
$M_s = rsl$
$M_t = Td_T$

Figure 6.33 Simple reinforced slope with driving and resisting moments.

shown in Figure 6.34. This slope has a single layer of reinforcement and $\phi = 0$. The factor of safety for the unreinforced slope is 0.91, thus indicating that the reinforcement is necessary to make the slope stable. Factors of safety were first computed by applying the factor of safety to shear strength only [Eq. (6.105)] and to both shear strength and reinforcement force equally [Eq. (6.104)]. A third factor of safety was computed by applying the factor of safety to only the reinforcement force (i.e., the shear strength was assumed to be fully mobilized and the reinforcement force was reduced by the factor of safety). The three different values for the factor of safety shown in Figure 6.34 range from approximately 1.3 to 4.8, a difference of over threefold. Clearly, the manner in which the factor of safety is defined will affect the computed value.

Although any of the foregoing definitions for factor of safety could be used to compute a factor of safety, only Eq. (6.105) is consistent with the definition of factor of safety generally used for slope stability analyses throughout this book. Instead of defining and computing a factor of safety that is applied equally to the reinforcement forces and soil strength [Eq. (6.104)] or to only the reinforcement force, it seems more appropriate first to apply a suitable factor of safety to the

reinforcement forces before any slope stability computations begin and then compute a separate factor of safety with respect to the shear strength of the soil. This approach is recommended and is discussed further in Chapter 8.

PORE WATER PRESSURE REPRESENTATION

Whenever the shear strength of one or more materials is expressed in terms of effective stresses, the pore water pressures must be determined and represented in the slope stability analysis. Several methods exist for doing this, depending on the seepage and groundwater conditions and the degree of rigor required. The various methods are described and discussed in this section.

Flow Net Solutions

When steady-state seepage conditions exist in a slope, a graphical flow net solution can be used to determine the pore water pressures. For most slopes this requires determining the location of a *line of seepage* representing the uppermost flow line, and then constructing families of curves representing flow and equipotential

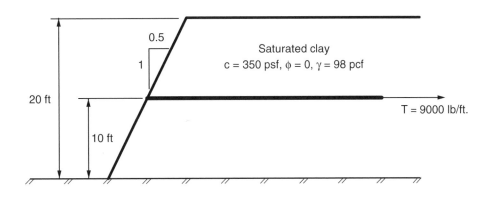

Case	Factor of Safety
Factor of safety applied to <u>shear strength</u> only.	1.51
Factor of safety applied to <u>reinforcement force</u> only.	4.82
Factor of safety applied to <u>both shear strength and reinforcement force</u>.	1.32

Figure 6.34 Reinforced slope with computed factors of safety defined (applied) in three different ways.

lines in the region of saturated flow, below the line of seepage (Casagrande, 1937). Once a correct flow net has been constructed, pore water pressures can be calculated at any point desired; pore water pressures are usually assumed to be zero above the line of seepage.

Although flow nets provide an accurate representation of pore water pressures, flow nets are difficult and tedious to use in slope stability computations. For each slice of each trial slip surface, pore water pressures must be calculated. This generally requires some interpolation between the equipotential lines in the flow net to get the pore water pressures at each slice. If done manually, the task is extremely tedious. It is also difficult to use a flow net in computer calculations. To do so, the pore water pressures must first be computed from the flow net at selected points. Usually, points corresponding to the intersections of the flow lines and equipotential lines are used. Once the pressures are computed, which must be done manually, the digitized values of pressure need to be entered into a computer program. In the computer program it is necessary to interpolate the values of pressure that were input to get pore water pressures at the center of the base of each slice. Although the interpolation process can be automated, much as it is when using pore water pressures calculated from finite element analyses, a substantial effort is still required to construct a flow net and then compute the pore water pressures for entry into the computer program. Also, for complex slopes it is impractical to construct a flow net by hand.

Numerical Solutions

Today, most analyses of seepage and groundwater flow, for any but the simplest conditions, are conducted using finite difference or finite element numerical solutions. Because of the great flexibility that it provides, most such analyses are done using the finite element method. Results of such analyses consist of values of pore water pressure at each of a number of nodal points in the finite element mesh.

Most of the earlier and even some of today's finite element modeling schemes model only the region of saturated flow, below the line of seepage. Essentially, these schemes mimic what is done with flow net solutions by establishing a line of seepage and assuming no flow above the line of seepage. Various schemes have been used to determine the location of the line of seepage; including adjusting the geometry of the finite-mesh and truncating the finite element mesh at the point where it intersects the line of seepage (zero pressure line). With these schemes pore water pressure are calculated only in the region of saturated flow, where

the pore water pressures are positive. Pore water pressures are assumed to be zero above the line of seepage.

Most current finite element modeling schemes model the entire cross section of a slope, including the region where the pore water pressures are negative and the soil may be unsaturated. These schemes employ finite elements everywhere there is soil, and the hydraulic conductivity is adjusted to reflect the pore water pressures and degree of saturation. Both positive and negative values of pore water pressure are calculated. However, for slope stability analyses the pore water pressures are usually assumed to be zero in the region where negative pressures have been calculated.

Regardless of the finite element scheme used, results of a finite element analysis consist of the value of pore water pressure at each nodal point. These values must then be used along with a suitable interpolation scheme to calculate the pore water pressures at the center of the base of individual slices along a slip surface.

Interpolation Schemes

Several interpolation schemes have been used to calculate the pore water pressures along the slip surface from the results of finite element analyses. Several of these schemes are described below.

Three- and four-point interpolation. One of the earliest schemes used to interpolate pore water pressures from gridded data for slope stability calculations was based on a three- or four-point interpolation function (Wright, 1974; Chugh, 1981). In this scheme the three or four points where pore water pressures were defined (e.g., nodal points) that are closest to the point where pore water pressures are to be calculated are located. If four points are used, the pore water pressures are then interpolated using an equation of the form

$$u = a_1 + a_2 x + a_3 y + a_4 xy \qquad (6.106)$$

where x and y are coordinates and a_1, a_2, a_3, and a_4 are four coefficients that are evaluated using the coordinates and pore water pressures at the four interpolation points (nodal points). Once the coefficients are determined, Eq. (6.106) is used to compute the pore water pressure at the point on the slip surface. If three points are used for interpolation, the form of the equation is

$$u = a_1 + a_2 x + a_3 y \qquad (6.107)$$

but otherwise, the procedure is the same as for four points. This three- or four-point interpolation scheme has problems that sometimes lead to erroneous values, especially when pore water pressures are interpolated (actually, extrapolated) outside the perimeter of the region formed by the three or four interpolation points. Some improvement was achieved in this scheme through an "averaging" scheme proposed by Chugh (1981); however, the scheme still can lead to erroneous values, especially when three or more of the interpolation points lie along a nearly straight line.

Spline interpolation. Two-dimensional interpolation schemes based on spline surfaces are more rigorous and overcome some of the limitations of the three- and four-point schemes described above. In the spline interpolation schemes a much larger number of points is used to interpolate the pore water pressures at any given point (Geo-Slope, 2002; RocScience, 2002). Also, a much larger system of equations must be solved for the interpolation coefficients. The procedures work well but can be relatively time consuming in terms of computational effort and computer memory requirements. These schemes may also result in values being extrapolated outside the actual range of the interpolation point data (i.e., the pore water pressures may exceed or be less than the respective highest and lowest values at points used for interpolation).

Finite element shape functions. Another scheme that should be quite accurate, but has not been widely used, employs the same shape functions that are used for the finite element formulation to interpolate pore water pressures once the finite element solution is completed. This scheme has been used by the writers as well as by Geo-Slope (2002). The scheme closely integrates the finite element solution with the subsequent interpolation of pore water pressures and thus should not introduce additional errors resulting from the interpolation scheme. With this scheme the element that contains the point where pore water pressures are to be interpolated is found first. The pore water pressures are then calculated using the values of the head at the adjoining nodal points and the finite element shape (interpolation) functions. This scheme is relatively complex computationally, but more important, the scheme is only well suited for interpolation of pore water pressures when the pore water pressures have been computed using the finite element method. The scheme requires more close integration of the finite element and slope stability analyses than most other schemes.

Triangulated irregular network scheme. One of the best interpolation schemes is based on use of a triangulated irregular network (TIN) for interpolation (Wright, 2002). The TIN consists of a set of triangles whose vertices coincide with the points where pore water pressures are defined and which cover without overlapping the entire region where pore water pressures are defined. A convenient scheme for creating the triangles is the *Delaunay triangulation scheme* (Figure 6.35). In this scheme the triangles are created such that no interpolation point lies within the circumcircle of any triangle (Lee and Schachter, 1980; Watson and Philip, 1984). A *circumcircle* is the circle passing through the three vertices of a triangle and contains the three vertices, but there are no points inside the circle.[11] Robust algorithms exist for creating a Delaunay triangulation for any series of discrete points.

Once a Delaunay triangulation is created interpolation consists of a two-step process for interpolating pressures at any point: First, the triangle that contains the point where pressures are to be interpolated is located. Then, the pore water pressures are interpolated linearly from the values of pore water pressure at the three vertices of the enclosing triangle. The interpolation equation is of the form of Eq. (6.107), but differs from the previous scheme in that the interpolation point never lies outside the perimeter of the triangle formed by the three points used for interpolation. Also by using the Delaunay triangulation scheme it is usually possible to avoid having the three points used for interpolation being located on essentially a straight line. One of the most important advantages of the TIN-based scheme is that algorithms exist for very quickly locating the appropriate triangle that contains the point where pressures are to be interpolated, and thus, the points to be used for interpolation are located quickly (Lee and Schachter, 1980; Mirante and Weingarten, 1982; Jones, 1990). The three- and four-point schemes described earlier as well as schemes using the finite element shape functions may involve a significant amount of time being spent on locating the appropriate points or finite element that are to be used for the interpolation. The TIN-based scheme also has the advantage of using a very simple linear interpolation function [Eq. (6.107)] which requires very little computational time and computer memory for storage.

Another advantage of the TIN-based interpolation scheme is that it is applicable to interpolation using any type of irregularly gridded data, not just the results from finite element analyses. For example, there may be cases where some pore water pressures are recorded in piezometers or by groundwater observations. These measured data may then be supplemented by additional data points based on judgment and interpretation to provide a grid of values. Such values can then easily

---

[11] It is possible for more than three points to lie on a circumcircle (e.g., if four points form a rectangle); however, no point will ever lie inside the circumcircles in a Delaunay triangulation.

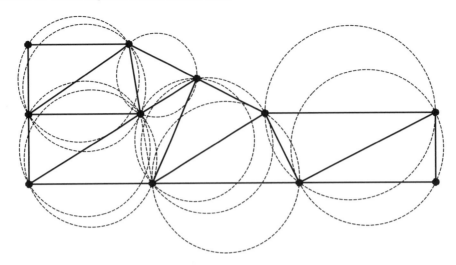

Figure 6.35 Delaunay triangulation of a series of interpolation points (e.g., nodal points) with accompanying circumcircles.

be used to compute values of pore water pressure at other points using the TIN-based interpolation scheme.

A related advantage of TIN-based interpolation schemes is that once such schemes are developed and implemented in computer codes, they can also be used for other forms of interpolation required in slope stability computations. For example, TIN-based schemes are useful for interpolating values of undrained shear strength. This is particularly useful when the undrained shear strengths vary both vertically and horizontally. Typical examples where such spatial variation in undrained shear strength occur include some mine tailings disposal structures and clay foundations beneath embankments built with staged construction techniques.

Phreatic Surface

Flow net and finite element seepage solutions are complex and time consuming to perform. In many cases it is more appropriate to use simple approximations of the seepage and pore water pressures in a slope. Considering that groundwater and seepage conditions are often not well known, the simple approximations may be more than justified. One approximation is to define the pore water pressures using a line that represents a phreatic surface. The phreatic surface corresponds to the line of seepage from a flow net [i.e., the phreatic surface is considered to be a flow line and a line of zero (atmospheric) pressure]. The line of zero pressure from a finite element solution is also often considered equivalent to the phreatic surface even though in this case the zero pressure line may not be a flow line.

When the pore water pressures are defined by a phreatic surface, the pore water pressure is equal to the product of the pressure head, h_p, and the unit weight of water, γ_w:

$$u = h_p \gamma_w \qquad (6.108)$$

If the phreatic surface is a straight line and the equipotential lines are also straight lines (Figure 6.36a), perpendicular to the phreatic surface, the pressure head is related to the vertical depth, z_p, below the phreatic surface by

$$h_p = z_p \cos^2 \beta \qquad (6.109)$$

where β is the slope of the phreatic surface. If, instead, the phreatic surface and equipotential lines are curved as shown in Figure 6.36b, the pore water pressure can be expressed by the following inequalities:

$$z_p \gamma_w \cos^2 \beta' < u < z_p \gamma_w \cos^2 \beta'' \qquad (6.110)$$

where β' is the slope of the phreatic surface at the point where the equipotential line intersects the phreatic surface and β'' is the slope of the phreatic surface directly above the point of interest. In this case ($\beta'' < \beta'$) it is conservative to express the pore water pressure as

$$u = z_p \gamma_w \cos^2 \beta'' \qquad (6.111)$$

where again β'' is the slope of the phreatic surface directly above the point of interest.

Several computer programs have used Eq. (6.111) or a similar form to approximate the pore water pressures from a phreatic surface. This seems to provide a

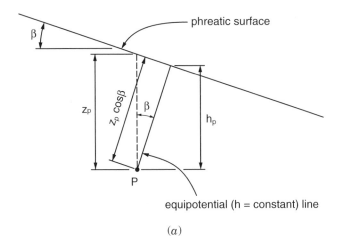

(a)

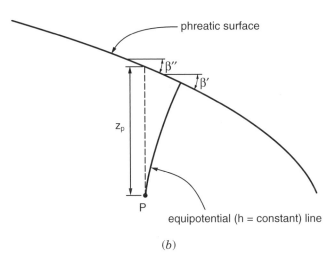

(b)

Figure 6.36 (a) Linear and (b) curved phreatic surfaces used to approximate pore water pressures.

reasonable approximation for many cases, but as shown later the approximation can also lead to unconservative results depending on seepage conditions.

Piezometric Line

As a further approximation and simplification of pore water pressures a piezometric line may be used. With a piezometric line the pore water pressures are computed by multiplying the vertical depth below the piezometric line by the unit weight of water. Thus,

$$u = z_p \gamma_w \qquad (6.112)$$

This representation is sometimes considered to be a conservative representation of the pore water pressures compared to the phreatic surface described previously. However, the differences between the two representations of pore water pressure are typically small because

the slope (β) for most surfaces is small. Furthermore, because the phreatic surface is at times unconservative (i.e., it can underestimate the pore water pressures), representation by a piezometric line may be just as suitable as a phreatic surface. It should be recognized that both represent approximations that may or may not be valid, depending on the particular seepage conditions in a slope.

Examples

To illustrate differences in the various representations of pore water pressure described above, analyses of two examples are presented. For each of these two examples a finite element steady-state seepage analysis was performed first using the GMS/SEEP2D software (EMRL, 2001) to calculate the pore water pressures. The seepage analyses were performed by modeling the entire soil cross section shown for each example and using appropriate values of saturated or unsaturated hydraulic conductivity depending on the pore water pressures. Pore water pressures calculated from the finite element seepage analyses were then used to interpolate pore water pressures along the slip surface for each trail slip surface and slice. Pore water pressures were interpolated using the TIN-based interpolation scheme described earlier. A phreatic surface was also established by locating the line (contour) corresponding to zero pore water pressure. Pore water pressures were calculated from the phreatic surface using Eq. (6.111). Finally, a piezometric line was defined from the line of zero pressure (i.e., the phreatic surface and piezometric line were the same). Pore water pressures were calculated from the piezometric line using Eq. (6.112).

Factors of safety were calculated for both examples using each of the pore water pressure representations described above. For each slope and representation of pore water pressure the critical circle with the minimum factor of safety was found.

Example 1. The slope for the first example is shown in Figure 6.37. The slope and foundation are homogeneous and composed of the same soil. Total heads, expressed relative to a datum at the bottom of the foundation, are 75 ft at the entrance to the cross section and 40 ft at the ground surface beyond the toe of the slope. The slope face is assumed to be a free discharge surface, and it is assumed that no infiltration occurs along the slope face or behind the crest of the slope.

The contour of zero pressure determined from the finite element seepage analysis is shown in Figure 6.38. In the finite element analysis, pore water pressures were negative above and positive below this line. In the subsequent slope stability analyses, the zero

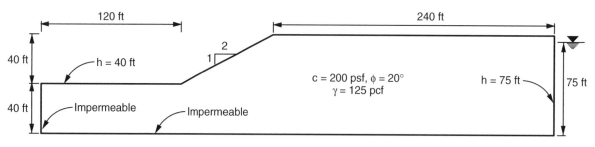

Figure 6.37 Homogeneous slope used to illustrate different schemes for representing pore water pressures in slope stability analyses.

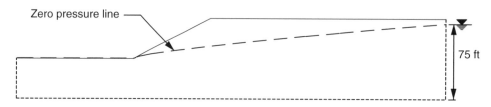

Figure 6.38 Zero-pressure line (contour) determined using finite element analysis, used to represent the phreatic surface and the piezometric line for a homogeneous slope.

pressure line was used to represent both the piezometric line and the phreatic surface.

Factors of safety calculated for each of the three representations of pore water pressure are summarized in Table 6.7. The three values shown for the different representations of pore water pressures are all extremely close; the values are shown to three decimal places to illustrate the differences. All three representations produced approximately the same factor of safety of approximately 1.14 to 1.15. The phreatic surface in this case produced a slightly higher factor of safety because of a slight upward component of flow beyond the toe of the slope, which causes the phreatic surface and Eq. (6.111) to underestimate pore water pressures slightly.

Table 6.7 Summary of Computed Factors of Safety for Three Different Representations of Pore Water Pressure: Homogeneous Slope and Foundation

Pore water pressure representation	Factor of safety
Finite element analysis with pore pressures interpolated from nodal points values using triangle-based interpolation scheme	1.138
Phreatic surface approximation	1.147
Piezometric line approximation	1.141

Example 2. The slope for the second example consists of the earth embankment dam resting on a layered soil foundation as shown in Figure 6.39. Soil properties are shown in Figure 6.39. The foundation consists of a low-permeability clay layer underlain by a more permeable layer of sand. A significant amount of underseepage occurs through the more permeable sand layer at depth and produces upward flow near the downstream portion of the dam. The zero-pressure line determined from the finite element analyses is shown in Figure 6.40. This line was used as a phreatic surface and piezometric line for the slope stability analyses.

Although the minimum factor of safety for this slope occurs for very shallow circles (sloughs) near the toe of the slope, deeper slip surfaces are likely to be of interest as well. Therefore, the overall minimum factor of safety (shallow circles) and the minimum factor of safety for circles tangent to elevation 197 (bottom of clay) were both calculated. The factors of safety are summarized in Table 6.8. Factors of safety calculated by both the phreatic surface and piezometric line representations of pore water pressures were essentially the same because of the relatively flat zero-pressure surface. However, both the piezometric line and phreatic surface representations of pore water pressures produced factors of safety that ranged from 14 to 19% greater than the factors of safety calculated from the finite element solution and interpolation. These relatively large differences are due to the upward flow of water through the clay foundation, which is not represented well by either a phreatic surface or a single piezometric line.

Material	k - ft/min	c' - psf	φ' - degrees	Unit wt. - pcf
Outer shell	1×10^{-2}	0	34	125
Clay core	1×10^{-6}	100	26	122
Foundation clay	1×10^{-5}	0	24	123
Foundation sand	1×10^{-3}	0	32	127

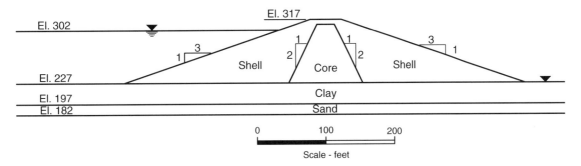

Figure 6.39 Embankment dam on layered foundation used to illustrate different schemes for representing pore water pressures in slope stability analyses.

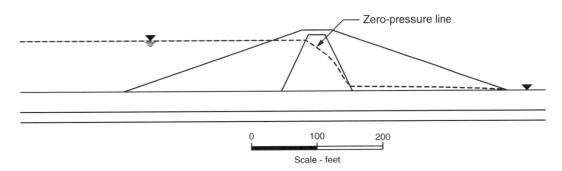

Figure 6.40 Zero-pressure line (contour) determined using finite element analysis, used to represent the phreatic surface and the piezometric line for an embankment dam.

Table 6.8 Summary of Computed Factors of Safety for Three Different Representations of Pore Water Pressure: Embankment Dam on Layered Soil Foundation

	Minimum factors of safety	
Pore water pressure representation	Overall critical circle	Critical circle tangent to elev. 197
Finite element analysis with pore pressures interpolated from nodal points values using triangle-based interpolation scheme	1.11	1.37
Phreatic surface approximation	1.32	1.57
Piezometric line approximation	1.30	1.57

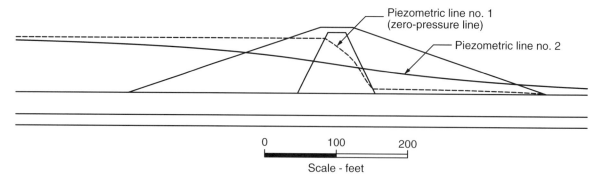

Figure 6.41   Two piezometric lines used to represent the pore water pressures for an embankment dam.

Results of the stability calculations for the second example can be improved by using more than one piezometric line. To illustrate this, a second piezometric line was established based on the pore water pressures at the bottom of the foundation clay layer (top of sand layer). The second piezometric line is shown in Figure 6.41 and better represents the pore water pressures (total head) in the sand and lower part of the clay. An additional set of slope stability calculations was performed using the second piezometric line to define the pore water pressures in the bottom half of the clay layer, and the original first piezometric line was used to define the pore water pressures in the upper half of the foundation clay as well as in the embankment. Calculations were performed for circles tangent to elevation 197, which is the bottom of the clay layer. The factor of safety computed using the two piezometric lines was 1.36, which is almost identical to the value (1.37) that was obtained when the pore water pressures were obtained by interpolation of the values from the finite element analysis. Thus, the use of multiple—in this case, two—piezometric lines can improve the solution obtained using piezometric lines, although the very close agreement between the results with two piezometric lines and interpolation may be somewhat fortuitous. Also, it is not always as easy to establish appropriate piezometric lines when multiple lines are to be used.

SUMMARY

When flow is predominately horizontal (vertical equipotential lines), both a phreatic surface and single piezometric line can be used to approximate the pore water pressures relatively well, with an error of only a few percent at most. There appears to be little difference between these two representations of pore water pressure (phreatic surface and piezometric line); the refinement of using a phreatic surface representation rather than a simpler piezometric line does not appear to produce a better representation. In fact, the phreatic surface may actually produce less accurate results than the piezometric line.

For cases where the flow is not predominately horizontal and as the component of flow and head loss in the vertical direction increases, neither a phreatic surface nor a single piezometric line represent pore water pressures well; both may result in errors on the unsafe side. In such cases, care must be exercised in selecting an appropriate representation of pore water pressures. It is probably better to use an appropriate seepage solution and interpolate pore water pressures. Currently, there are several good finite element software programs available for seepage analyses. Use of pore water pressures from a finite element analysis requires some effort to interpolate pore water pressures from nodal point values to points along a slip surface (base of slices), but robust and efficient interpolation schemes are available for this as well.

CHAPTER 7

Methods of Analyzing Slope Stability

Methods for analyzing stability of slopes include simple equations, charts, spreadsheet software, and slope stability computer programs. In many cases more than one method can be used to evaluate the stability for a particular slope. For example, simple equations or charts may be used to make a preliminary estimate of slope stability, and later, a computer program may be used for detailed analyses. Also, if a computer program is used, another computer program, slope stability charts, or a spreadsheet should be used to verify results. The various methods used to compute a factor of safety are presented in this chapter.

SIMPLE METHODS OF ANALYSIS

The simplest methods of analysis employ a single simple algebraic equation to compute the factor of safety. These equations require at most a hand calculator to solve. Such simple equations exist for computing the stability of a vertical slope in purely cohesive soil, of an embankment on a much weaker, deep foundation, and of an infinite slope. Some of these methods, such as the method for computing the stability of an infinite slope, may provide a rigorous solution, whereas others, such as the equations used to estimate the stability of a vertical slope, represent some degree of approximation. Several simple methods are described below.

Vertical Slope in Cohesive Soil

For a vertical slope in cohesive soil a simple expression for the factor of safety is obtained based on a planar slip surface like the one shown in Figure 7.1. The average shear stress, τ, along the slip plane is expressed as

$$\tau = \frac{W \sin \alpha}{l} = \frac{W \sin \alpha}{H/\sin \alpha} = \frac{W \sin^2\alpha}{H} \qquad (7.1)$$

where α is the inclination of the slip plane, H is the slope height, and W is the weight of the soil mass. The weight, W, is expressed as

$$W = \frac{1}{2} \frac{\gamma H^2}{\tan \alpha} \qquad (7.2)$$

which when substituted into Eq. (7.2) and rearranged gives

$$\tau = \tfrac{1}{2} \gamma H \sin \alpha \cos \alpha \qquad (7.3)$$

For a cohesive soil ($\phi = 0$) the factor of safety is expressed as

$$F = \frac{c}{\tau} = \frac{2c}{\gamma H \sin \alpha \cos \alpha} \qquad (7.4)$$

To find the minimum factor of safety, the inclination of the slip plane is varied. The minimum factor of safety is found for $\alpha = 45°$. Substituting this value for α (45°) into Eq. (7.4) gives

$$F = \frac{4c}{\gamma H} \qquad (7.5)$$

Equation (7.5) gives the factor of safety for a vertical slope in cohesive soil, assuming a plane slip surface. Circular slip surfaces give a slightly lower value for the factor of safety ($F = 3.83c/\gamma h$); however, the difference between the factors of safety based on a plane and a circular slip surface is small for a vertical slope in cohesive soil and can be ignored.

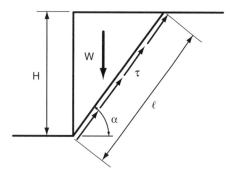

Figure 7.1 Vertical slope and plane slip surface.

Equation (7.5) can also be rearranged to calculate the critical height of a vertical slope (i.e., the height of a slope that has a factor of safety of unity). The critical height of a vertical slope in cohesive soil is

$$H_{\text{critical}} = \frac{4c}{\gamma} \qquad (7.6)$$

Bearing Capacity Equations

The equations used to calculate the bearing capacity of foundations can also be used to estimate the stability of embankments on deep deposits of saturated clay. For a saturated clay and undrained loading ($\phi = 0$), the ultimate bearing capacity, q_{ult}, based on a circular slip surface is[1]

$$q_{\text{ult}} = 5.53c \qquad (7.7)$$

Equating the ultimate bearing capacity to the load, $q = \gamma H$, produced by an embankment of height, H, gives

$$\gamma H = 5.53c \qquad (7.8)$$

where γ is the unit weight of the soil in the embankment; γh represents the maximum vertical stress produced by the embankment. Equation (7.8) is an equilibrium equation corresponding to ultimate conditions (i.e., with the shear strength of the soil fully developed). If, instead, only some fraction of the shear strength is developed (i.e., the factor of safety is

greater than unity), a factor of safety can be introduced into the equilibrium equation (7.8) and we can write

$$\gamma H = 5.53 \frac{c}{F} \qquad (7.9)$$

In this equation F is the factor of safety with respect to shear strength; the term c/F represents the developed cohesion, c_d. Equation (7.9) can be rearranged to give

$$F = 5.53 \frac{c}{\gamma H} \qquad (7.10)$$

Equation (7.10) can be used to estimate the factor of safety against a deep-seated failure of an embankment on soft clay.

Equation (7.10) gives a conservative estimate of the factor of safety of an embankment because it ignores the strength of the embankment and the depth of the foundation in comparison with the embankment width. Alternative bearing capacity equations that are applicable to reinforced embankments on thin clay foundations are presented in Chapter 8.

Infinite Slope

In Chapter 6 the equations for an infinite slope were presented. For these equations to be applicable, the depth of the slip surface must be small compared to the lateral extent of the slope. However, in the case of cohesionless soils, the factor of safety does not depend on the depth of the slip surface. It is possible for a slip surface to form at a small enough depth that the requirements for an infinite slope are met, regardless of the extent of the slope. Therefore, an infinite slope analysis is rigorous and valid for cohesionless slopes. The infinite slope analysis procedure is also applicable to other cases where the slip surface is parallel to the face of the slope and the depth of the slip surface is small compared to the lateral extent of the slope. This condition may exist where there is a stronger layer of soil at shallow depth: for example, where a layer of weathered soil exists near the surface of the slope and is underlain by stronger, unweathered material.

The general equation for the factor of safety for an infinite slope with the shear strength expressed in terms of total stresses is

$$F = \cot \beta \tan \phi + (\cot \beta + \tan \beta) \frac{c}{\gamma z} \qquad (7.11)$$

---

[1] Although Prandtl's solution of $q_{\text{ult}} = 5.14c$ is commonly used for bearing capacity, it is more appropriate to use the solution based on circles, which gives a somewhat higher bearing capacity and offsets some of the inherent conservatism introduced when bearing capacity equations are applied to slope stability.

where z is the vertical depth of the slip surface below the face of the slope. For shear strengths expressed by effective stresses the equation for the factor of safety can be written as

$$F = \left[\cot \beta - \frac{u}{\gamma z}(\cot \beta + \tan \beta)\right] \tan \phi'$$
$$+ (\cot \beta + \tan \beta)\frac{c'}{\gamma z} \qquad (7.12)$$

where u is the pore water pressure at the depth of the slip surface.

For effective stress analyses, Eq. (7.12) can also be written as

$$F = [\cot \beta - r_u (\cot \beta + \tan \beta)] \tan \phi'$$
$$+ (\cot \beta + \tan \beta)\frac{c'}{\gamma z} \qquad (7.13)$$

where r_u is the pore pressure ratio defined by Bishop and Morgenstern (1960) as

$$r_u = \frac{u}{\gamma z} \qquad (7.14)$$

Values of r_u can be determined for specific seepage conditions. For example, for seepage parallel to the slope, the pore pressure ratio, r_u, is given by

$$r_u = \frac{\gamma_w}{\gamma} \frac{h_w}{z} \cos^2\beta \qquad (7.15)$$

where h_w is the height of the free water surface vertically above the slip surface (Figure 7.2a). If the seepage exits the slope face at an angle (Figure 7.2b), the value of r_u is given by

$$r_u = \frac{\gamma_w}{\gamma} \frac{1}{1 + \tan \beta \tan \theta} \qquad (7.16)$$

where θ is the angle between the direction of seepage (flow lines) and the horizontal. For the special case of horizontal seepage ($\theta = 0$), the expression for r_u reduces to

$$r_u = \frac{\gamma_w}{\gamma} \qquad (7.17)$$

Recapitulation

- Simple equations can be used to compute the factor of safety for several slope and shear strength conditions, including a vertical slope in cohesive soil, an embankment on a deep deposit of saturated clay, and an infinite slope.
- Depending on the particular slope conditions and equations used, the accuracy ranges from excellent, (e.g., for a homogeneous slope in cohesionless soil) to relatively crude (e.g., for bearing capacity of an embankment on saturated clay).

SLOPE STABILITY CHARTS

The stability of many relatively homogeneous slopes can be calculated using slope stability charts based on one of the analysis procedures presented in Chapter 6. Fellenius (1936) was one of the first to recognize that factors of safety could be expressed by charts. His work was followed by the work of Taylor (1937) and Janbu (1954b). Since the pioneering work of these authors, numerous others have developed charts for computing the stability of slopes. However, the early charts of Janbu are still some of the most useful for many conditions, and these are described in further detail in the Appendix. The charts cover a range in slope and soil conditions and they are quite easy to use. In addition, the charts provide the minimum factor of safety and eliminate the need to search for a critical slip surface.

Stability charts rely on dimensionless relationships that exist between the factor of safety and other parameters that describe the slope geometry, soil shear strengths, and pore water pressures. For example, the infinite slope equation for effective stresses presented earlier [Eq. (7.13)] can be written as

$$F = [1 - r_u(1 + \tan^2\beta)] \frac{\tan \phi'}{\tan \beta} + (1 + \tan^2\beta) \frac{c'}{\gamma z} \qquad (7.18)$$

or

$$F = A \frac{\tan \phi'}{\tan \beta} + B \frac{c'}{\gamma z} \qquad (7.19)$$

where

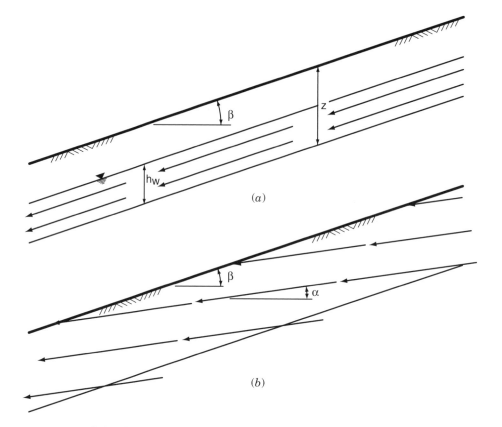

Figure 7.2   Infinite slope with seepage: (*a*) parallel to slope face; (*b*) exiting the slope face.

$$A = 1 - r_u(1 + \tan^2\beta) \qquad (7.20)$$

$$B = 1 + \tan^2\beta \qquad (7.21)$$

A and B are dimensionless parameters (*stability numbers*) that depend only on the slope angle, and in the case of A, the dimensionless pore water pressure coefficient, r_u. Simple charts for A and B as functions of the slope angle and pore water pressure coefficient, r_u, are presented in the Appendix.

For purely cohesive ($\phi = 0$) soils and homogeneous slopes, the factor of safety can be expressed as

$$F = N_0 \frac{c}{\gamma H} \qquad (7.22)$$

where N_0 is a stability number that depends on the slope angle, and in the case of slopes flatter than about 1:1, on the depth of the foundation below the slope. For vertical slopes the value of N_0 according to the Swedish slip circle method is 3.83. This value (3.83) is slightly less than the value of 4 shown in Eq. (7.5) based on a plane slip surface. In general, circular slip surfaces give a lower factor of safety than a plane, especially for flat slopes. Therefore, circles are gener-

ally used for analysis of most slopes in cohesive soils. A complete set of charts for cohesive slopes of various inclinations and foundation depths is presented in the Appendix. Procedures are also presented for using average shear strengths with the charts when the shear strength varies.

For slopes with both cohesion and friction, additional dimensionless parameters are introduced. Janbu (1954) showed that the factor of safety could be expressed as

$$F = N_{cf} \frac{c'}{\gamma H} \qquad (7.23)$$

where N_{cf} is a dimensionless stability number. The stability number depends on the slope angle, β, the pore water pressures, u, and the dimensionless parameter, $\lambda_{c\phi}$, which is defined as

$$\lambda_{c\phi} = \frac{\gamma H \tan\phi'}{c'} \qquad (7.24)$$

Stability charts employing $\lambda_{c\phi}$ and Eq. (7.23) to calculate the factor of safety are presented in the Appendix. These charts can be used for soils with cohesion and friction as well as a variety of pore water pressure and external surcharge conditions.

Although all slope stability charts are based on the assumption of constant shear strength (c, c' and ϕ, ϕ' are constant) or else a simple variation in undrained shear strength (e.g., c varies linearly with depth), the charts can be used for many cases where the shear strength varies. Procedures for using the charts for cases where the shear strength varies are described in the Appendix. Examples for using the charts are also presented in the Appendix.

Recapitulation

- Slope stability charts exist for computing the factor of safety for a variety of slopes and soil conditions.

SPREADSHEET SOFTWARE

Detailed computations for the procedures of slices can be performed in tabular form using a table where each row represents a particular slice and each column represents the variables and terms in the equations presented in Chapter 6. For example, for the case where $\phi = 0$ and the slip surface is a circle, the factor of safety is expressed as

$$F = \frac{\sum c\,\Delta l}{\sum W \sin \alpha} \qquad (7.25)$$

A simple table for computing the factor of safety using Eq. (7.25) is shown in Figure 7.3. For the Ordinary Method of Slices with the shear strength expressed in terms of effective stresses, the preferred equation for computing the factor of safety is

$$F = \frac{\sum [c'\,\Delta l + (W \cos \alpha - u\,\Delta l \cos^2\alpha)\tan \phi']}{\sum W \sin \alpha} \qquad (7.26)$$

A table for computing the factor of safety using this form of the Ordinary Method of Slices equation is illustrated in Figure 7.4. Tables such as the ones shown in Figures 7.3 and 7.4 are easily represented and implemented in computer spreadsheet software. In fact, more sophisticated tables and spreadsheets can be developed for computing the factor of safety using procedures of slices such as the Simplified Bishop, force equilibrium, and even Chen and Morgenstern's procedures (Low et al., 1998).

The number of different computer spreadsheets that have been developed and used to compute factors of safety is undoubtedly very large. This attests to the usefulness of spreadsheets for slope stability analyses, but at the same time presents several important problems: First, because such a large number of different spreadsheets are used and because each spreadsheet is often used only once or twice, it is difficult to validate spreadsheets for correctness. Also, because one person may write a spreadsheet, use it for some computations and then discard the spreadsheet, results are often poorly archived and difficult for someone else to interpret or to understand later. Electronic copies of the spreadsheet may have been discarded. Even if an electronic copy is maintained, the software that was used to create the spreadsheet may no longer be available or the software may have been updated such that the old spreadsheet cannot be accessed. Hard copies of numerical tabulations from the spreadsheet may have been saved, but unless the underlying equations, formulas, and logic that were used to create the numerical values are also clearly documented, it may be difficult to resolve inconsistencies or check for errors.

Recapitulation

- Spreadsheets provide a useful way of performing calculations by the procedures of slices.
- Spreadsheet calculations can be difficult to check and archive.

COMPUTER PROGRAMS

For more sophisticated analyses and complex slope, soil, and loading conditions, computer programs are generally used to perform the computations. Computer programs are available that can handle a wide variety of slope geometries, soil stratigraphies, soil shear strength, pore water pressure conditions, external loads, and internal soil reinforcement. Most programs also have capabilities for automatically searching for the most critical slip surface with the lowest factor of safety and can handle slip surfaces of both circular and noncircular shapes. Most programs also have graphics capabilities for displaying the input data and the results of the slope stability computations.

Types of Computer Programs

Two types of computer programs are available for slope stability analyses: The first type of computer program allows the user to specify as input data the slope geometry, soil properties, pore water pressure condi-

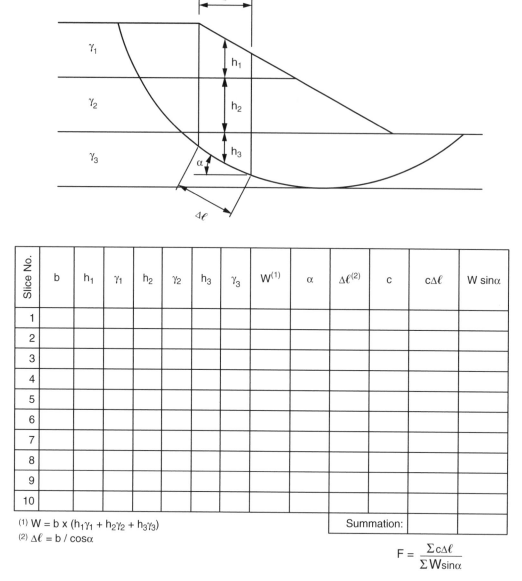

Slice No.	b	h_1	γ_1	h_2	γ_2	h_3	γ_3	$W^{(1)}$	α	$\Delta\ell^{(2)}$	c	$c\Delta\ell$	$W\sin\alpha$
1													
2													
3													
4													
5													
6													
7													
8													
9													
10													

(1) $W = b \times (h_1\gamma_1 + h_2\gamma_2 + h_3\gamma_3)$

(2) $\Delta\ell = b / \cos\alpha$

Summation:

$$F = \frac{\Sigma c\Delta\ell}{\Sigma W\sin\alpha}$$

Figure 7.3 Sample table for manual calculations using the Swedish circle ($\phi = 0$) procedure.

tions, external loads, and soil reinforcement, and computes a factor of safety for the prescribed set of conditions. These programs are referred to as *analysis programs*. They represent the more general type of slope stability computer program and are almost always based on one or more of the procedures of slices.

The second type of computer program is the *design program*. These programs are intended to determine what slope conditions are required to provide one or more factors of safety that the user specifies. Many of the computer programs used for reinforced slopes and other types of reinforced soil structures such as soil nailed walls are of this type. These programs allow the user to specify as input data general information about the slope geometry, such as slope height and external loads, along with the soil properties. The programs may also receive input on candidate reinforcement materials such as either the tensile strength of the reinforcement or even a particular manufacturer's product number along with various factors of safety to be achieved. The computer programs then determine what type and extent of reinforcement are required to produce suitable factors of safety. The design programs may be based on either procedures of slices or single-free-body procedures. For example, the logarithmic spiral procedure has been used in several computer

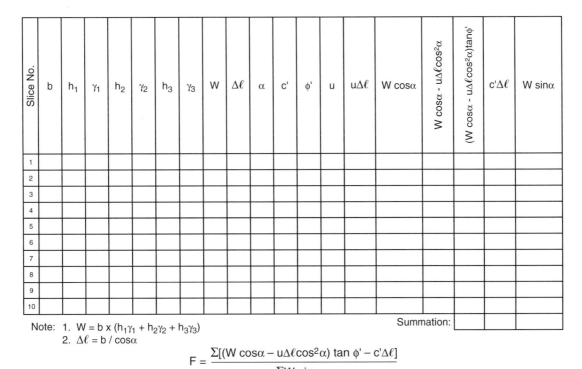

Slice No.	b	h_1	γ_1	h_2	γ_2	h_3	γ_3	W	$\Delta\ell$	α	c'	ϕ'	u	$u\Delta\ell$	W cosα	W cosα - $u\Delta\ell$cos$^2\alpha$	(W cosα - $u\Delta\ell$cos$^2\alpha$)tanϕ'	c'$\Delta\ell$	W sinα
1																			
2																			
3																			
4																			
5																			
6																			
7																			
8																			
9																			
10																			

Note: 1. W = b × $(h_1\gamma_1 + h_2\gamma_2 + h_3\gamma_3)$
2. $\Delta\ell = b / \cos\alpha$

Summation:

$$F = \frac{\Sigma[(W\cos\alpha - u\Delta\ell\cos^2\alpha)\tan\phi' - c'\Delta\ell]}{\Sigma W\sin\alpha}$$

Figure 7.4 Sample table for manual calculations using the Ordinary Method of Slices and effective stresses.

programs for both geogrid and soil nail design (Leshchinsky, 1997; Byrne, 2003[2]). The logarithmic spiral procedure is very well suited for such applications where only one soil type may be considered in the cross section.

Design programs are especially useful for design of reinforced slopes using a specific type of reinforcement (e.g., geogrids or soil nails) and can eliminate much of the manual trial-and-error effort required. However, the design programs are usually restricted in the range of conditions that can be handled and they often make simplifying assumptions about the potential failure mechanisms. Most analysis program can handle a much wider range of slope and soil conditions.

Automatic Searches for Critical Slip Surface

Almost all computer programs employ one or more schemes for searching for a critical slip surface with the minimum factor of safety. Searches can be performed using both circular and noncircular slip surfaces. Usually, different schemes are used depending

on the shape (circular vs. noncircular) of slip surface used. Many different search schemes have been used, and it is beyond the scope of this chapter to discuss these in detail. Nevertheless, several recommendations and guidelines can be offered for searching for a critical slip surface:

1. *Start with circles.* It is almost always preferable to begin searching for a critical slip surface using circles. Very robust schemes exist for searching with circles, and it is possible to examine a large number of possible locations for a slip surface with relatively little effort on the part of the user.

2. *Let stratigraphy guide the search.* For both circular and noncircular slip surfaces, the stratigraphy often suggests where the critical slip surface will be located. In particular, if a relatively weak zone exists, the critical slip surface is likely to pass through it. Similarly, if the weak zone is relatively thin and linear, the slip surface may follow the weak layer and is more likely to be noncircular than circular.

3. *Try multiple starting locations.* Almost all automatic searches begin with a slip surface that the user specifies in some way. Multiple starting locations should be tried to determine if one location leads to a lower factor of safety than another.

[2]Byrne has utilized the log spiral procedure in an unreleased version of the GoldNail software. One of the authors (Wright) has also used the log spiral successfully for this purpose in unreleased software for analyzing soil nail walls.

4. *Be aware of multiple minima.* Many search schemes are essentially optimization schemes that seek to find a single slip surface with the lowest factor of safety. However, there may be more than one "local" minimum and the search scheme may not necessarily find the local minimum that produces the lowest factor of safety overall. This is one of the reasons why it is important to use multiple starting locations for the search.

5. *Vary the search constraints and other parameters.* Most search schemes require one or more parameters that control how the search is performed. For example, some of the parameters that may be specified include:
   • The incremental distances that the slip surface is moved during the search
   • The maximum depth for the slip surface
   • The maximum lateral extent of the slip surface or search
   • The minimum depth or weight of soil mass above the slip surface
   • The maximum steepness of the slip surface where it exits the slope
   • The lowest coordinate allowed for the center of a circle (e.g., to prevent inversion of the circle)
   Input data should be varied to determine how these parameters affect the outcome of the search and the minimum factor of safety.

A relatively large number of examples and benchmarks can be found in the literature for the factor of safety for a particular slip surface. However, many fewer examples can be found to confirm the location of the most critical slip surface (lowest factor of safety), even though this may be the more important aspect of verification. For complex slopes, much more effort is usually spent in a slope stability analysis to verify that the most critical slip surface is found than is spent to verify that the factor of safety for a given slip surface has been computed correctly.

Restricting the Critical Slip Surfaces of Interest

In general, all areas of a slope should be searched to find the critical slip surface with the minimum factor of safety. However, is some cases it may be desirable to search only a certain area of the slope by restricting the location of trial slip surfaces. There are two common cases where this is appropriate. One case is where there are insignificant modes of failure that lead to low factors of safety, but the consequences of failure are small. The other case is where the slope geometry is

such that a circle with a given center point and radius does not define a unique slip surface and slide mass. These two cases are described and discussed further below.

Insignificant modes of failure. For cohesionless slopes it has been shown that the critical slip surface is a very shallow plane, essentially coincident with the face of the slope. However, the consequences of a slide where only a thin layer of soil is involved may be very low and of little significance. This is particularly the case for some mine tailings disposal dams. In such cases it is desirable to investigate only slip surfaces that have some minimum size and extent. This can be done in several ways, depending on the particular computer program being used:

   • The slip surfaces investigated can be required to have a minimum depth.
   • The slip surfaces investigated can be forced to pass through a specific point at some depth below the surface of the slope.
   • The soil mass above the slip surface can be required to have a minimum weight.
   • An artificially high shear strength, typically expressed by a high value of cohesion, can be assigned to a zone of soil near the face of the slope so that shallow slip surfaces are prevented. In doing so, care must be exercised to ensure that slip surfaces are not unduly restricted from exiting in the toe area of the slope.

Ambiguities in slip surface location. In some cases it is possible to have a circle where more than one segment of the circle intersects the slope (Figure 7.5). In such cases there is not just a single soil mass above the slip surface, but rather there are multiple, disassociated soil masses, probably with different factors of safety. To avoid ambiguities in this case, it is necessary to be able to designate that only a particular portion of the slope is to be analyzed.

Recapitulation

• Computer programs can be categorized as design programs and analysis programs. Design programs are useful for design of simple reinforced slopes, while analysis programs generally can handle a much wider range of slope and soil conditions.
• Searches to locate a critical slip surface with a minimum factor of safety should begin with circles.

- Multiple searches with different starting points and different values for the other parameters that affect the search should be performed to ensure that the most critical slip surface is found.
- In some case it is appropriate to restrict the region where a search is conducted; however, care must be taken to ensure that an important slip surface is not overlooked.

VERIFICATION OF ANALYSES

Most slope stability analyses are performed using general-purpose computer programs. The computer programs offer a number of features and may involve tens of thousands, and sometimes millions, of lines of computer code with many possible paths through the logic, depending on the problem being solved. Forester and Morrison (1994) point out the difficulty of checking even simple computer programs with multiple

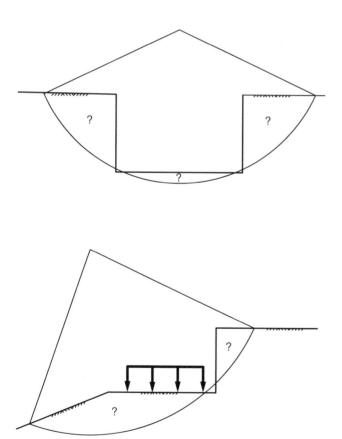

Figure 7.5 Cases where the slide mass defined by a circular slip surface is ambiguous and may require selective restriction.

combinations of paths through the software. Consider, for example, a comprehensive computer program for slope stability analysis that contains the features listed in Table 7.1. Most of the more sophisticated computer programs probably contain at least the number of options or features listed in this table. Although some programs will not contain all of the options listed, they may contain others. A total of 40 different features and options is listed in Table 7.1. If we consider just two different possibilities for the input values for each option or feature, there will be a total of over 1×10^{12} ($= 2^{40}$) possible combinations and paths through the software. If we could create, run, and verify problems to test each possible combination at the rate of one test problem every 10 minutes, over 20 million years would be required to test all possible combinations, working 24 hours a day, 7 days a week. Clearly, it is not possible to test sophisticated computer programs for all possible combinations of data, or even a reasonably small fraction, say 1 of 1000, of the possible combinations. Consequently, there is a significant possibility that any computer program being used has not been tested for the precise combination of paths involved in a particular problem.

Because it is very possible that any computer program has not been verified for the particular combination of conditions the program is being used for, some form of independent check should be made of the results. This is also true for other methods of calculation. For example, spreadsheets are just another form of computer program, and the difficulty of verifying spreadsheet programs was discussed earlier. It is also possible to make errors in using slope stability charts and even in using simple equations. Furthermore, the simple equations generally are based on approximations that can lead to important errors for some applications. Consequently, regardless of how slope stability computations are performed, some independent check should be made of the results. A number of examples of slope stability analyses and checks that can be made are presented in the next section.

Recapitulation

- Because of the large number of possible paths through most computer programs, it is likely most programs have not been tested for the precise combination of paths involved in any particular analysis.
- Some check should be made of the results of slope stability calculations, regardless of how the calculations are performed.

Table 7.1   Possible Options and Features for a Comprehensive Slope Stability Computer Program

Soil shear strength	Soil profile lines—stratigraphy
	c–ϕ soil—total stresses
	c'–σ' soil—effective stresses
	Curved Mohr failure envelope—total stresses
	Curved Mohr failure envelope—effective stresses
	Undrained shear strength varies with depth below horizontal datum
	Undrained shear strength defined by contour lines or interpolation
	Shear strength defined by a c/p ratio
	Anisotropic strength variation—undrained strength and total stresses
	Anisotropic strength variation—drained strength and effective stresses
	Consolidated–undrained shear strength (e.g., for rapid drawdown—linear strength envelopes)
	Consolidated–undrained shear strength (e.g., for rapid drawdown—curved strength envelopes)
	Structural materials (e.g., steel, concrete, timber)
Pore water pressure	Constant pore water pressure
	Constant pore pressure coefficient, r_u
	Piezometric line
	Phreatic surface
	Interpolated values of pore water pressure (e.g., from finite element analyses)
	Interpolated values of pore water pressure coefficient, r_u
	Slope geometry
	Left vs. right face of slope analyzed
	Distributed surface loads (e.g., water)
	Line loads
Reinforcement	Geotextiles
	Geogrids
	Soil nails
	Tieback anchors
	Piles
	Piers
Slip surface(s)	Individual circle
	Individual noncircular slip surface
	Systematic search with circles
	Random search with circles
	Systematic search with noncircular slip surfaces
	Random search with noncircular slip surfaces
Procedure of analysis	Simplified Bishop procedure
	Spencer's procedure
	Corps of Engineers' Modified Swedish procedure
	Simplified Janbu procedure
	Chen and Morgenstern's procedure

EXAMPLES FOR VERIFICATION OF STABILITY COMPUTATIONS

Ten example slopes were selected for the slope stability analyses presented in this section. These examples were selected with two purposes in mind: First, to illustrate the different methods for computing the factor of safety that were discussed in the preceding sections of this chapter, and second, to illustrate several impor-

tant details and features of slope stability analyses. For example, one problem addresses the use of submerged unit weights. Several other problems illustrate the differences among various procedures of slices. Some of these and other examples illustrate the importance of locating the critical slip surface. Most of the examples are presented with enough detail that they can be used as benchmarks for verifying results of calculations using other means (e.g., with other computer programs).

The 10 example problems selected for analysis are summarized in Table 7.2. Each example is described briefly and the methods of calculation (simple equations, charts, spreadsheets, and computer programs) are indicated. Any calculations presented using computer programs were performed with the UTEXAS4 software (Wright, 1999) unless otherwise stated. The summary also indicates whether analyses were performed for short- or long-term stability conditions. Additional features illustrated by each example are indicated in the last column of Table 7.2. The 10 cases listed in this table provide a useful collection of problems for computer program verification.

Example 1: Unbraced Vertical Cut in Clay

Tschebotarioff (1973) describes the failure of a vertical excavated slope that was made for a two-story basement in varved clay. The excavation was made, without bracing, to a depth of 22 ft on one side and 31.5 ft on the other side. The average unconfined compressive strength of the clay from an investigation nearby was reported to be 1.05 tons/ft^2 and the unit weight of the clay was 120 lb/ft^3. Factors of safety were calculated for the deeper of the two cuts (Figure 7.6) using the equation for a vertical slope with a plane slip surface, and using the slope stability charts presented in the Appendix. Calculations were also performed using a computer program. For an undrained shear strength, S_u of 1050 psf ($= q_u/2$), the factor of safety for a plane slip surface is calculated as

$$F = \frac{4c}{\gamma H} = \frac{(4)(1050)}{(120)(31.5)} = 1.11 \quad (7.27)$$

Using Janbu's charts for $\phi = 0$ presented in the Appendix, the factor of safety is calculated as

$$F = N_0 \frac{c}{\gamma H} = (3.83)\frac{1050}{(120)(31.5)} = 1.06 \quad (7.28)$$

Calculations with circles using the computer program resulted in a factor of safety of 1.06. The calculations with the charts confirm the results with the computer program, and both show that circular slip surfaces give a slightly lower factor of safety than plane slip surfaces.

Although the foregoing calculations are in close agreement, they may not correctly reflect the true factor of safety of the slope. Terzaghi (1943) pointed out that the upper part of the soil adjacent to a vertical slope is in tension. If the soil cannot withstand tension, cracks will form and the factor of safety will be reduced. Terzaghi showed that if one conservatively estimates that a crack will form to a depth equal to one-half the slope height, the equation for the factor of safety (assuming a planar slip surface) becomes

$$F = 2.67\frac{c}{\gamma H} \quad (7.29)$$

Thus, for the slope described above,

$$F = \frac{(2.67)(1050)}{(120)(31.5)} = 0.74 \quad (7.30)$$

which would clearly indicate that the slope was not stable. A computed factor of safety less than 1.0 for this case seems reasonable, because the slope failed and the unconfined compression tests that were used to measure the shear strength would be expected to underestimate strength due to sample disturbance.

In the first calculations with the computer program, tension was observed on the bottoms of several of the slices near the upper part of the slope. Subsequently, a series of slope stability calculations was performed in which vertical tension cracks were introduced, beginning with a crack depth of 1 ft, and successively increasing the crack depth in 1-ft increments until there was no longer tension. The assumed crack depths, corresponding factors of safety, and minimum normal stresses on the base of slices are summarized in Table 7.3. If we take the factor of safety as being the value where the tensile stresses are first eliminated, we would conclude that the factor of safety is less than 1 (between 0.96 and 0.99).

For this example the stability calculations support the behavior observed quite well. However, the closeness of the factor of safety to unity may be due in part to compensating errors caused by factors that were not considered. The shear strengths used were based on unconfined compression tests, which typically underestimate the shear strength. Thus, it is likely that the undrained shear strength of the clay was actually greater than what was assumed. At the same time, because the slope was excavated, the unloading due to excavation would cause the soil to swell gradually and lose strength with time. Also, it is possible that vertical cracks may have opened to substantial depths. It is possible to imagine that the undrained strength measured in more appropriate UU tests would have been considerably higher than the shear strength used, while losses of strength due to swell and the development of deep tension cracks could have reduced the stability by a substantial amount. These offsetting factors could have affected the stability of the slope significantly, and it can be seen that the failure may have taken place

Table 7.2 Summary of Example Problems for Verification of Slope Stability Analyses

No.	Description	Short or long term	Simple equations			Charts				Spreadsheets			Computer program		Additional features
			Vertical slope—plane	Bearing capacity	Infinite slope	$\phi = 0$: Janbu	$\phi = 0$: Hunter and Schuster	c, ϕ Soil: Janbu	$\phi = 0$	OMS	Simplified Bishop	Force equilibrium	UTEXAS4	Other	
1	Unbraced vertical cut in saturated clay (after Tschebotarioff); including effects of tension crack	S	Y			Y							Y		Effects of tension and a tension crack.
2	LASH terminal: submerged slope excavated in saturated, nearly normally consolidated clay	S					Y						Y		Use of total unit weights and pore water pressures vs. submerged unit weights.
3	Bradwell slip—excavated slope in stiff-fissured clay	S							Y				Y		Application of Janbu correction factor in simplified Janbu procedure. Slope may fail even with high factor of safety.
4	Hypothetical example of cohesionless slope ($c = 0$) on saturated clay ($\phi = 0$) foundation	S		Y						Y			Y		Application of Janbu correction factor in simplified Janbu procedure. Relatively large differences in F by various procedures.
5	Oroville Dam—high rockfill dam	L			Y								Y		Stability computations with a curved Mohr shear strength envelope.

114

No.	Example							Purpose
6	James Bay dike—embankments constructed on soft clay foundation	S		Y	Y			Importance of finding critical slip surface.
7	Homogeneous earth dam with steady-state seepage	L			Y		Y	Effects of how pore water pressures are represented (by flow net, piezometric line, phreatic surface). Illustrates effects of pore pressure in Ordinary Method of Slices.
8	Zoned (or clay core) earth dam with steady-state seepage	L	Y		Y			Effects of how pore water pressures are represented (by flow net, piezometric line, phreatic surface).
9	Reinforced slope (1): embankment on a soft clay foundation	S		Y	Y	Y		Reinforced slope analysis; influence of location of critical circle.
10	STABGM reinforced slope (2): steep reinforced slope	L		Y	Y	Y		Reinforced slope analysis; influence of location of critical circle.

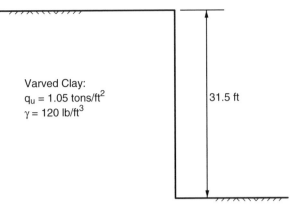

Varved Clay:
$q_u = 1.05$ tons/ft^2
$\gamma = 120$ lb/ft^3

31.5 ft

Figure 7.6 Unbraced vertical cut in clay described by Tschebotarioff (1973).

Table 7.3 Variation in the Factor of Safety and Minimum Normal Stress on the Slip Surface with the Assumed Depth of Tension Crack

Assumed crack depth (ft)	Minimum normal stress on slip surface (base of slices) (psf)	Calculated factor of safety
0	−241	1.06
1	−160	1.04
2	−67	1.01
3	−62	0.99
4	40	0.96

under conditions quite different from what was assumed in the stability calculations.

Recapitulation

- Slope stability charts, the computer program, and the simple equation for stability of a vertical cut based on plane slip surfaces all gave nearly identical values for the factor of safety.
- Plane slip surfaces, compared to circles, give similar but slightly higher values for the factor of safety of a vertical slope.
- Tensile stresses may develop behind the crest of steep slopes in clay and may lead to cracking that will substantially reduce the stability of the slope.
- Close agreement between computed and actual factors of safety may be fortuitous and a result of multiple large errors that compensate.

Example 2: Underwater Slope in Soft Clay

Duncan and Buchignani (1973) described the failure of a slope excavated underwater in San Francisco Bay. The slope was part of a temporary excavation and was designed with an unusually low factor of safety to minimize construction costs. During construction a portion of the excavated slope failed. A drawing of the slope cross section is shown in Figure 7.7. The undrained shear strength profile is presented in Figure 7.8. The original design factor of safety based on undrained shear strengths was reported by Duncan and Buchignani to be 1.17.

Recently (2003), new slope stability calculations were performed by the writers, first using a computer program with Spencer's procedure of slices. The minimum factor of safety calculated was 1.17. Because the undrained shear strength for the clay in the slope increases linearly with depth, Hunter and Schuster's (1968) slope stability charts described in the Appendix can also be used to compute the factor of safety. The factor of safety computed using these charts is 1.18.

The slope stability calculations described above were performed using submerged (buoyant) unit weights to account for the slope being fully submerged. Submerged unit weights are convenient to use when the computations are being performed with either slope stability charts or by hand using a spreadsheet. Submerged unit weights can be used for this example because there was no seepage force (no flow of water). However, in general when using computer programs it is preferable to use total unit weights and to specify external and internal water pressures. Computer calculations were repeated for this slope using total unit weights and distributed loads on the surface of the slope to represent the water pressures. The factor of safety was again found to be 1.17. This not only confirms what is expected but provides a useful check on the calculations of the weights of slices and the forces due to external distributed loads calculated by the computer program.

A simple and useful check of any computer program is to perform separate sets of slope stability calculations for a submerged slope (with no flow) using (1) submerged unit weights and (2) total unit weights with water pressures. If the computer program is working properly and being used properly, it should give the same result for both sets of calculations.[3]

---

[3]This may not be true with force equilibrium procedures with inclined interslice forces. Similar results may not be obtained with submerged unit weights and total unit weights plus water pressures when the interslice forces are total forces, due to both earth and water pressures, as described in Chapter 6. In this case the differences in factors of safety calculated using submerged unit weights and total unit weights plus water pressures may be large.

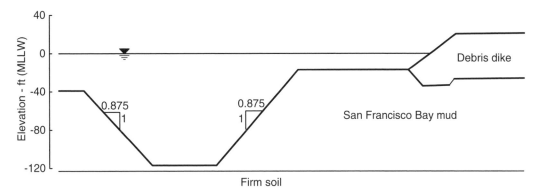

Figure 7.7 Underwater slope in San Francisco Bay mud described by Duncan and Buchignani (1973) and Duncan (2000).

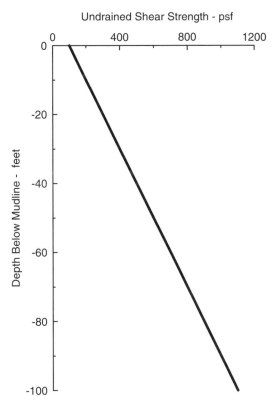

Figure 7.8 Undrained shear strength profile for underwater slope in San Francisco Bay mud. (From Duncan, 2000.)

the probability of failure was almost 20%. This probability of failure is consistent with the fact that about 20% of the length of the slope actually failed. Given the accuracy with which such analyses can be made, the close agreement between the probability of failure and the fraction of the slope that failed is probably fortuitous.

Because this slope was only temporary, it was appropriate to compute the stability using undrained shear strengths. However, if the slope was permanent, much lower drained shear strengths would apply. As the soil swells due to unloading by excavation, the shear strength would gradually be reduced. Eventually, the fully drained shear strength would become applicable. Representative values of the drained (effective stress) shear strength parameters for San Francisco Bay mud are $c' = 0$, $\phi' = 34.5°$ (Duncan and Seed, 1966b). For a fully submerged slope and $c' = 0$, the factor of safety can be calculated using the equation for an infinite slope as

$$F = \frac{\tan \phi'}{\tan \beta} = \frac{\tan 34.5°}{1/0.875} = 0.60 \qquad (7.31)$$

Clearly, this factor of safety (0.60) is much less that the factor of safety (1.17) based on undrained shear strengths, indicating that a substantial reduction in factor of safety would have occurred if the excavated trench had not been filled with sand.

Recapitulation

- Identical values for the factor of safety were obtained using a computer program and a slope stability chart.
- Either submerged unit weights or total unit weights and water pressures may be used to compute the stability of a submerged slope when there is no flow.

Although the calculations presented above confirm the factor of safety calculated by Duncan and Buchignani (1973) and indicate that the slope would be expected to be stable, a portion of the slope failed, as noted earlier. Duncan and Buchignani (1973) showed that the effects of sustained loading (creep) under undrained conditions was probably sufficient to reduce the shear strength and cause the failure. More recent reliability analyses by Duncan (2000) have shown that

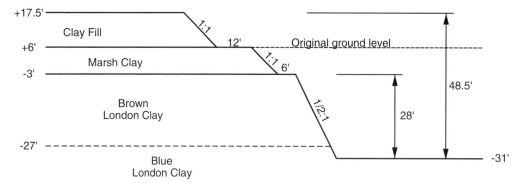

Figure 7.9 Cross section of excavated slope for reactor 1 at Bradwell. (From Skempton and LaRochelle, 1965.)

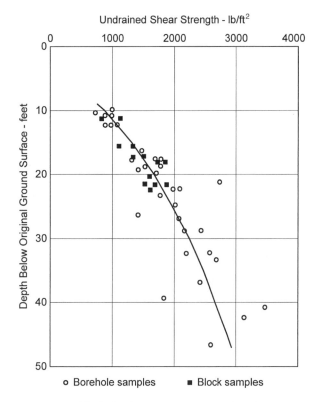

Undrained Shear Strength - lb/ft²

o Borehole samples   ■ Block samples

Figure 7.10 Undrained shear strength profile for reactor 1 excavation slope at Bradwell. (From Skempton and La-Rochelle, 1965.)

- Even though the calculated factor of safety was greater than unity (1.17), the slope failed due to creep strength loss.
- For an excavated slope, the short-term factor of safety based on undrained conditions may be much higher than the long-term factor of safety based on drained conditions.

Example 3: Excavated Slope in Stiff-Fissured Clay

Skempton and LaRochelle (1965) describe a deep excavation in the London Clay at Bradwell. A cross section of the excavation for reactor 1 is shown in Figure 7.9. The excavation is 48.5 ft deep. The lower 28 ft of the excavation is in London Clay and is inclined at $\frac{1}{2}$(horizontal):1(vertical). The London Clay is overlain by 9 ft of Marsh clay where the excavation slope was inclined at 1:1 (45°). Approximately 11.5 ft of clay from the excavation was placed at the top of the excavation, over the marsh clay. The clay fill was also inclined at 1:1.

Short-term stability analyses were performed for the slope using undrained shear strengths. The marsh clay was reported to have an average undrained shear strength of 300 psf and a total unit weight of 105 pcf. The clay fill was assumed to crack to the full depth of the fill (11.5 ft), and thus its strength was ignored. Skempton and LaRochelle reported a total unit weight of 110 pcf for the fill. The undrained shear strength profile for the London Clay is shown in Figure 7.10. The undrained shear strength increases at a decreasing

Table 7.4 Summary of Short-Term Slope Stability Analyses for an Excavated Slope in Stiff-Fissured Clay: The Bradwell Slip

Procedure of slices	Factor of safety
Spencer	1.76
Simplified Bishop	1.76
Corps of Engineers' Modified Swedish	1.80
Simplified Janbu—no correction	1.63
Simplified Janbu—with correction, f_0	1.74

rate with depth. A representative unit weight for the London Clay at the site is 120 pcf.

Stability computations were first performed for this example using a computer program and several procedures of slices. The resulting factors of safety are summarized in Table 7.4. The values for the factor of safety are as expected: Spencer's procedure and the Simplified Bishop procedure give identical values because they both satisfy moment equilibrium; there is only one value for the factor of safety that will satisfy moment equilibrium for a circular slip surface. The Corps of Engineers' Modified Swedish procedure, a force equilibrium procedure, overestimates the factor of safety compared to procedures that satisfy complete equilibrium, as is commonly the case. The Simplified Janbu procedure (force equilibrium with horizontal interslice forces) without Janbu et al.'s (1956) correction factor underestimates the factor of safety, as is also typically the case. The correction factor, f_0, for the Simplified Janbu procedure was calculated from the following equation presented by Abramson et al., (2002):

$$f_0 = 1 + b_1 \left[\frac{d}{L} - 1.4 \left(\frac{d}{L} \right)^2 \right] \qquad (7.32)$$

where b_1 is a factor that depends on the soil type (c and ϕ) and d/L represents the slide depth-to-length

ratio. For $\phi = 0$, b_1 is 0.69 and the depth-to-length ratio for the critical circle found for the Simplified Janbu procedure is 0.13. The resulting correction factor calculated from Eq. (7.32) is 1.07 and the corrected factor of safety is 1.74 (= 1.07 × 1.63). This corrected value (1.74) for the factor of safety by the Simplified Janbu procedure agrees well with the value (1.76) calculated by procedures that satisfy moment equilibrium.

The factor of safety was also calculated manually using a spreadsheet program based on the Ordinary Method of Slices. Because ϕ is zero for this problem and the Ordinary Method of Slices satisfies moment equilibrium, the Ordinary Method of Slices should give the same value for the factor of safety as Spencer's and the Simplified Bishop procedures. There is no need to use a more complex procedure than the Ordinary Method of Slices for this case. The calculations for the Ordinary Method of Slices are shown in Figure 7.11. As expected, the factor of safety is 1.76, which is the same as the value shown previously for Spencer's and the Simplified Bishop procedures.

Although the factor of safety calculated for this slope is almost 1.8, the slope failed approximately 5 days after excavation was completed. Skempton and LaRochelle (1965) discuss the probable causes of failure. These include overestimates of the shear strength due to testing of samples of small size, strength losses due to sustained loading (creep), and the presence of fissures. Skempton and LaRochelle concluded that the

Slice No.	b (ft)	h_{fill} (ft)	γ_{fill} (pcf)	h_{marsh} (ft)	γ_{marsh} (pcf)	h_{clay} (ft)	γ_{clay} (pcf)	W (pounds)	α (deg)	$\Delta\ell$ (ft)	c (psf)	$c\Delta\ell$	$W \sin\alpha$
1	10.8	11.5	110	4.5	105	-	-	18,748	39.8	14.1	300	4,215	12,008
2	9.1	11.5	110	9.0	105	3.2	120	23,637	35.1	11.1	1069	11,908	13,602
3	11.5	5.8	110	9.0	105	9.8	120	31,665	30.5	13.3	1585	21,157	16,082
4	12.0	-	-	9.0	105	16.1	120	34,466	25.6	13.3	1968	26,183	14,873
5	9.0	-	-	4.5	105	20.7	120	26,857	21.3	9.7	2222	21,461	9,636
6	6.0	-	-	-	-	23.4	120	16,862	18.3	6.3	2349	14,841	5,284
7	5.5	-	-	-	-	19.7	120	13,002	16.0	5.7	2429	13,900	3,590
8	8.5	-	-	-	-	7.5	120	7,645	13.3	8.7	2503	21,861	1,759
										Summations:		135,525	76,835

$$F = \frac{135,525}{76,835} = 1.76$$

Figure 7.11 Manual calculations by the Ordinary Method of Slices for short-term stability of the slope at Bradwell.

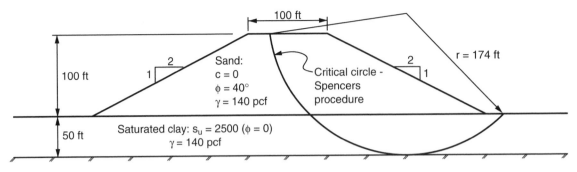

Figure 7.12 Cohesionless fill slope on saturated clay foundation.

opening of fissures and a lower, residual strength along the fissures were probable causes of failure of the slope, even though the factor of safety computed based on undrained shear strengths was relatively high.

Recapitulation

• Spencer's, the Simplified Bishop, and Ordinary Method of Slices procedures all gave the same value for the factor of safety for circular slip surfaces because $\phi = 0$, and all these procedures satisfy moment equilibrium.

• The computer solution and the manual solution using a spreadsheet gave the same value for the factor of safety.

• The Corps of Engineers' Modified Swedish procedure overestimated the factor of safety for this case by a small amount (2%).

• The Simplified Janbu procedure without the correction factor applied underestimated the factor of safety by about 7%.

• The corrected factor of safety by the Simplified Janbu procedure agrees within 1% with the value of the factor of safety calculated using methods that satisfy moment equilibrium.

• Although the factor of safety for short-term stability was much greater than 1, the slope failed approximately five days after construction, due to several factors that influenced the shear strength.

Example 4: Cohesionless Slope on Saturated Clay Foundation

The fourth example is for a hypothetical embankment constructed of cohesionless granular material resting on a saturated clay ($\phi = 0$) foundation, as shown in Figure 7.12. The embankment is assumed to drain almost instantaneously, and thus its strength will not change over time. The clay in the foundation is expected to consolidate with time and its strength is ex-

pected to increase with time. Therefore, the critical period (lowest factor of safety) for the embankment should be immediately after construction.

Soil shear strength and unit weight properties are shown in Figure 7.12. Drained (effective stress) shear strength parameters are shown for the embankment, and undrained shear strengths are shown for the clay foundation. Stability computations were first performed using a computer program and several procedures of slices. The minimum factors of safety for various procedures are summarized in Table 7.5, and the critical slip surface by Spencer's procedure is shown in Figure 7.12.

As expected, the Simplified Bishop procedure gives a value for the factor of safety that is very close to the one calculated by Spencer's procedure. The Simplified Janbu procedure without the correction factor applied gives a factor of safety that is approximately 10% lower, but the corrected value (1.16) agrees closely with the values by the Simplified Bishop and Spencer's procedures. The Corps of Engineers' Modified Swedish procedure produced a factor of safety about 25% higher than the value by Spencer's procedure. The much higher value clearly demonstrates the potentially

Table 7.5 Summary of Slope Stability Analyses for a Cohesionless Embankment Supported by a Saturated Clay Foundation

Procedure of slices	Factor of safety
Spencer	1.19
Simplified Bishop	1.22
Corps of Engineers' Modified Swedish	1.54
Simplified Janbu—no correction	1.07
Simplified Janbu—with correction $f_0{}^a$	1.16

[a]Correction based on Eq. (7.32) with $b_1 = 0.5$ and $d/L = 0.34$; $f_0 = 1.09$.

unconservative nature of the Modified Swedish force equilibrium procedure.

The factor of safety was also computed using the Ordinary Method of Slices with a spreadsheet program for the critical circle found by Spencer's procedure. The computations are shown in Figure 7.13. The computed factor of safety is 1.08, approximately 10% less than the value calculated using Spencer's procedure. Differences of this order (10%) are typical for cases like this one where c and ϕ vary significantly along the slip surface.

As an additional, approximate check on the stability of the embankment, the bearing capacity equation [Eq. (7.10)] was used to calculate a factor of safety. This gave

$$F = 5.53 \frac{2500}{(140)(100)} = 0.99 \qquad (7.33)$$

Although the bearing capacity solution represented by Eq. (7.33) underestimates the stability of the embankment in this example, it provides a simple and convenient way of preliminary screening for potential problems. In general, if the factor of safety for bearing capacity is near or below 1, the factor of safety is likely to be marginal and additional, more detailed analyses are probably warranted.

Recapitulation

- Spencer's procedure and the Simplified Bishop procedure give very similar values for the factor of safety.
- The Corps of Engineers' Modified Swedish procedure can substantially overestimate the factor of safety.
- The Simplified Janbu procedure without the correction factor applied underestimated the factor of safety, but the value is improved by applying the correction.
- The Ordinary Method of Slices underestimates the factor of safety but provides a convenient way of checking a computer solution using more accurate methods.
- The simple equation for bearing capacity on a saturated clay foundation gives a conservative estimate of stability but provides a useful tool for screening for stability problems.

Example 5: Cohesionless Embankment (Oroville Dam)—Curved Mohr–Coulomb Envelope

The next example is of the Oroville Dam, in particular, the stability of the downstream slope (Figure 7.14).

The downstream slope is composed of rockfill (amphibolite gravel). As for most granular materials, the Mohr failure envelope is curved. Due to the great height of the Oroville Dam (778 ft) and the large variation in the pressures from the top to the bottom of the embankment, the curved Mohr failure envelope requires special consideration for the slope stability computations.

Curved (nonlinear) Mohr failure envelope. For this example the shear strength of the downstream shell material is characterized by a *secant friction angle* (i.e., $\tan \phi' = \tau_f/\sigma_f'$), which represents the slope of a line drawn from the origin of the Mohr diagram to a point on the Mohr failure envelope. As discussed in Chapter 5, the secant friction angle varies with confining pressure and can be related to the minor principal stress, σ_3' by

$$\phi' = \phi_0 - \Delta\phi \log_{10} \frac{\sigma_3'}{p_a} \qquad (7.34)$$

where ϕ_0 is the friction angle for a confining pressure (σ_3') of 1 atm, $\Delta\phi$ is the change in friction angle per log-cycle (10-fold) change in confining pressure, and p_a is atmospheric pressure. Duncan et al. (1989) summarize shear strength data for the Oroville dam and report values of $\phi_0 = 51°$ and $\Delta\phi = 6°$ for the shell material.

For slope stability computations the shear strength needs to be defined by a Mohr failure envelope that expresses the shear strength, τ, as a function of the normal stress, σ or σ', depending on whether total or effective stress analyses are being performed. The normal stress, σ', should be the normal stress on the failure plane at failure, σ_{ff}'. The relationship between σ_{ff}' and confining pressure, σ_3', depends on the shear strength parameters. For a cohesionless soil ($c' = 0$) the relationship is expressed as

$$\sigma_{ff}' = \sigma_{3f}' \frac{\cos^2\phi'}{1 - \sin\phi'} \qquad (7.35)$$

Ratios of σ_{ff}' to σ_{3f}' are tabulated in Table 7.6 for a range in friction angles that are representative for granular materials. The ratios shown in Table 7.6 vary with friction angle; however, it is convenient to assume that the ratio, $\sigma_{ff}'/\sigma_{3f}'$ is constant and equal to 1.5 for any value of ϕ'. This assumption has very little effect on the Mohr failure envelope that is subsequently computed. To illustrate this, consider a material with a value of ϕ_0 of 40° and $\Delta\phi$ of 10°. Computation of the friction angles, ϕ', for values of effective normal stress, σ' (= σ_{ff}'), of 100, 1000, and 10,000 psf are shown in Table 7.7. For each value of the effective

Slice No.	b (ft)	h_{fill} (ft)	γ_{fill} (pcf)	h_{clay} (ft)	γ_{clay} (pcf)	W (lb)	$\Delta\ell$	α	c (psf)	ϕ (deg)	W cosα	(W cosα) tanφ'	c'Δℓ	W sinα
1	10.7	19.2	140	0.0	125	28706	40	74.5	0	40	7666	6432	0	27664
2	19.0	56.0	140	0.0	125	148726	40	61.6	0	40	70653	59285	0	130872
3	22.5	86.8	140	0.0	125	273040	35	49.6	0	40	176932	148464	0	207956
4	22.7	100.0	140	9.3	125	343995	29	39.3	2500	0	266206	0	73301	217869
5	35.2	91.2	140	28.0	125	572546	40	28.1	2500	0	504909	0	99788	269954
6	38.5	72.8	140	42.6	125	597628	40	15.3	2500	0	576542	0	99796	157348
7	27.3	56.3	140	49.0	125	382166	27	4.4	2500	0	381040	0	68427	29322
8	39.7	39.6	140	47.8	125	456562	40	-6.4	2500	0	453677	0	99781	-51247
9	37.7	20.3	140	38.9	125	290072	40	-19.3	2500	0	273752	0	99789	-95926
10	21.7	5.4	140	26.1	125	87202	25	-29.8	2500	0	75702	0	62405	-43283
11	24.7	0.0	140	10.0	125	30754	32	-38.9	2500	0	23931	0	79325	-19317
										Summation:	214181	682612		831211

$$F = \frac{\Sigma[(W\cos\alpha)\tan\phi + c\Delta\ell]}{\Sigma W\sin\alpha} = \frac{214181 + 682612}{831211} = 1.08$$

Figure 7.13 Manual calculations for stability of cohesionless fill slope on saturated clay foundation using the Ordinary Method of Slices and the critical circular slip surface found using Spencer's procedure.

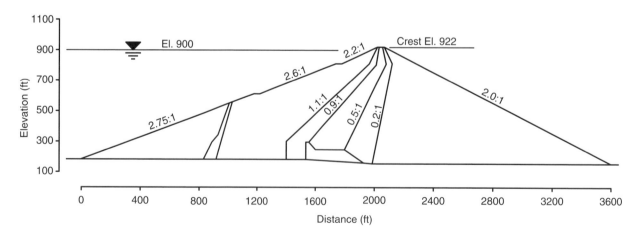

Figure 7.14 Cross section of Oroville Dam.

Table 7.6   Relationship Between the Ratio $\sigma'_{ff}/\sigma'_{3f}$ and the Friction Angle

ϕ' (deg)	$\dfrac{\sigma'_{ff}}{\sigma'_{3f}}$
30	1.50
40	1.64
50	1.77

normal stress, the confining pressure, σ'_{3f}, was computed assuming values of 1.5, 1.65, and 1.8 for the ratio $\sigma'_{ff}/\sigma'_{3f}$. The resulting confining pressures were then used to compute the friction angles from Eq. (7.34). The assumed value for the ratio, $\sigma'_{ff}/\sigma'_{3f}$, can be seen to have very little effect on the friction angle computed for a given confining pressure. The maximum difference between the friction angles computed assuming 1.5 and 1.8 for the ratio, $\sigma'_{ff}/\sigma'_{3f}$, was only 0.8°. The shear stress, τ_{ff}, defining the Mohr failure envelope for a given normal stress, σ'_{ff}, is computed by multiplying the normal stress by the tangent of the friction angle (i.e., $\tau_{ff} = \sigma'_{ff} \tan \phi'$). A 0.8° difference in the friction angle corresponds to a difference of no more than 3% in the shear stress. If it is assumed that the ratio, $\sigma'_{ff}/\sigma'_{3f}$, is 1.5 when the value may actually be somewhat higher, the resulting value for the friction angle will be estimated slightly conservatively.

Based on the preceding discussion, a nonlinear Mohr failure envelope was determined for the Oroville Dam shell material using values of $\phi_0 = 51°$ and $\Delta\phi = 6°$, and assuming that $\sigma'_{ff}/\sigma'_{3f} = 1.5$. The envelope was defined by a series of discrete points computed using the following steps:

1. A range in values of normal stress was established to encompass the maximum range expected for Oroville Dam. The minimum normal stress of interest was considered to be the normal stress at a depth of 1 ft, which for a total unit weight of 150 pcf was taken to be 150 psf. The maximum normal stress was estimated based on the height of the dam. For a height of 770 ft and a unit weight of 150 pcf, the maximum stress is approximately 115,000 psf.

2. Specific values of normal stress, σ', ranging from 150 psf to the maximum were selected for computing points defining the Mohr–Coulomb failure envelope. Beginning with the minimum stress of 150 psf, a geometric progression of values was used (e.g., 150, 300, 600, 1200 psf). Particular attention was paid to selecting points at low stresses because it was anticipated that the critical slip surface would be relatively shallow due to the cohesionless nature of the Oroville Dam shell material.

3. For each value of normal stress the corresponding value of the confining pressure was computed as

Table 7.8   Points Calculated to Define the Nonlinear Mohr Failure Envelope for the Oroville Dam Shell Material

σ'	σ'_3	ϕ'	τ
150	100	59.0	250
300	200	57.1	465
600	400	55.3	870
1,200	800	53.5	1,625
2,400	1,600	51.7	3,040
4,800	3,200	49.9	5,705
9,600	6,400	48.1	10,705
19,200	12,800	46.3	20,100
38,400	25,600	44.5	37,740
76,800	51,200	42.7	70,865
153,600	102,400	40.9	133,015

Table 7.7   Computed Secant Friction Angles for Different Confining Pressures and Various Assumed Values for the Ratio σ'_{ff}/σ'_3

σ'_{ff} (psf)	$\dfrac{\sigma'_{ff}}{\sigma'_{3f}} = 1.5$		$\dfrac{\sigma'_{ff}}{\sigma'_{3f}} = 1.65$		$\dfrac{\sigma'_{ff}}{\sigma'_{3f}} = 1.8$	
	σ'_3 (psf)	ϕ' (deg)	σ'_3 (psf)	ϕ' (deg)	σ'_3 (psf)	ϕ' (deg)
100	67	55.0	61	55.4	56	55.8
1,000	667	45.0	606	45.4	556	45.8
10,000	6,667	35.0	6,061	35.4	5556	35.8

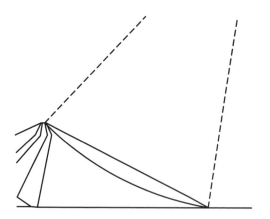

Figure 7.15 Critical circular slip surface for downstream slope of Oroville Dam.

$$\sigma_3' = \frac{\sigma'}{1.5} \qquad (7.36)$$

4. Secant values of the friction angle were computed for each value of σ_3' from Eq. (7.34) as

$$\phi' = 51° - 6° \log_{10} \frac{\sigma_3'}{p_a} \qquad (7.37)$$

where p_a (atmospheric pressure) is 2116 psf for units of pounds and feet that were used.

5. Shear stresses, τ, were calculated for each value of normal stress from

$$\tau = \sigma' \tan \phi' \qquad (7.38)$$

The values calculated for the Oroville Dam shell material are summarized in Table 7.8. The pairs of values of σ' and τ were used as points defining a nonlinear Mohr failure envelope for the slope stability

computations. Although not shown in this table, an additional point representing the origin ($\sigma' = 0$, $\tau = 0$) was included in the data defining the envelope for the slope stability computations. Nonlinear Mohr failure envelopes were also defined for the transition zone and the core of Oroville Dam; however, because the critical slip surface did not pass significantly through these zones, the shear strength data are not included here.

Slope stability computations. Slope stability computations were performed using the computer program and the nonlinear Mohr failure envelopes discussed earlier. Computations were performed using Spencer's procedure and circular slip surfaces. The critical slip surface is shown in Figure 7.15, and the minimum factor of safety is 2.28.

One way of checking the computer solution is to calculate the factor of safety manually using a procedure of slices such as the Ordinary Method of Slices or Simplified Bishop procedure. The friction angle could be varied for each slice depending on the normal stress, σ'. This is easiest to do with the Ordinary Method of Slices because the normal stress can be calculated independently of the shear strength using the following equation

$$\sigma' = \frac{W \cos \alpha - u \, \Delta l \cos^2\alpha}{\Delta l} \qquad (7.39)$$

With the Simplified Bishop procedure, the normal stress depends on the friction angle (i.e., the normal stress is part of the solution for the unknowns). Therefore, the normal stress must first be estimated to compute the friction angle, and then trial and error is used until the estimated and calculated values are in reasonable agreement. To estimate the friction angle initially for the Simplified Bishop procedure, either the normal stress can be estimated from the vertical overburden pressure or the normal stress can be calculated from Eq. (7.39) from the Ordinary Method of Slices.

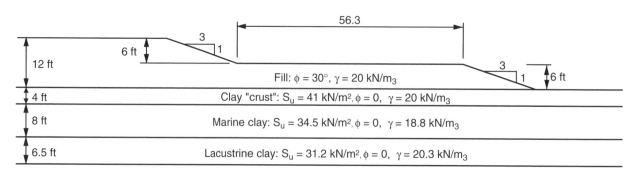

Figure 7.16 Cross section of James Bay dike.

Because the critical slip surface for the downstream slope was relatively shallow for this case, the infinite slope procedure was used to check the results of the computer solution. To do this the average normal stress was calculated for the critical slip surface found from the computer solution. The average normal stress was calculated using the equation

$$\sigma'_{av} = \frac{\sum \sigma_i \, \Delta l_i}{\sum \Delta l_i} \qquad (7.40)$$

where the summations were performed for all slices. The average normal stress calculated for the critical slip surface was 12,375 psf. From the nonlinear Mohr failure envelope in Table 7.8, the corresponding shear stress, τ, is 13,421 psf and the equivalent secant friction angle is 47.3°. The factor of safety based on an infinite slope is then

$$F = \frac{\tan \phi'}{\tan \beta} = \frac{\tan 47.3°}{1/2.0} = 2.17 \qquad (7.41)$$

This value (2.17) from the infinite slope analysis is within 5% of the value of 2.28 obtained from the computer solution with circular slip surfaces.

Recapitulation

- When the friction angle depends on the confining stress, the friction angle is expressed conveniently by a secant angle, which is a function of confining pressure, σ_3. This requires additional steps to determine an equivalent nonlinear Mohr failure envelope for slope stability analyses.

- To relate confining pressure, σ_3, to normal stress, σ, for a nonlinear Mohr failure envelope, the confining pressure can be assumed to be equal to two-thirds ($= 1/1.5$) the normal stress. This facilitates defining points on the Mohr failure envelope when the friction angle is defined in terms of confining pressure, σ_3.

- For shallow slides in cohesionless soils, stability computations can be checked with an infinite slope analysis, even when the Mohr failure envelope is nonlinear. When the Mohr failure envelope is nonlinear, the average normal stress on the slip surface from a computer solution can be used to define an equivalent secant friction angle that is then used in the infinite slope analysis.

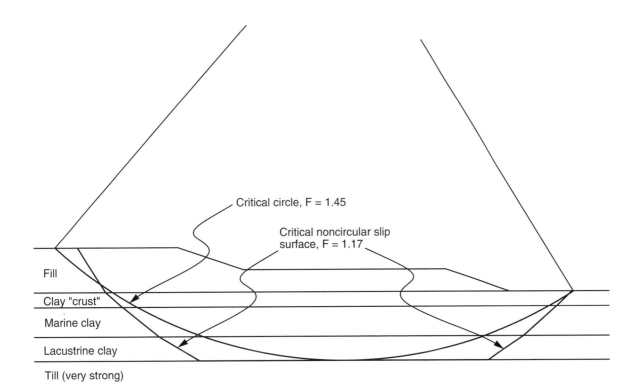

Fill

Clay "crust"

Marine clay

Lacustrine clay

Till (very strong)

Figure 7.17 Critical circular and noncircular slip surfaces for James Bay dike.

Example 6: James Bay Dike

The James Bay hydroelectric project involved the design of dikes that were to be constructed on soft and sensitive clays (Christian et al., 1994). A typical cross section of one of the planned dikes is shown in Figure 7.16. Soil properties for the materials in the dike and its foundation are summarized in this figure.

An analysis was first performed using circular slip surfaces, with a computer program and Spencer's procedure. The minimum factor of safety was calculated to be 1.45 for the critical circle, which is shown in Figure 7.17. This value of 1.45 for the factor of safety is the same as the value (1.453) that Christian et al. reported for the slope.

Additional analyses were performed using noncircular slip surfaces and an automatic search. The automatic search was started with the critical circle from the previous analyses. Ten points were used to define the slip surface. These points were shifted systematically using the search routine implemented in the computer program until a minimum factor of safety was found. The corresponding critical noncircular slip surface was then adjusted by adding some points and removing others. Points were adjusted so that there was a point located at the interfaces between soil layers, and these points were shifted in the horizontal direction until the minimum factor of safety was again found. The most critical noncircular slip surface found after searching is shown in Figure 7.17. The corresponding minimum factor of safety is 1.17, which is approximately 20% less than the minimum value computed using circles.

Christian et al. (1994) discussed the effects of variations and uncertainties in shear strength on the computed factors of safety for the James Bay dikes. They showed that the variation in shear strength could have an important effect on the evaluation of stability. The results presented in the preceding paragraph show that the effect of using noncircular slip surfaces is of comparable magnitude, thus illustrating the importance of locating the critical slip surface accurately.

To verify the computations with noncircular slip surfaces, additional computations were performed using a force equilibrium procedure and a computer spreadsheet program. For these computations the interslice forces were assumed to be parallel; the interslice force inclination was assumed to be the same as the inclination (2.7 degrees) determined from the computer solution with Spencer's procedure. The spreadsheet computations are presented in Figure 7.18. The computed factor of safety was 1.17, thus verifying the value calculated using the computer program.

Recapitulation

- Noncircular slip surfaces may give significantly lower factors of safety than circles.
- The critical circle provides a good starting point for searching for the critical noncircular slip surface.
- A force equilibrium solution using a spreadsheet with the interslice force inclination from Spencer's procedure provides a good method for checking a computer solution with Spencer's procedure and is applicable to circular and noncircular slip surfaces.

Example 7: Homogeneous Earth Dam with Steady-State Seepage

A homogeneous earth embankment resting on a relatively impervious foundation is illustrated in Figure 7.19. The embankment impounds water on one side, and steady-state seepage is assumed to have developed. Stability computations were performed for this embankment to evaluate the long-term stability of the downstream slope. Drained effective stress shear strength parameters were used and are shown on the cross section of the embankment.

Pore water pressures. Finite element seepage analysis was performed for the embankment to calculate pore water pressures. The GMS/SEEP2D software was used for this purpose (Tracy, 1991; EMRL, 2001). The entire cross section of the embankment was modeled with finite elements, and appropriate saturated or unsaturated hydraulic conductivities were assigned depending on the pore water pressure. The hydraulic conductivity for the saturated soil (positive pressures) was 1×10^{-5} ft/min. The hydraulic conductivity for the unsaturated soil was assumed to decrease sharply to a residual hydraulic conductivity equal to 0.1% of the value for saturated conditions as the water pressures decreased below atmospheric. This essentially restricted almost all flow to the saturated (positive water pressure) zone of the cross section. The finite element mesh used for this problem contained 589 node points and 1044 elements.

Finite element seepage analyses produced values of pore water pressure at each node point. These values of pore water pressure were then used to interpolate values of pore water pressure along each slip surface in the slope stability computations. Interpolation was performed using the triangle-based interpolation scheme described by Wright (2002). This scheme is very efficient and introduces negligible error caused by

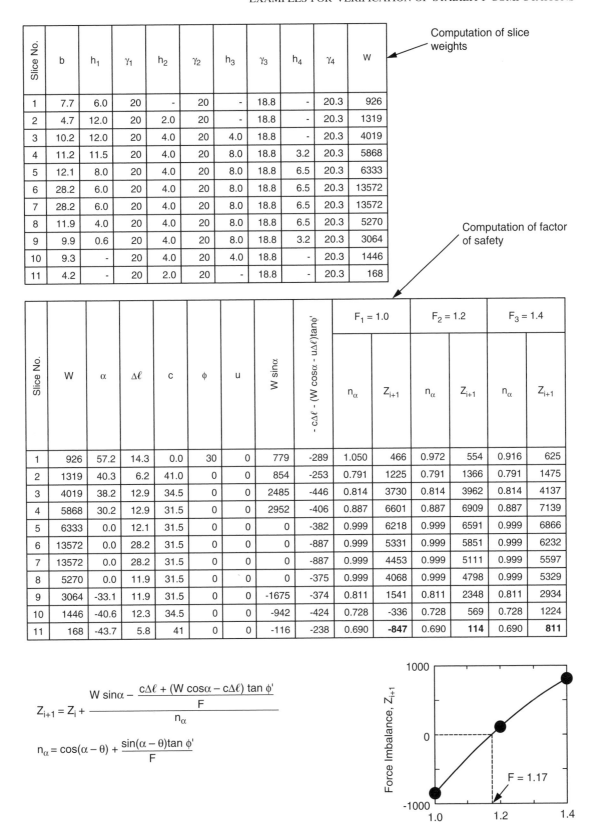

Computation of slice weights

Slice No.	b	h_1	γ_1	h_2	γ_2	h_3	γ_3	h_4	γ_4	W
1	7.7	6.0	20	-	20	-	18.8	-	20.3	926
2	4.7	12.0	20	2.0	20	-	18.8	-	20.3	1319
3	10.2	12.0	20	4.0	20	4.0	18.8	-	20.3	4019
4	11.2	11.5	20	4.0	20	8.0	18.8	3.2	20.3	5868
5	12.1	8.0	20	4.0	20	8.0	18.8	6.5	20.3	6333
6	28.2	6.0	20	4.0	20	8.0	18.8	6.5	20.3	13572
7	28.2	6.0	20	4.0	20	8.0	18.8	6.5	20.3	13572
8	11.9	4.0	20	4.0	20	8.0	18.8	6.5	20.3	5270
9	9.9	0.6	20	4.0	20	8.0	18.8	3.2	20.3	3064
10	9.3	-	20	4.0	20	4.0	18.8	-	20.3	1446
11	4.2	-	20	2.0	20	-	18.8	-	20.3	168

Computation of factor of safety

Slice No.	W	α	$\Delta\ell$	c	ϕ	u	W sinα	$-c\Delta\ell - (W\cos\alpha - u\Delta\ell)\tan\phi'$	$F_1 = 1.0$		$F_2 = 1.2$		$F_3 = 1.4$	
									n_α	Z_{i+1}	n_α	Z_{i+1}	n_α	Z_{i+1}
1	926	57.2	14.3	0.0	30	0	779	-289	1.050	466	0.972	554	0.916	625
2	1319	40.3	6.2	41.0	0	0	854	-253	0.791	1225	0.791	1366	0.791	1475
3	4019	38.2	12.9	34.5	0	0	2485	-446	0.814	3730	0.814	3962	0.814	4137
4	5868	30.2	12.9	31.5	0	0	2952	-406	0.887	6601	0.887	6909	0.887	7139
5	6333	0.0	12.1	31.5	0	0	0	-382	0.999	6218	0.999	6591	0.999	6866
6	13572	0.0	28.2	31.5	0	0	0	-887	0.999	5331	0.999	5851	0.999	6232
7	13572	0.0	28.2	31.5	0	0	0	-887	0.999	4453	0.999	5111	0.999	5597
8	5270	0.0	11.9	31.5	0	0	0	-375	0.999	4068	0.999	4798	0.999	5329
9	3064	-33.1	11.9	31.5	0	0	-1675	-374	0.811	1541	0.811	2348	0.811	2934
10	1446	-40.6	12.3	34.5	0	0	-942	-424	0.728	-336	0.728	569	0.728	1224
11	168	-43.7	5.8	41	0	0	-116	-238	0.690	**-847**	0.690	**114**	0.690	**811**

$$Z_{i+1} = Z_i + \frac{W\sin\alpha - \dfrac{c\Delta\ell + (W\cos\alpha - c\Delta\ell)\tan\phi'}{F}}{n_\alpha}$$

$$n_\alpha = \cos(\alpha - \theta) + \frac{\sin(\alpha - \theta)\tan\phi'}{F}$$

Figure 7.18 Manual calculations for James Bay dike using force equilibrium procedure.

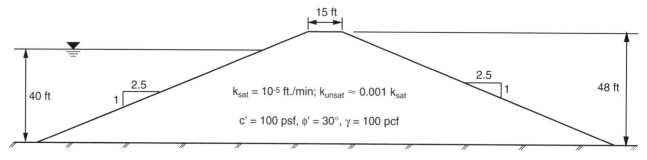

Figure 7.19 Cross section for homogeneous embankment with steady-state seepage.

interpolation. In fact, many fewer nodes and elements could probably have been used for the finite element seepage analysis with no loss in accuracy in the computed factor of safety (Wright, 2002).

The finite element seepage analysis was also used to determine a position for a phreatic surface and a piezometric line (Figure 7.20). The phreatic surface and piezometric line were assumed to be the same as the line of zero pore water pressure determined from contours of pore water pressure obtained from the finite element seepage analysis. Experience with a number of finite element seepage analyses where both saturated and unsaturated flow has been modeled has shown that the contour of zero pore water pressure corresponds very closely to the classical line of seepage described by Casagrande (1937) for saturated flow.

For the slope stability computations, pore water pressures were calculated from the phreatic surface and piezometric line using the procedures discussed in Chapter 6. The pore water pressures calculated from the phreatic surface were always as small as or smaller than those based on the piezometric line and were usually slightly larger than those based on the actual finite element seepage analysis.

Although the pore water pressures calculated in the finite element analyses were negative in the uppermost part of the flow region, above the piezometric line and phreatic surface, negative pore water pressures were neglected in all of the slope stability calculations. Negative pore water pressures probably would exist and

would contribute slightly to stability, but their effect would be small for this problem and it seems reasonable to neglect them. Only when negative pore water pressures can be sustained throughout the life of the slope should they be counted on for stability. Sustainable negative pore water pressures seem unlikely for most slopes.

Stability analyses. Slope stability calculations were first performed using a computer program and each of the three representations of pore water pressure discussed above. Spencer's procedure was used for all of the calculations. In each case an automatic search was conducted to locate the most critical circle for each representation of pore water pressures. The minimum factors of safety are summarized in Table 7.9. All three

Table 7.9 Summary of Factors of Safety from Slope Stability Computations for Homogeneous Embankment Subjected to Steady-State Seepage (Spencer's Procedure and Circular Clip Surfaces)

Procedure of slices	Factor of safety
Finite element seepage analysis— pore water pressures interpolated	1.19
Piezometric line	1.16
Phreatic surface	1.24

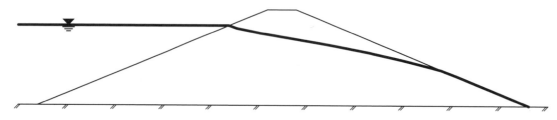

Figure 7.20 Zero-pressure line (contour) used as phreatic surface and piezometric line for homogeneous embankment.

representations of pore water pressure produced similar values for the factor of safety, with the differences between the rigorous finite element solution and the approximate representations being 4% or less. As expected, the piezometric line produced lower factors of safety than the phreatic surface, although the differences were small. The phreatic surface did, however, result in a higher factor of safety than the rigorous finite element solution, and thus this approximation errs on the unsafe side for these conditions.

A manual solution using a computer spreadsheet program and the Ordinary Method of Slices was performed as verification of the computer solutions. Calculations were performed using both the preferred and alternative methods of handling pore water pressures in the Ordinary Method of Slices that were discussed in Chapter 6. For both sets of calculations the pore water pressures were calculated using the piezometric line.

The calculations for the Ordinary Method of Slices are presented in Figures 7.21 and 7.22 for the preferred and alternative methods, respectively. Using the pre-

ferred method of handling pore water pressures, the factor of safety was 1.19 (Figure 7.21). This value is identical to the value calculated using the finite element seepage solution and slightly higher than the value with the piezometric line based on the computer solutions with Spencer's procedure. The other, alternative method of handling pore water pressures in the Ordinary Method of Slices resulted in a lower factor of safety of 1.08, which is approximately 10% lower than the other solutions. This is consistent with what was shown in Chapter 6 for the Ordinary Method of Slices and illustrates why the form of the method illustrated in Figure 7.21 is preferred.

The final set of calculations was performed using the charts in the Appendix and are summarized in Figure 7.23. The factor of safety from the chart solution is 1.08. This value is slightly lower than the values from the more accurate computer solutions. The slightly lower value from the chart solution probably reflects use of the Ordinary Method of Slices to develop the charts as well as other approximations that are made regarding seepage and pore water pressures.

Slice No.	b (ft)	h_{soil} (ft)	γ_{soil} (pcf)	W (lb)	$\Delta\ell$	α (deg)	c' (psf)	ϕ (deg)	$h_{piezometric}$[1]	u[2] (psf)	$u\Delta\ell\cos^2\alpha$	$W\cos\alpha$	$W\cos\alpha - u\Delta\ell\cos^2\alpha$	$(W\cos\alpha - u\Delta\ell\cos^2\alpha)\tan\phi'$	$c'\Delta\ell$	$W\sin\alpha$
1	7.3	2.6	125	2392	11.0	48.1	100	30	0.0	0	0	1597	1597	922	1096	1780
2	11.1	7.6	125	10580	14.4	39.8	100	30	3.4	213	1818	8128	6310	3643	1443	6772
3	7.3	10.9	125	9866	8.6	32.2	100	30	8.2	515	3160	8345	5185	2994	858	5264
4	7.5	12.1	125	11383	8.4	26.7	100	30	10.6	663	4462	10174	5711	3297	843	5107
5	5.1	12.5	125	7886	5.5	22.1	100	30	11.9	745	3488	7307	3819	2205	545	2966
6	7.3	12.2	125	11153	7.7	17.8	100	30	12.1	756	5253	10617	5364	3097	767	3418
7	11.6	10.8	125	15671	11.9	11.4	100	30	10.8	673	7669	15360	7691	4440	1186	3106
8	11.5	7.7	125	11098	11.5	3.8	100	30	7.7	482	5529	11073	5545	3201	1153	732
9	12.3	2.9	125	4466	12.4	-4.0	100	30	2.9	181	2225	4455	2230	1287	1235	-315
												Summation:		25087	9126	28832

[1] $h_{piezometric}$ = depth below piezometric line to slip surface.
[2] $u = \gamma_w \times h_{piezometric}$

$$F = \frac{\Sigma[(W\cos\alpha - u\Delta\ell\cos^2\alpha)\tan\phi - c\Delta\ell]}{\Sigma W\sin\alpha} = \frac{25087 + 9126}{28832} = 1.19$$

Figure 7.21 Manual calculations for stability of embankment with steady-state seepage using the Ordinary Method of Slices (the preferred method of representing pore water pressures).

Slice No.	b (ft)	h_{soil} (ft)	γ_{soil} (pcf)	W (lb)	$\Delta\ell$	α (deg)	c' (psf)	ϕ (deg)	$h_{piezometric}$[1]	u[2] (psf)	$u\,\Delta\ell$	W cosα	W cosα − u $\Delta\ell$	(W cosα − u $\Delta\ell$) tanϕ'	c'$\Delta\ell$	W sinα
1	7.3	2.6	125	2392	11.0	48.1	100	30	0.0	0	0	1597	1597	922	1096	1780
2	11.1	7.6	125	10580	14.4	39.8	100	30	3.4	213	3080	8128	5048	2915	1443	6772
3	7.3	10.9	125	9866	8.6	32.2	100	30	8.2	515	4417	8345	3928	2268	858	5264
4	7.5	12.1	125	11383	8.4	26.7	100	30	10.6	663	5587	10174	4587	2648	843	5107
5	5.1	12.5	125	7886	5.5	22.1	100	30	11.9	745	4062	7307	3245	1873	545	2966
6	7.3	12.2	125	11153	7.7	17.8	100	30	12.1	756	5798	10617	4819	2782	767	3418
7	11.6	10.8	125	15671	11.9	11.4	100	30	10.8	673	7983	15360	7377	4259	1186	3106
8	11.5	7.7	125	11098	11.5	3.8	100	30	7.7	482	5553	11073	5521	3187	1153	732
9	12.3	2.9	125	4466	12.4	−4.0	100	30	2.9	181	2236	4455	2219	1281	1235	−315
												Summation:		22136	9126	28832

[1] $h_{piezometric}$ = depth below piezometric line to slip surface.
[2] $u = \gamma_w \times h_{piezometric}$

$$F = \frac{\Sigma[(W\cos\alpha - u\Delta\ell\cos^2\alpha)\tan\phi - c\Delta\ell]}{\Sigma W\sin\alpha} = \frac{22136 + 9126}{28832} = 1.08$$

Figure 7.22 Manual calculations for stability of embankment with steady-state seepage using the Ordinary Method of Slices (a poor method of representing pore water pressures).

Nevertheless, the chart solution confirms the validity of the other more accurate solutions for this example and is conservative.

Recapitulation

- The zero-pressure line from finite element seepage analysis can be used to define an equivalent piezometric line and phreatic surface.
- Pore water pressures determined by interpolation of pressures from finite element seepage analysis and pore water pressures represented by an equivalent piezometric line and a phreatic surface all produced similar factors of safety, with pore water pressures from a piezometric line producing the lowest factors of safety.
- The form of the Ordinary Method of Slices recommended for effective stress analyses in Chapter 6 produces much better agreement with complete equilibrium procedures than the alternative form.

Example 8: Earth Dam with Thick Core—Steady-State Seepage

A series of stability computations similar to those for the preceding example was performed for the earth dam shown in Figure 7.24. Soil properties for the core and shell of the dam are shown in the figure. The primary purposes of this example were to illustrate with additional computations the differences among various methods for representing pore water pressures in a slope stability analysis and to present results of a spreadsheet solution using the Simplified Bishop procedure for effective stress analyses.

Pore water pressures were calculated from a finite element analysis with GMS/SEEP2D software. A piezometric line and phreatic surface were determined from the finite element analysis by locating the zero-pressure line as described for the preceding example. Slope stability computations were performed for the three representations of pore water pressure using the computer program and Spencer's procedure with circular slip surfaces. Results of these calculations are

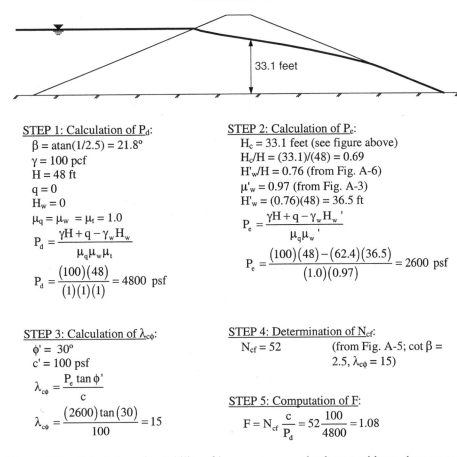

STEP 1: Calculation of P_d:
$\beta = \text{atan}(1/2.5) = 21.8°$
$\gamma = 100$ pcf
$H = 48$ ft
$q = 0$
$H_w = 0$
$\mu_q = \mu_w = \mu_t = 1.0$
$$P_d = \frac{\gamma H + q - \gamma_w H_w}{\mu_q \mu_w \mu_t}$$
$$P_d = \frac{(100)(48)}{(1)(1)(1)} = 4800 \text{ psf}$$

STEP 2: Calculation of P_e:
$H_c = 33.1$ feet (see figure above)
$H_c/H = (33.1)/(48) = 0.69$
$H'_w/H = 0.76$ (from Fig. A-6)
$\mu'_w = 0.97$ (from Fig. A-3)
$H'_w = (0.76)(48) = 36.5$ ft
$$P_e = \frac{\gamma H + q - \gamma_w H_w{}'}{\mu_q \mu_w{}'}$$
$$P_e = \frac{(100)(48) - (62.4)(36.5)}{(1.0)(0.97)} = 2600 \text{ psf}$$

STEP 3: Calculation of $\lambda_{c\phi}$:
$\phi' = 30°$
$c' = 100$ psf
$$\lambda_{c\phi} = \frac{P_e \tan\phi'}{c}$$
$$\lambda_{c\phi} = \frac{(2600)\tan(30)}{100} = 15$$

STEP 4: Determination of N_{cf}:
$N_{cf} = 52$ (from Fig. A-5; cot $\beta =$
                    2.5, $\lambda_{c\phi} = 15$)

STEP 5: Computation of F:
$$F = N_{cf}\frac{c}{P_d} = 52\frac{100}{4800} = 1.08$$

Figure 7.23 Calculations for stability of homogeneous embankment with steady-state seepage using Janbu's slope stability charts.

Material Properties		
Zone	c' - psf	ϕ' - deg	k - ft/min
Core	0	20	1 x 10^{-5}
Shell	0	38	1 x 10^{-3}

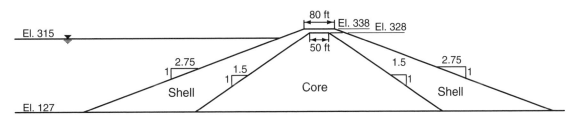

Figure 7.24 Cross section and soil properties for earth dam with thick clay core.

summarized in Table 7.10. The results shown in this table are very similar to those shown for the homogeneous dam in the preceding example: The piezometric line produces a slightly lower factor of safety that the more rigorous interpolation of pore water pressures, while the phreatic surface approximation produces a slightly higher value for the factor of safety. All three representations in this case produce very similar values for the factor of safety.

Additional computations were performed using the computer program with the Simplified Bishop procedure and the piezometric line to represent pore water pressures. The factor of safety calculated by the Simplified Bishop procedure was 1.61, which is approximately 4% less than the corresponding value (1.67) by Spencer's procedure. One of the purposes of calculating the factor of safety by the Simplified Bishop procedure was to be able to compare the value from the computer solution with a manual solution using a spreadsheet. Results of the spreadsheet solution for the critical circle found with the Simplified Bishop procedure are summarized in Figure 7.25. The factor of safety calculated by the spreadsheet solution (1.61) is the same as the value calculated with the computer program.

Recapitulation

- Conclusions similar to those reached for the homogeneous dam can be drawn regarding the representations of pore water pressures: All three representations of pore water pressure produced similar results, with the piezometric line producing the lowest factors of safety.
- The Simplified Bishop procedure provides an excellent check of a computer solution for effective stress analyses with circular slip surfaces.

Example 9: Reinforced Slope (1)

This and the next example were selected to demonstrate how results obtained using one computer program may be verified using another computer program. The two examples chosen are part of the user's documentation for the STABGM 2.0 slope stability software (Duncan et al., 1998). The first example involves an embankment resting on a soft, saturated clay foundation with a single layer of reinforcement at the base of the embankment (Figure 7.26). Because the clay in the foundation will consolidate with time and become stronger under the weight of the overlying embankment, the short-term stability condition is most critical and was selected for analyses. The undrained shear strength profile for the foundation is shown along with the other soil properties in Figure 7.26.

Stability computations were performed with the UTEXAS4 computer program using the Simplified Bishop procedure. An automatic search was conducted to find the most critical slip surface, giving a minimum factor of safety of 1.13. This value is approximately 5% less than the value of 1.19 that is reported in the STABGM user's documentation. Differences of this magnitude (5%) are larger than normally expected when two different computer programs are used, provided that both programs use the same input data and the same procedures for computation. In the present case the relatively large difference is due to differences in the degree of refinement of the search used to locate a critical slip surface. The critical circle reported using STABGM is shown in Figure 7.27 along with the critical circle found using UTEXAS4. With STABGM, circles were only investigated at depths corresponding to the bottom of each soil layer shown in Figure 7.27. With UTEXAS4, circles were also investigated at intermediate depths, and a more critical circle was found. STABGM would probably have given the same factor of safety as UTEXAS4 had a more refined search been used in the STABGM analyses.

An additional set of stability calculations was performed with UTEXAS4 for the critical circle from STABGM. The computed factor of safety for this circle with UTEXAS4 was 1.19, which is identical to the value reported by STABGM. This confirms that the differences in factor of safety reported earlier were due to the input data for the automatic search rather than to differences between the two programs. The close agreement in results when comparable conditions are used verifies the correctness of the results for this case.

Recapitulation

- One computer program may be used to verify results obtained by another computer program.
- When comparing results from two different computer programs, it is important to compare results for the same conditions, including the same slope geometry, soil properties, and slip surface.

Table 7.10 Summary of Slope Stability Calculations for Dam with Clay Core (Spencer's Procedure and Circular Clip Surfaces)

Procedure of slices	Factor of safety
Finite element seepage analysis— pore water pressures interpolated	1.69
Piezometric line	1.67
Phreatic surface	1.70

Slice No.	b (ft)	h_{shell} (ft)	γ_{shell} (pcf)	h_{core} (ft)	γ_{core} (pcf)	W (lb)
1	20.0	13.1	140	0.0	120	36659
2	35.2	20.8	140	26.6	120	215174
3	65.0	10.0	140	76.0	120	683527
4	62.5	29.5	140	84.9	120	894190
5	89.9	52.5	140	69.1	120	1406585
6	105.7	82.2	140	34.2	120	1648280
7	21.2	101.4	140	5.6	120	315386
8	81.8	92.2	140	0.0	120	1055716
9	114.7	54.1	140	0.0	120	867907
10	49.8	14.2	140	0.0	120	98993

(a)

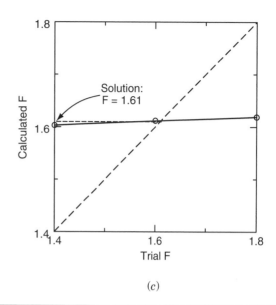

(c)

Slice No.	W (lbs)	α (degs)	$W \sin\alpha$	c' (psf)	ϕ' (degs)	$h_{piezometric}$[1]	u[2] (psf)	$cb + (W - ub)\tan\phi'$	Trial F = 1.4		Trial F = 1.6		Trial F = 1.8	
									m_α	$[cb + (W - ub)\tan\phi'] \div m_\alpha$	m_α	$[cb + (W - ub)\tan\phi'] \div m_\alpha$	m_α	$[cb + (W - ub)\tan\phi'] \div m_\alpha$
1	36659	43.2	25110	0	38	0.0	0	28641	1.11	28641	1.06	26942	1.03	27918
2	215174	40.2	138831	0	20	23.2	1450	59724	0.93	59724	0.91	65575	0.89	666770
3	683527	35.0	391572	0	20	38.3	2392	192187	0.97	192187	0.95	202309	0.94	205441
4	894190	28.7	429214	0	20	55.3	3449	247055	1.00	247055	0.99	250447	0.97	253565
5	1406585	21.7	519784	0	20	59.9	3741	389600	1.03	389600	1.01	384494	1.00	388071
6	1648280	13.1	374514	0	20	34.2	2131	517970	1.03	517970	1.03	505074	1.02	507919
7	315386	7.7	42435	0	20	11.5	719	109240	1.03	109240	1.02	106939	1.02	107296
8	1055716	3.4	62854	0	38	12.9	805	773349	1.03	773349	1.03	752799	1.02	755174
9	867907	-4.8	-72569	0	38	7.2	452	637567	0.95	671239	0.96	667142	0.96	663990
10	98993	-11.7	-20099	0	38	0.0	0	77342	0.87	77342	0.88	87886	0.89	86799
		Sum:	1891646						Sum:	3033119	Sum:	3049608	Sum:	3062944

[1] $h_{piezometric}$ = depth below piezometric line to slip surface.

[2] $u = \gamma_w \times h_{piezometric}$

$$F = \frac{\left[\dfrac{c'b + (W - ub)\tan\phi}{m_\alpha} \right]}{\Sigma W \sin\alpha}$$

$$m_\alpha = \cos\alpha + \frac{\sin\alpha + \tan\phi}{F}$$

$$F_1 = \frac{3033119}{1891646} = 1.603 \qquad F_2 = \frac{3049608}{1891646} = 1.612 \qquad F_3 = \frac{3062944}{1891646} = 1.619$$

(b)

Figure 7.25 Manual calculations for stability of clay core dam with steady-state seepage using the Simplified Bishop procedure: (a) calculation of slice weights; (b) calculations for factor of safety; (c) summary of trial-and-error solution for F.

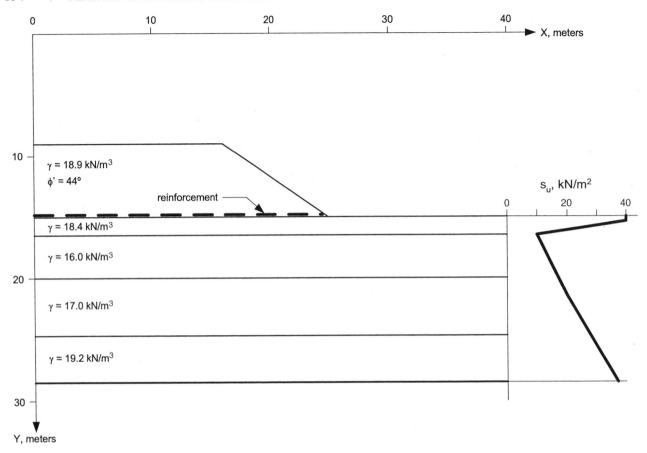

Figure 7.26 Soil properties and slope geometry for reinforced embankment on a soft clay foundation from the STABGM user's documentation.

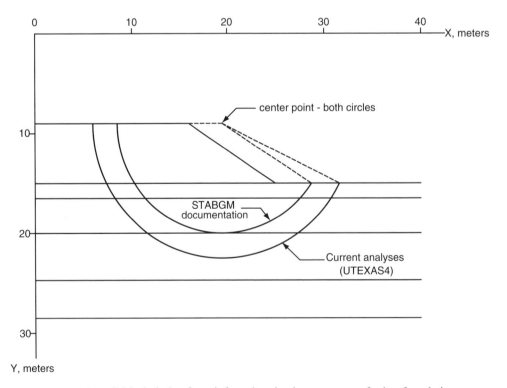

Figure 7.27 Critical circles for reinforced embankment on a soft clay foundation.

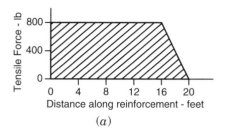

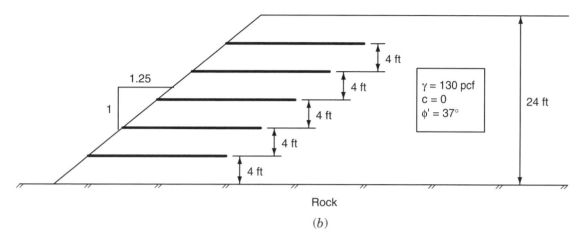

Figure 7.28 Reinforced slope on a rock foundation foundation from the STABGM user's documentation: (*a*) distribution of longitudinal forces along each reinforcing element; (*b*) slope and reinforcement geometry with soil properties.

Example 10: Reinforced Slope (2)

This example is for a reinforced fill slope resting on a much stronger soil foundation and is also described in the STABGM 2.0 user's manual (Duncan et al., 1998). The slope and soil properties are presented in Figure 7.28. The soil in the slope is assumed to drain freely, and thus long-term stability computations were performed using shear strengths expressed in terms of effective stresses. The slope contains six layers of reinforcement, beginning at the bottom of the slope and spaced 4 ft apart vertically. Each reinforcement layer is 20 ft long, with a tensile force of 800 lbs per lineal foot, decreasing linearly to zero over the final 4 ft of embedded length.

Slope stability analyses were performed using UTEXAS4 with circular slip surfaces and Spencer's procedure. The minimum factor of safety calculated was 1.61. This value is less that the value of 1.71 that is reported by Duncan et al. using STABGM 2.0, representing a difference of approximately 6% in the factor of safety. As discussed for the preceding example, differences of this magnitude (5 to 6%) are larger than expected for two computer programs if the same con-

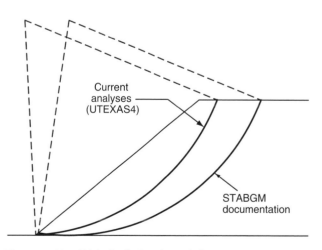

Figure 7.29 Critical circles for reinforced slope on rock foundation.

ditions are analyzed. Subsequently, an analysis was performed using UTEXAS4 and the critical circle that was found using SSTABGM 2.0. The factor of safety was then calculated to be 1.71, which is identical to the value reported by Duncan et al. for the critical circle found using STABGM. The critical circles found using UTEXAS4 and STABGM 2.0 are plotted in Figure 7.29. It can be seen that the two circles are different. If a more refined search had been conducted with STABGM 2.0, it is likely that the minimum factor of safety would have been similar to the minimum factor of safety (1.61) found with UTEXAS4.

CHAPTER 8

Reinforced Slopes and Embankments

Reinforcement can be used to improve the stability of slopes and embankments, making it possible to construct slopes and embankments steeper and higher than would otherwise be possible. Reinforcement has been used in four distinct types of applications:

1. *Reinforced slopes.* Multiple layers of reinforcement at various elevations within fill slopes have been used to increase the factor of safety for slip surfaces that cut through the reinforcement, making it possible to construct slopes steeper than would be possible without reinforcement.

2. *Reinforced embankments on weak foundations.* Reinforcement at the bottom of an embankment on a weak foundation can increase the factor of safety for slip surfaces passing through the embankment, making it possible to construct the embankment higher than would be possible without reinforcement.

3. *Reinforced soil walls or mechanically stabilized earth walls.* Several different proprietary systems have been developed for reinforced soil walls, which are used as alternatives to conventional retaining walls. Most of the companies that market MSE walls have developed proprietary design procedures. The stability of MSE walls can also be evaluated using the methods described in this chapter.

4. *Anchored walls.* Vertical soldier pile walls or slurry trench concrete walls can be "tied back" or anchored at one or more levels to provide vertical support for excavations or fills. Anchored walls have been used in both temporary and permanent applications. The methods described in this chapter can be used to evaluate the stability of anchored walls.

LIMIT EQUILIBRIUM ANALYSES WITH REINFORCING FORCES

Reinforced slopes can be analyzed using the procedures described in Chapter 6 by including the reinforcement forces in the analyses as known forces. Zornberg et al. (1998a,b) have shown through centrifuge tests that limit equilibrium analyses provide valid indications of factor of safety and failure mechanisms for reinforced slopes. Their analyses, which agreed well with the results of their tests, were performed using peak values of ϕ' rather than the lower critical-state friction angle of the backfill soil.

The amount of force required to achieve a target value of factor of safety can be determined using repeated trials, varying the magnitude of the force until the factor of safety computed is the one desired. Some computer programs can perform this operation automatically—the input is the desired factor of safety, and the output is the required reinforcement force. This type of program is better adapted to design of reinforced slopes, since there is no need for repeated analyses.

FACTORS OF SAFETY FOR REINFORCING FORCES AND SOIL STRENGTHS

Two methods have been used for limit equilibrium analyses of reinforced slopes.

- *Method A.* The reinforcement forces used in the analysis are *allowable* forces and *are not divided* by the factor of safety calculated during the slope stability analysis. Only the soil strength is divided by the factor of safety calculated in the slope stability analysis.

• *Method B.* The reinforcement forces used in the analysis are *ultimate* forces, and *are divided* by the factor of safety calculated in the slope stability analysis. Both the reinforcing force and the soil strength are divided by the factor of safety calculated in the slope stability analysis.

Method A is preferable, because the soil strength and the reinforcement forces have different sources of uncertainty, and they therefore involve different amounts of uncertainty. Factoring them separately makes it possible to reflect these differences.

When a computer program is used to analyze reinforced slopes, it is essential to understand which of these methods is being used within the program, so that the appropriate measure of reinforcing force (allowable force or ultimate force) can be specified in the input for the analysis.

If the documentation of a computer program does not specify whether the reinforcement force should be allowable or ultimate, this can be deduced from the equations employed to compute the factor of safety.

Method A Equations

If the factor of safety for circular slip surfaces is defined by an equation of the form

$$F = \frac{\text{soil resisting moment}}{\text{overturning moment} - \text{reinforcement moment}}$$

(8.1)

or, more generally, if the factor of safety is defined by an equation of the form

$$F = \frac{\text{shear strength}}{\substack{\text{shear stress required for equilibrium} \\ - \text{ reinforcement resistance}}}$$

(8.2)

the program uses method A, and the reinforcement forces specified in the input should be allowable forces, denoted here as P_{all}.

Method B Equations

If the factor of safety for circular slip surfaces is defined by an equation of the form

$$F = \frac{\text{soil resisting moment} + \text{reinforcement moment}}{\text{overturning moment}}$$

(8.3)

or, more generally, by an equation of the form

$$F = \frac{\text{shear strength} + \text{reinforcement resistance}}{\text{shear stress required for equilibrium}}$$

(8.4)

the program uses method B, and the reinforcement forces specified in the input should be the unfactored long-term load capacity of the reinforcement, denoted here as P_{lim}.

If it is not clear which method is used by a computer program, this can be determined by analyzing the reinforced slope problem shown in Figure 8.1. This slope is 20 ft high and is inclined at 2.0 vertical on 1.0 horizontal. The soil within the slope is uniform, with $\gamma = 100$ pcf, $\phi = 0$, and $c = 500$ psf. There is a firm layer beneath the toe. A reinforcing force of 10,000 lb/ft acts horizontally at midheight, 10 ft above the toe of the slope. Results for two analyses are shown in Figure 8.1. The analyses considered only circular slip surfaces tangent to the top of the firm layer. The critical circles, located as shown in Figure 8.1, exit slightly above the toe of the slope.

The method used by any computer program can be determined based on the computed factor of safety:

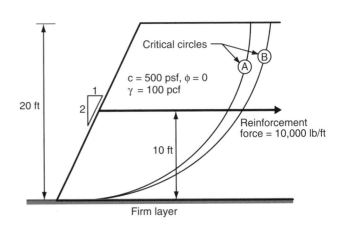

Method A: The reinforcement force used in the analysis is an allowable force, and is not divided by the factor of safety calculated during the slope stability analysis.
The factor of safety computed using method A is F = 2.19.

Method B: The reinforcement force used in the analysis is an ultimate force, and is divided by the factor of safety calculated during the slope stability analysis.
The factor of safety computed using method B is F = 1.72.

Figure 8.1 Check problem for determining whether a computer program is using method A or method B for analysis of reinforced slopes.

- If the program computes $F = 2.19$, the program uses method A. Allowable reinforcement forces should be used with the program.

- If the program computes $F = 1.72$, the program uses method B. Ultimate reinforcement forces should be used with the program.

Slight deviations from $F = 2.19$ or $F = 1.72$ can be expected, depending on the number of slices used by the program, and the method used to locate the critical circle. The differences should be no more than 1 or 2%, however.

TYPES OF REINFORCEMENT

The principal types of reinforcing materials that have been used for slopes and embankments are geotextile fabrics, geogrids, steel strips, steel grids, and high-strength steel tendons. *Geotextiles* are manufactured by weaving polymeric fibers into a fabric or by matting the fibers together to form a continuous nonwoven fabric. Woven geotextiles are stiffer and stronger than nonwoven geotextiles and more useful for reinforced slope applications. *Geogrids* are manufactured by stretching sheets of polymer plastic in one or both directions to form a high-strength grid. Stretching the polymeric materials makes them stiffer and stronger. *Galvanized* or *epoxy-coated steel strips* have been used for slope reinforcement. The strips usually have raised ribs to increase their pullout resistance. Welded mild steel mats or grids have also been used for reinforcing slopes and embankments.

Key sources of information about geosynthetic reinforcing for slopes are the book by Koerner (1998), which contains a great deal of information on the fundamental characteristics and properties of polymers, geotextiles, and geogrids, and the FHWA (2000) publication entitled *Mechanically Stabilized Earth Walls and Reinforced Soil Slopes: Design and Construction Guidelines*, which covers a wide range of subjects related to geotextiles, geogrids, steel strips, and steel grids, and their use in reinforced walls and slopes.

REINFORCEMENT FORCES

The long-term capacity of reinforcement, denoted here as T_{lim}, depends on the following factors:

- *Tensile strength.* For steel, the tensile strength is the yield strength. For geosynthetics, the tensile strength is measured using short-term wide-width tensile tests.

- *Creep characteristics.* Steel does not creep appreciably, but geosynthetic materials do. The tensile loads used for design of geotextile- and geogrid-reinforced walls must be reduced to values lower than those measured in short-term tensile tests, to stresses that are low enough so that little or no creep deformation will occur over the design life of the structure.

- *Installation damage.* Geotextiles and geogrids are subject to damage during installation that results in holes and tears in the material. Epoxy-coated and PVC coatings on steel are subject to damage during installation, and galvanization is therefore preferred for corrosion protection.

- *Durability.* The mechanical properties of geosynthetics are subject to deterioration during service as a result of attack by chemical and biological agents. Steel is subject to corrosion.

- *Pullout resistance.* Near the ends of the reinforcement, capacity is limited by the resistance to pullout, or slip between the reinforcement and the soil within which it is embedded.

- *Reinforcement stiffness and tolerable strain within the slope.* To be useful for slope reinforcement, the reinforcing material must have stiffness as well as strength. A very strong but easily extensible rubber band would not provide effective reinforcement, because it would have to stretch so much to mobilize its tensile capacity that it would not be able to limit the deformation of the slope.

Values of T_{lim}, the long-term capacity of reinforcing materials, must satisfy the following three criteria:

1. $T_{lim} \leq$ capacity determined by short-term tensile strength, creep, installation damage, and deterioration of properties over time.
2. $T_{lim} \leq$ capacity determined by pullout resistance.
3. $T_{lim} \leq$ capacity determined by stiffness and tolerable strain.

Methods of applying these requirements to geosynthetics and steel reinforcing are described in the following sections.

Criterion 1: Creep, Installation Damage, and Deterioration in Properties over Time

Geotextiles and geogrids. The effects of creep, installation damage, and long-term deterioration on geosynthetic materials can be evaluated using the expression

$$T_{lim} = \frac{T_{ult}}{(RF_{CR})(RF_{ID})(RF_D)} \quad (8.5)$$

where T_{lim} is the long-term limit load (F/L); T_{ult} the short-term ultimate strength, measured in a wide-strip tension test (F/L); RF_{CR} the strength reduction factor to allow for creep under long-term load; RF_{ID} the strength reduction factor to allow for installation damage; RF_D the strength reduction factor to allow for deterioration in service. Values of RF_{CR}, RF_{ID}, and RF_D recommended by the FHWA are given in Table 8.1. The units of T_{lim} and T_{ult} are force per unit length of reinforced slope.

Steel reinforcement. Steel reinforcing does not creep appreciably and is not subject to installation damage. Epoxy and PVC coatings are subject to installation damage, but galvanized coating is not. The effects of long-term deterioration of steel due to corrosion can be evaluated using the expression

$$T_{lim} \leq A_c f_y \quad (8.6)$$

where T_{lim} is the allowable long-term reinforcement tension load (F/L); A_c the cross-sectional area of reinforcement after corrosion, calculated by reducing the metal thickness by the loss expected during the life of the installation [A_c is the area per unit length of slope (L^2/L)]; and f_y the yield strength of steel (F/L^2). Corrosion rates for steel in mildly corrosive backfill materials are given in Table 8.2.

Table 8.1 Reduction Factors for Tensile Strengths of Geotextiles and Geogrids for Use in Eq. (8.5)

Reduction for:	Factor	Polymer	Range of values[a]
Creep	RF_{CR}	Polyester	1.6–2.5
		Polypropylene	4.0–5.0
		Polyethylene	2.6–5.0
Installation damage	RF_{ID}	Any polymer	1.1–3.0
Deterioration in service	RF_D	Any polymer	1.1–2.0

[a]These values (from FHWA, 2000) are applicable to reinforcement in granular soils with maximum particle sizes up to 19 mm, values of pH from 4.5 to 9.0, and in-service temperatures below 30°C. Geotextiles weighing less than 270 g/m^2 are subject to greater damage during installation and should not be used for reinforcement.

Table 8.2 Corrosion Rates for Steel Reinforcement in Mildly Corrosive Backfill

Material corroding	Period of time	Corrosion rate (μm/yr)	Corrosion rate[a] (in./yr)
Zinc	First two years	15	5.9×10^{-4}
Zinc	Thereafter	4	1.6×10^{-4}
Carbon steel	Thereafter	12	4.7×10^{-4}

[a]These corrosion rates are applicable for steel reinforcement in backfill with these electrochemical properties: Resistivity greater than 3000 $\Omega \cdot$ cm, pH between 5 and 10, chlorides less than 100 ppm, sulfates less than 200 ppm, organic content less than 1%.
Source: After FHWA (2000).

Criterion 2: Pullout Resistance

To develop tensile capacity, reinforcement must be restrained sufficiently by friction in the soil. The maximum possible resistance (T_{po}) is proportional to the effective overburden pressure. T_{po} begins from zero at the end of the reinforcement, where the embedded length is zero and increases with distance from the end, as shown in Figure 8.2. The slope of the curve representing the variation of T_{po} with distance can be expressed as

$$\frac{dT_{po}}{dL} = 2\gamma z \alpha F^* \quad (8.7)$$

where T_{po} is the pullout resistance (F/L); L the length of embedment, or distance from the end of the reinforcement (L); γ the unit weight of fill above the reinforcement (F/L^3); z the depth of fill above the reinforcement (L); α the adjustment factor for extensible reinforcement (dimensionless); and F^* the pullout resistance factor (dimensionless).

Values of α and F^* recommended by the FHWA (2000) are listed in Table 8.3. These values are conservative estimates. Larger values may be applicable and can be used if they are supported by tests performed on the specific soil and reinforcing material.

Equation (8.7) gives the slope of the pullout resistance curve at any location. If the thickness of fill above the reinforcement is constant, the slope of the pullout curve is constant, and the pullout resistance can be expressed as

$$T_{po} = 2\gamma z \alpha F^* L_e \quad (8.8)$$

where L_e is the distance from the end of the reinforcement or length of embedment (L).

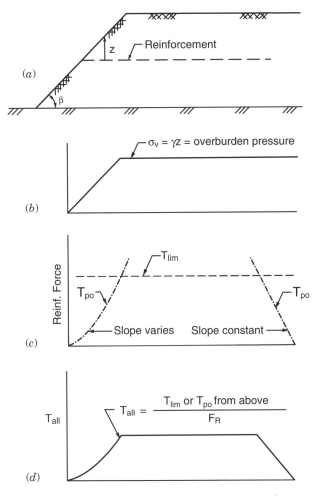

Figure 8.2 Variation of T_{lim} and T_{all} with distance along reinforcement.

Table 8.3 Pullout Resistance Factors α and F^* for Use in Eq. (8.7)

Pullout resistance factor[a]	Type of reinforcement	Resistance factor value
α	Geotextiles	0.6
	Geogrids	0.8
	Steel strips and steel grids	1.0
F^*	Geotextiles	0.67 tan ϕ
	Geogrids	0.8 tan ϕ
	Steel strips and steel grids	1.0 tan ϕ

[a]Higher values of F^* generally apply at depths shallower than 6 m. Larger values of both α and F^* may be applicable and can be used if they are supported by tests performed on the specific soil and reinforcing material.
Source: FHWA (2000).

If the overburden pressure increases with distance from the end of the reinforcement, as it does beneath the slope on the left in Figure 8.2, the slope of the T_{po} curve also increases with distance, and the pullout resistance diagram is a curve. In this case the pullout resistance can be expressed as

$$T_{\text{po}} = \gamma \tan \beta \, \alpha F^*(L_e)^2 \qquad (8.9)$$

where β is the slope angle in degrees, as shown in Figure 8.2.

Criterion 3: Reinforcement Stiffness

Reinforcing materials must be stiff enough so that reinforcement forces can be mobilized without excessive strain. The value of T_{lim} should not exceed the product of the long-term secant modulus of the reinforcement multiplied by the tolerable strain for the slope:

$$T_{\text{lim}} \leq E_{\text{secant}}\varepsilon_{\text{tolerable}} \qquad (8.10)$$

where E_{secant} is the secant modulus of reinforcing at axial strain = $\varepsilon_{\text{tolerable}}$ (F/L) and $\varepsilon_{\text{tolerable}}$ is the strain within the slope at the location of the reinforcing that can be tolerated without excessive slope deformation or failure (dimensionless).

Steel reinforcing is often described as *inextensible* because it stiffness is very high compared to its yield strength. With a modulus equal to 500 to 1000 times its yield strength, the yield strength of steel is mobilized at a strain of only 0.1 to 0.2%, far less than the tolerable strains for soil reinforcing applications. As a result, criterion 3 never governs the value of T_{lim} for steel reinforcing. The stiffness of geosynthetic materials, on the other hand, may be low enough so that criterion 3 governs the value of T_{lim} for applications where the tolerable strain is small.

As shown in Figure 8.3, E_{secant} is the slope of a line extending from the origin to the point on the $T–\varepsilon$ curve where the strain is equal to $\varepsilon_{\text{tolerable}}$. Note that the units of E_{secant}, like the units of T_{lim}, are force per unit length.

Values of tolerable strain are based on the results of finite element analyses (Rowe and Soderman, 1985); and on experience (Fowler, 1982; Christopher and Holtz, 1985; Haliburton et al., 1982; Bonaparte et al., 1987). A summary of published recommendations is given in Table 8.4.

ALLOWABLE REINFORCEMENT FORCES AND FACTORS OF SAFETY

The preceding section is concerned with the long-term capacity of reinforcement (T_{lim}). These values of T_{lim}

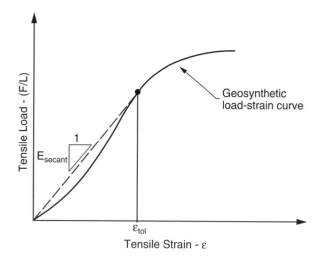

Figure 8.3 Definition of E_{secant} for geosynthetic reinforcement.

Table 8.4   Tolerable Strains for Reinforced Slopes and Embankments

Application	ε_{tol} (%)
Reinforced soil walls	10
Reinforced slopes of embankments on firm foundations	10
Reinforced embankments on nonsensitive clay, moderate crest deformations tolerable	10
Reinforced embankments on nonsensitive clay, moderate crest deformations not tolerable	5–6
Reinforced embankments on highly sensitive clay	2–3

Source: Compiled from Fowler (1982), Christopher and Holtz (1985), Haliburton et al. (1982), Rowe and Soderman (1985), and Bonaparte et al. (1987).

reflect consideration of long-term loading, installation damage, deterioration in properties over time, pullout resistance, and tolerable strains, but they do not include a factor of safety.

The allowable load assigned to reinforcing materials should include a factor of safety, as indicated by

$$T_{all} = \frac{T_{lim}}{F_R} \qquad (8.11)$$

where T_{all} is the allowable long-term reinforcement force (F/L) and F_R is the factor of safety for reinforce-

ment force. The value of F_R should reflect (1) the degree of uncertainty involved in estimating the value of T_{lim}, (2) the degree of uncertainty involved in estimating the load that the reinforcement must carry, and (3) the consequences of failure. Recommended values of F_R are given in Table 8.5.

ORIENTATION OF REINFORCEMENT FORCES

Various orientations of reinforcement forces have been suggested (Schmertmann et al., 1987; Leshchinsky and Boedeker, 1989; Koerner, 1998; FHWA, 2000). The extremes are (1) reinforcement forces that are aligned with the original orientation of the reinforcement, and (2) reinforcement forces that are parallel to the slip surface. The latter assumption, which results in larger factors of safety, has been justified by the concept that the reinforcement will be realigned where the slip surface crosses the reinforcement. This is more likely if the reinforcement is very flexible. The assumption that the orientation of the reinforcement force is the same as the orientation of the reinforcement is more conservative, is supported by the findings of Zornberg et al. (1998a), and is the more logical, reliable choice. This is the approach recommended here.

REINFORCED SLOPES ON FIRM FOUNDATIONS

Reinforcement in embankments can be used to construct slopes steeper than would be possible without reinforcing. Usually, several layers of reinforcing are used, spaced more closely near the base of the slope and farther apart near the top, as shown in Figure 8.4. Secondary shorter lengths of lower-strength reinforcement, between the primary reinforcement layers, can be used to improve surficial stability (FHWA, 2000). Zornberg et al. (1998a) showed that such layers near the bottom of the slope significantly enhance stability if they are wrapped around at the face and overlap the adjacent layers. In this configuration they are anchored firmly and not subject to pullout failure.

Table 8.5   Recommended Values of F_R

Consequences of failure	Uncertainties in T_{lim} and load in reinforcement	Appropriate value of F_R
Minimal	Small	1.5
Minimal	Large	2.0
Great	Small	2.0

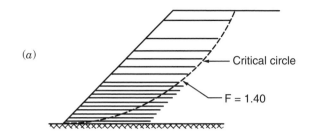

(a)

Critical circle

F = 1.40

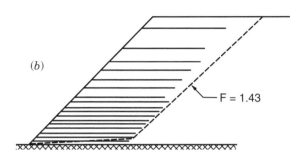

(b)

F = 1.43

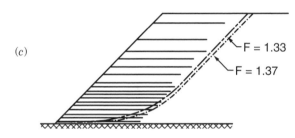

(c)

F = 1.33

F = 1.37

Figure 8.4 Limit equilibrium analyses of a reinforced slope using circular, wedge, and smooth noncircular slip surfaces: (After Wright and Duncan, 1991). (*a*) critical circular slip surface; (*b*) critical two-part wedge; (*c*) critical noncircular slip surfaces.

The stability of reinforced slopes can be evaluated using the procedures outlined in Chapter 6. An example is shown in Figure 8.4. It can be seen that the factor of safety varies slightly with the shape of the slip surface, from $F = 1.43$ for the most critical two-part wedge slip surface, to $F = 1.33$ for the most critical noncircular slip surface, a difference of about 7%. The most critical circle gives a factor of safety $F = 1.40$, which is sufficiently accurate for practical purposes.

Schmertmann et al. (1987) Charts

Before an analysis of the type illustrated in Figure 8.4 can be performed, the strength, length, and spacing of the reinforcement must be estimated. Determining these by trial and error can be time consuming because many trials can be required to determine strength, length, and spacing.

Designing reinforced slopes is facilitated greatly by slope stability charts of the type developed by Schmertmann et al. (1987), which are shown in Figure 8.5*a* and *b*. Figure 8.5*a* can be used to determine the total reinforcing force, and Figure 8.5*b* can be used to determine the length of reinforcing required for a given factor of safety. The terminology used in these charts is:

$$K = \frac{2T_{all}}{\gamma(H')^2} = \text{dimensionless force coefficient}$$

$$(8.12)$$

$$H' = H + \frac{q}{\gamma} = \text{effective height, including effect of surcharge (L)} \quad (8.13)$$

$$\phi'_f = \arctan\frac{\tan\phi'}{F} = \text{factored friction angle (degrees)} \quad (8.14)$$

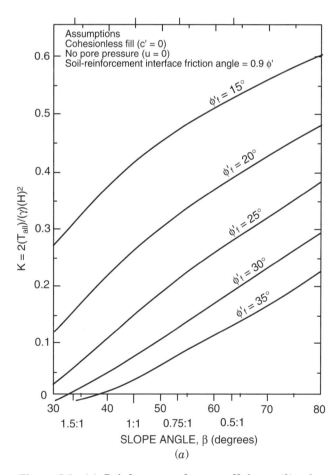

Figure 8.5 (*a*) Reinforcement force coefficients; (*b*) reinforcement length coefficients. (After Schmertmann et al., 1987.)

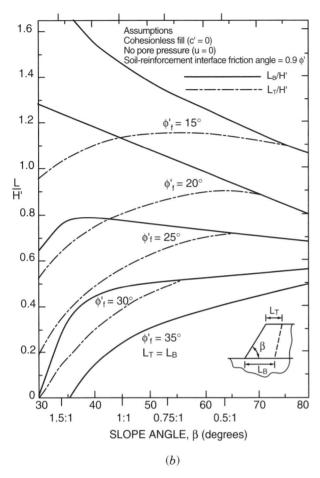

(b)

Figure 8.5 (*Continued*)

the STABGM user's manual and Chapter 7 is $F = 1.71$.

The charts in Figure 8.5a and b can be used to determine the total reinforcement force and the length of reinforcement required for a factor of safety $F = 1.71$ using the following steps:

1. Compute $\phi'_f = \arctan \tan \phi'/F = \arctan 0.75/ 1.71 = 24°$.
2. Compute $H' = H + q/\gamma = 24 \text{ ft} + 0/130 \text{ pcf} = 24 \text{ ft}$.
3. From Figure 8.5a, determine $K = 0.11$.
4. Compute $T_{all} = (0.11)(\frac{1}{2})(130 \text{ pcf})(24 \text{ ft})^2 = 4100$ lb/ft.

In Figure 7.27 there are five active layers of reinforcement. The sixth layer shown in the STABGM manual and Figure 7.27, at the elevation of the toe, does not cut across any of the slip surfaces shown in Figure 7.28 and therefore does not influence the factor of safety. The factors of safety computed using STABGM and UTEXAS4 are the same whether this bottom layer of reinforcement is included or not. The total reinforcement force from step 4, 4100 lb/ft, agrees well with the total of 4000 lb/ft provided by the five active layers of reinforcement.

L_B = required length of reinforcement at the bottom of the slope (L)

L_T = required length of reinforcement at the top of the slope (L)

β = slope angle (degrees)

γ = unit weight of soil (F/L³)

q = surcharge pressure (F/L²)

u = pore pressure, assumed to be zero throughout the slope

Example. As an example of the use of these charts, consider the slope shown in Figure 7.27. Pertinent parameter values from Figure 7.27 are $H = 24$ ft, $\gamma = 130$ pcf, $\phi = 37°$, $\beta = \arctan(1.25) = 39°$, $q = 0$, $c = 0$, and $u = 0$. The factor of safety computed in

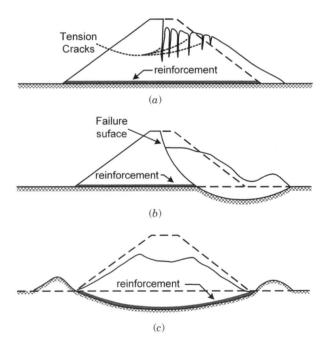

Figure 8.6 Potential modes of failure of reinforced embankments: (*a*) block sliding outward along reinforcement with slumping of the crest; (*b*) foundation failure with rotational sliding through embankment; (*c*) excessive elongation of reinforcement. (Modified from Haliburton et al., 1978).

5. From Figure 8.5b determine $L_B/H' = 0.88$, $L_T/H' = 0.55$.
6. Compute $L_B = (0.88)(24$ ft$) = 21$ ft and $L_T = (0.55)(24$ ft$) = 13$ ft.

In Figure 7.27, $L_T = L_B = 20$ ft. The results from Figure 8.5b indicate that the reinforcement could be somewhat shorter at the top of the slope.

EMBANKMENTS ON WEAK FOUNDATIONS

Reinforcement near the base of an embankment can be used to improve stability with regard to spreading of the embankment and with regard to shear failure through the embankment and foundation. With reinforcement at the bottom of the embankment, the slopes can be made as steep as for an embankment con-structed on a firm foundation. The volume of the embankment and the total load it imposes on the foundation can be reduced and its height can be increased.

Reinforced embankments have been used at a number of sites where weak foundations posed difficult stability problems, including Almere in the Netherlands (Rowe and Soderman, 1985); Mohicanville Dike 2 in Ohio (Duncan et al., 1988; Franks et al., 1988, 1991); St. Alban in Canada (Busbridge et al., 1985; Schaefer and Duncan, 1986, 1987); Hubrey Road in Ontario, Canada (Rowe and Mylleville, 1996); and Sackville, New Brunswick, Canada (Rowe et al., 1996).

Modes of failure. Potential modes of failure of reinforced embankments on weak foundations have been discussed by Haliburton et al. (1978) and by Bonaparte and Christopher (1987). Three possible modes of failure are shown in Figure 8.6.

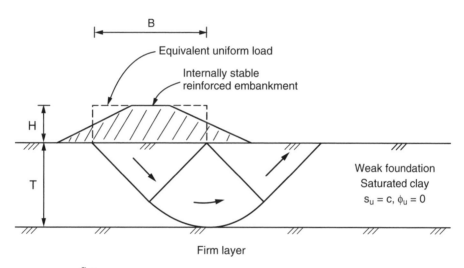

$$F = \frac{q_{ult}}{q}$$

$q_{ult} = cN_C$ (F/L^2)

c = average undrained strength of foundation (F/L^2)

$q = \gamma H$ (F/L^2)

γ = total unit weight of embankment fill (F/L^6)

H = embankment height (L)

Ratio B/T	Value of N_c (NAVFAC, 1986)
≤ 1.4	5.1
2.0	6.0
3.0	7.0
5.0	10.0
10.0	17.0
20.0	30.0

Figure 8.7 Bearing capacity mode of failure of a strongly reinforced embankment on a weak foundation.

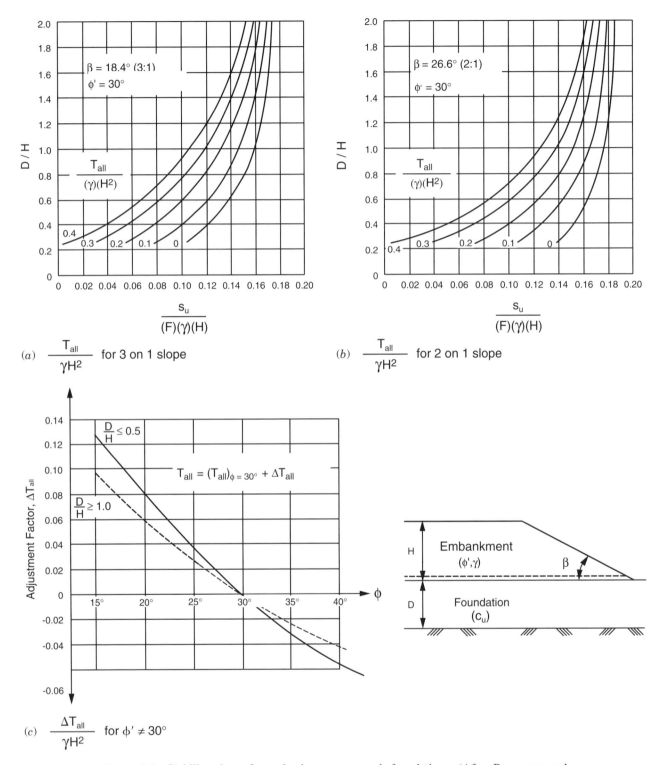

(a) $\dfrac{T_{all}}{\gamma H^2}$ for 3 on 1 slope

(b) $\dfrac{T_{all}}{\gamma H^2}$ for 2 on 1 slope

(c) $\dfrac{\Delta T_{all}}{\gamma H^2}$ for $\phi' \neq 30°$

Figure 8.8 Stability charts for embankments on weak foundations. (After Bonaparte and Christopher, 1987.)

Figure 8.6*a* shows the embankment sliding across the top of the reinforcing. This mode of failure is most likely if the interface friction angle between the embankment and the reinforcement is low, as it may be with geotextile reinforcement. A wedge analysis can be used to assess the safety of the embankment with regard to this mode of failure.

Figure 8.6*b* shows a shear surface cutting across the reinforcement and into the weak foundation. This mode of failure can occur only if the reinforcement ruptures or pulls out. Safety with regard to this mode of failure can be evaluated using circular, wedge, or noncircular slip surfaces, including reinforcement forces in the analysis as discussed previously.

Figure 8.6*c* shows large settlement of the embankment resulting from excessive elongation of the reinforcement. This mode of failure can occur if the strain in the reinforcement required to mobilize the reinforcement load is too large. Satisfying limit load criterion 3, discussed previously, will prevent this type of failure.

Even if an embankment is completely stable internally, it may still be subject to bearing capacity failure, as shown in Figure 8.7. This mode of failure can be analyzed using bearing capacity theory. If the foundation thickness (T) is small compared to the width of the equivalent uniform embankment load (B), the value of the bearing capacity factor N_c increases, as shown in the tabulated values in Figure 8.7 and the factor of

safety with respect to bearing capacity failure also increases. Therefore, the shallower is the weak foundation, the less likely is the bearing capacity mode of failure.

Bonaparte and Christopher (1987) charts. Preliminary estimates of the reinforcement force required for a given factor of safety can be made using the stability charts shown in Figure 8.8, which were developed by Bonaparte and Christopher (1987). The terminology used in these charts is:

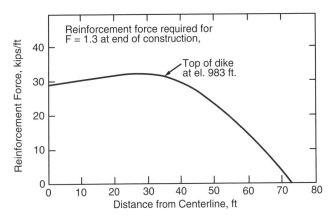

Figure 8.10 Required reinforcement force for Mohicanville Dike 2. (After Fowler et al., 1983.)

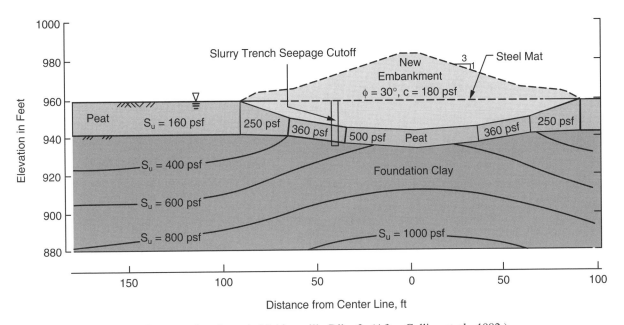

Figure 8.9 Cross section through Mohicanville Dike 2. (After Collins et al., 1982.)

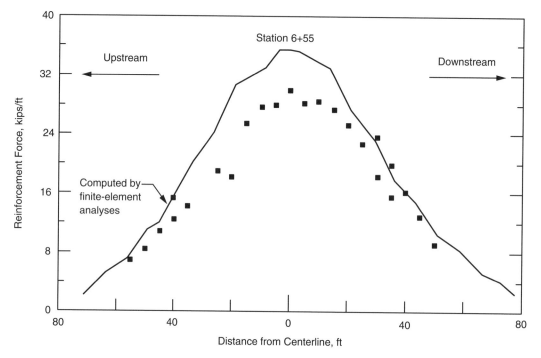

Figure 8.11 Reinforcement force distribution about the centerline for Station 6+55. (After Duncan et al., 1988.)

$\dfrac{T_{all}}{\gamma H^2}$ = dimensionless reinforcement force coefficient

(8.15)

$\dfrac{s_u}{F\gamma H}$ = dimensionless factored stability coefficient

(8.16)

T_{all} = allowable reinforcement force (F/L)

ΔT_{all} = change in T_{all} for embankment $\phi \neq 30°$

γ = total unit weight of embankment (F/L^3)

H = height of embankment (L)

s_u = undrained shear strength of foundation soil (foundation $\phi = 0$)

D = foundation depth (L)

β = slope angle (degrees)

Example. As an example of the use of these charts, consider the Mohicanville Dike No. 2 embankment shown in Figure 8.9. The Mohicanville project is described in more detail in the next section. Pertinent parameter values for the embankment and foundation are $H = 24$ ft, $\gamma = 136$ pcf, $\phi' = 32°$ for the embankment fill, embankment slope angle $\beta = \arctan$ (0.33) = 18°, average foundation shear strength = 700 psf, $D = 80$ ft, $s_u/F\gamma H = 700/(1.3)(136)(24) = 0.16$ for factor of safety on soil shear strength $F = 1.3$.

The cohesion of the embankment fill ($c = 200$ psf) is neglected because the charts in Figure 8.8 were developed for cohesionless embankment fill. The charts assume that the reinforcement is placed at the bottom of the embankment, where it is most effective in improving stability. The charts can be used to determine the magnitude of the reinforcement force required for a given factor of safety using the following steps:

1. Compute $D/H = (80 \text{ ft})/(24 \text{ ft}) = 3.3$.
2. Compute

$$\frac{s_u}{F\gamma H} = \frac{700 \text{ psf}}{(1.3)(136 \text{ pcf})(24 \text{ ft})} = 0.16$$

3. From Figure 8.8a, estimate $T_{all}/\gamma H^2 = 0.4$ (the value of D/H is above the top of the chart, and $T_{all}/\gamma H^2 = 0.4$ was estimated by extrapolation).
4. From Figure 8.8c, determine $\Delta T_{all}/\gamma H^2 = -0.01$.
5. Compute $T_{all}/\gamma H^2 = 0.40 - 0.01 = 0.39$.

6. Compute $T_{all} = (0.39)(136)(24)^2 = 31,000$ lb/ft.
7. Compute $T_{lim} = T_{all}F_R = (31,000$ lb/ft$)(1.5) = 47,000$ lb/ft for $F_R = 1.5$.

As shown in the next section, this result is in reasonable agreement with the results of detailed studies, indicating that the charts in Figure 8.8 can be used for preliminary assessment of reinforcement force.

Case history. Mohicanville Dike No. 2 is a rim dike on the Mohicanville Reservoir in Wayne County, Ohio (Duncan et al., 1988; Franks et al., 1988, 1991). Constructed on a weak peat and clay foundation, the dike failed during construction, and for many years the crest was 22 ft below its design elevation. A cross section through the dike is shown in Figure 8.9.

After evaluation of a number of alternatives for raising the dike to its design height, it was concluded that construction of a reinforced embankment afforded the best combination of cost and reliability. Limit equilibrium analyses and finite element analyses were preformed to determine the reinforcing force required for stability of the embankment. It was found that to achieve a factor of safety $F = 1.3$, a reinforcing force of 30,000 lb/ft was required. The results of equilibrium analyses are shown in Figure 8.10.

A heavy steel mesh was selected for reinforcement. This mesh has No. 3 mild steel bars spaced 2 in. apart perpendicular to the axis of the dike, welded into a mesh with No. 2 bars spaced 6 in. apart parallel to the axis of the dike. This mesh provided a cross-sectional area of about 1 in^2 of steel per foot of embankment length and a yield force (T_{lim}) equal to 48,000 lb/ft. This provides a factor of safety on reinforcement capacity, $F_R = T_{lim}/T_{all} = 48,000/30,000 = 1.6$.

The steel mesh was rolled up after fabrication into rolls containing strips 8 ft wide and 320 ft long. The steel yielded in bending, deformed plastically as it was rolled up, and stayed rolled up without restraint. The rolled strips of mesh were transported on trucks and were unrolled at the project site using the same equipment as that used to roll up the mesh in the fabricating plant. Each strip was cut into two 160-ft-long pieces that reached across the full width of the embankment, from upstream to downstream. The strips were dragged into position on the embankment using a front-end loader and a bulldozer. They were laid on, and were covered by, about 1 ft of clean sand.

The reinforcing mat was placed at elevation 960 ft, approximately 4 ft above the original ground elevation. In most areas, about 6 to 8 ft of old embankment fill was excavated to reach elevation 960 ft. In one 100-ft-long section of the embankment, where the foundation soils were thought to be exceptionally weak, a second layer of reinforcing was placed at elevation 961 ft.

The steel mat was not galvanized or otherwise protected against corrosion. Although the steel reinforcement will probably corrode in time, it is needed only for the first few years of the embankment's life. After the foundation gains strength through consolidation, the reinforcement will no longer be required for stability.

The embankment was designed using limit equilibrium analyses and finite element analyses that modeled consolidation of the foundation soils as well as interaction between the embankment and the steel reinforcing. The embankment was instrumented to measure reinforcement forces, settlements, horizontal movements, and pore pressures. Computed and measured reinforcement forces at the end of construction are shown in Figure 8.11. It can be seen that the calculated values agree quite well with the measured values. It is worthwhile to note that the finite element analyses were performed before the embankment was constructed, and the results shown in Figure 8.11 therefore constitute a true prediction of performance, not an after-the-fact matching of analytical results and field measurements.

This case history indicates that both limit equilibrium analyses and finite element analyses can be used to design reinforced embankments on weak foundations and to anticipate their performance. In most cases limit equilibrium analyses can be used as the sole design tool. However, in precedent-setting cases, as Mohicanville Dike No. 2 was in the mid-1980s, it is prudent to perform more thorough analyses using the finite element method.

Recapitulation

- Reinforcement can be used to improve the stability of slopes and embankments, making it possible to construct slopes and embankments steeper and higher than would otherwise be possible.

- Reinforced slopes and embankments can be analyzed using the procedures described in Chapter 6 by including the reinforcement forces as known forces in the analyses. The amount of force required to achieve a target value of factor of safety can be determined using repeated trials.

- Two methods have been used for limit equilibrium analyses of reinforced slopes: *method A,* in which allowable reinforcing forces are specified, and *method B,* in which *ultimate* reinforcing forces are specified. Method A is preferable, because it provides a means of applying different factors of safety to soil strength and reinforcing force, which have different sources of uncertainty and different amounts of uncertainty associated with their values.

- The principal types of reinforcing materials that have been used for slopes and embankments are geotextile fabrics, geogrids, steel strips, steel grids, and high-strength steel tendons.
- The long-term capacity of reinforcement, denoted here as T_{lim}, depends on tensile strength, creep characteristics, installation damage, durability, pullout resistance, and stiffness.
- The allowable load assigned to reinforcing materials should include a factor of safety, as indicated by the expression $T_{all} = T_{lim}/F_R$, where T_{all} is the allowable force, T_{lim} is the capacity of the reinforcement to carry long-term loads, and F_R is the reinforcement factor of safety. The value of F_R should reflect the level of uncertainty in the analyses and the consequences of failure.
- Designing reinforced slopes is facilitated greatly by slope stability charts of the type developed by Schmertmann et al. (1987), which are shown in Figure 8.5a and b.

- Potential modes of failure of reinforced embankments on weak foundations include sliding across the top of the reinforcing, shear through the reinforcement and into the weak foundation, large settlement of the embankment resulting from excessive elongation of the reinforcement, and bearing capacity failure.
- Preliminary estimates of the reinforcement force required for a given factor of safety can be made using the stability charts shown in Figure 8.8.
- The Mohicanville Dike No. 2 case history shows that both limit equilibrium analyses and finite element analyses can be used to design reinforced embankments on weak foundations and to anticipate their performance. Finite element analyses should be performed for precedent-setting applications.

CHAPTER 9

Analyses for Rapid Drawdown

Rapid drawdown takes place when the water level outside a slope drops so quickly that impermeable soils within the slope do not have sufficient time to drain. As the water level drops, the stabilizing effect of the water outside the slope is removed, and the shear stresses for equilibrium increase. The shear stresses within the slope are resisted by undrained strength in zones of low permeability and by drained strength within zones of higher permeability. This is a severe loading condition that can cause failure of slopes that are stable before drawdown.

Whether a soil zone drains or not can be estimated by calculating the value for the dimensionless time factor, T, given by

$$T = \frac{c_v\,t}{D^2} \qquad (9.1)$$

where c_v is the coefficient of consolidation, t the time for drawdown, and D the drainage distance. Typical values of c_v for various soils are shown in Table 9.1. If the calculated value of T is equal to 3 or more, the dissipation of pore water pressures induced by the drawdown exceeds 98%, and it is reasonable to treat the material as drained. Most soils with coefficients of permeability of 10^{-4} cm/s or more can be assumed to drain under normal rates of drawdown, and drained shear strengths can be used for these zones.

Rapid drawdown may occur at any time during the life of a slope, including during construction if the slope is built in water or water rises next to the slope during construction. Therefore, it may be necessary to analyze rapid drawdown for during and end-of-construction stability as well as for long-term conditions. The approaches and shear strengths used for the two cases (during and end of construction; long term)

are different and are described in the following sections.

DRAWDOWN DURING AND AT THE END OF CONSTRUCTION

If drawdown occurs during or immediately after construction, the appropriate shear strengths to be used in stability computations are the same as those used when no drawdown occurs: For soils that drain freely, the shear strengths are expressed in terms of effective stresses and appropriate pore water pressures are used. For soils that do not drain freely, undrained shear strengths, determined from the results of unconsolidated–undrained shear tests, are used. Stability computations are performed using effective stresses for soils that drain and total stresses for soils that do not drain.

DRAWDOWN FOR LONG-TERM CONDITIONS

If drawdown occurs long after construction, the soils within the slope will have had time to come to equilibrium under a new effective stress regime. The undrained shear strengths of low-permeability soils during drawdown are governed by the consolidation stresses in the equilibrium state before drawdown. The drained shear strengths of high-permeability soils are governed by the water pressures after drawdown.

Stability at the end of rapid drawdown has been analyzed in two basically different ways: (1) using effective stress methods, and (2) using total stress methods, in which the undrained shear strengths of low-permeability soils are related to effective consolidation pressures in the slope prior to drawdown. Both methods treat free-draining materials in the same way. The

Table 9.1   Typical Values of the Coefficient of Consolidation, c_v

Type of soil	Values of c_v (ft^2/day)
Coarse sand	>10,000
Fine sand	100–10,000
Silty sand	10–1000
Silt	0.5–100
Compacted clay	0.05–5
Soft clay	<0.2

strengths of free-draining materials are expressed in terms of effective stresses, and the pore water pressures are estimated assuming either steady seepage or hydrostatic conditions depending on the particular slope.

Effective Stress Methods

The advantage of effective stress analyses is that it is relatively easy to evaluate the required shear strength parameters. The effective stress shear strength parameters for soils are readily determined by means of isotropically consolidated undrained (IC-U) triaxial compression tests with pore pressure measurements. This type of test is well within the capability of most soil mechanics laboratories.

The disadvantage of effective stress analyses is that it is difficult to estimate the pore water pressures that will exist within low-permeability soils during drawdown. The pore pressure changes during drawdown depend on the changes in stress that result from the changing water loads on the slope and the undrained response of the soils within the slope to these changes in load. While the changes in stress can be estimated with reasonable accuracy, particularly at shallow depths beneath the surface of the slope, the undrained response of the soil is much harder to estimate. The changes in pore pressure are considerably different for materials that tend to dilate during shear and those that do not. Although in principle it is possible to estimate these pore water pressures, for example, by using Skempton's pore water pressure coefficients (Skempton, 1954), in practice this is difficult and the results are uncertain.

Most effective stress analyses of stability during rapid drawdown have used the assumptions regarding pore water pressures that were suggested by Bishop (1954) and later used by Morgenstern (1963). These assumptions have been justified on the basis of the fact that they are conservative in most cases. They have been found to result in reasonable values of factor of

safety for dams that suffered rapid drawdown failures: Wong et al. (1983) found that the values of safety factor calculated using Morgenstern's assumption were $F = 1.2$ for Pilarcitos Dam and $F = 1.0$ for Walter Bouldin Dam, both of which failed.

It seems likely that Bishop and Morgenstern's assumptions for pore water pressures during drawdown may be more accurate for soils that do not tend to dilate during shear than for those that do tend to dilate. Thus, although these assumptions may show reasonable correspondence with failures of slopes in materials that are not densely compacted and do not tend to dilate during shear, they are likely to be unduly conservative for better-compacted materials that do tend to dilate during shear. Use of effective stress analyses based on the Bishop and Morgenstern assumptions would treat all fill materials alike with respect to pore water pressures during drawdown, regardless of how well they are compacted or how strongly they might tend to dilate during shear.

Terzaghi and Peck (1967) suggested that pore water pressures during drawdown in well-compacted silty sands could be estimated using flow nets. Several investigators (Browzin, 1961; Brahma and Harr, 1963; Newlin and Rossier, 1967; Desai and Sherman, 1971; Desai, 1972, 1977) used theoretical methods to analyze the nonsteady flow conditions following drawdown. Like the Bishop and Morgenstern pore pressure assumptions, these methods do not consider the behavior of the soil with regard to dilatancy and are thus not capable of representing all of the important factors that control the pore water pressures during drawdown.

Svano and Nordal (1987) and Wright and Duncan (1987) used procedures for estimating pore water pressures during drawdown that reflect the influence of dilatancy on the pore water pressure changes. Svano and Nordal used two-stage stability analyses and iterated to achieve consistency between the calculated factor of safety and the values of the pore water pressures. Wright and Duncan used finite element analyses to estimate the stress changes during drawdown and Skempton's pore pressure parameters to calculate the pore water pressures. These studies indicate that it is possible to estimate realistic pore water pressures for effective stress analyses, but it is more cumbersome and difficult than using total stress analyses, as described in the following sections.

Terzaghi and Peck (1967) summarized the problems in estimating pore water pressures during drawdown in these words:

> [I]n order to determine the pore pressure conditions for the drawdown state, all the following factors need to be known: The location of the boundaries between materials

with significantly different properties; the permeability and consolidation characteristics of each of these materials; and the anticipated maximum rate of drawdown. In addition, the pore pressures induced by the changes in the shearing stresses themselves . . . need to be taken into consideration. In engineering practice, few of these factors can be reliably determined. The gaps in the available information must be filled by the most unfavorable assumptions compatible with the known facts.

By using undrained strengths in low-permeability zones, as described in the following sections, many of the problems associated with estimating pore water pressures for effective stress analyses can be avoided, and the accuracy of rapid drawdown stability analyses can be improved very considerably.

Total Stress Methods

Total stress methods are based on undrained shear strengths in low-permeability zones. The undrained shear strengths are estimated based on the effective stresses that exist in the slope prior to drawdown. Some zones within the slope may consolidate with time following construction, and their undrained strengths will increase. Portions of the same soils at lower stresses (near the surface of the slope) may expand following construction, and their undrained strengths will decrease with time.

Several total stress analysis methods have been suggested and used for sudden-drawdown analyses. These include the U.S. Army Corps of Engineer's (1970) method, and Lowe and Karafiath's (1959) method. Duncan et al. (1990) reviewed both of these methods and suggested an alternative *three-stage analysis* procedure that is described in the following paragraphs. The three-stage procedure combines the best features of both the Corps of Engineers' and Lowe and Karafiath's methods. Following along the lines of the Corps of Engineers' method, the three-stage procedure accounts for the effect of drainage and the fact that the drained strength may be less than the undrained strength. It differs from the Corps of Engineers' procedure in the way that the undrained strength is evaluated and the way that the drained strength is taken into account. Following Lowe and Karafiath's suggestion, the three-stage procedure accounts for the effects of anisotropic consolidation, which can result in significantly higher undrained shear strength.

Each stage in the three-stage stability analysis procedure involves a separate set of slope stability computations for each trial slip surface. For free-draining materials, effective stresses are used for all three stages, with different pore water pressures based on water levels and seepage conditions. The effective stress shear strength parameters are the same for all three stages. For low-permeability zones effective stresses are used for the first stage, before drawdown, and total stresses and undrained strengths are used for the second stage, after drawdown. For the third stage the lower of the drained and undrained strengths is used, whichever is lower, to be conservative.

First-stage computations. The first-stage stability computations are performed for conditions prior to drawdown. The purpose of the computations is to estimate effective stresses along the slip surface prior to drawdown. Effective stress shear strength parameters are used with pore water pressures based on estimated groundwater and seepage conditions. Steady-state seepage is assumed. Although the first-stage stability computations are performed in the same way that long-term stability computations are performed, the purpose is not to calculate the factor of safety; only the effective stresses on the slip surface are of interest.

The first-stage stability computations are used to calculate the shear stress and effective normal stress on the slip surface. The effective normal stress (σ') is computed for the base of each slice from the total normal force (N) on the bottom of each slice and the corresponding pore water pressure:

$$\sigma'_{fc} = \frac{N}{\Delta l} - u \qquad (9.2)$$

This normal stress represents the effective stress on a potential *failure* plane that exists at the time of *consolidation*. Accordingly, this normal stress and the corresponding shear stress are designated with the subscript *fc*. The corresponding shear stress is computed from the Mohr–Coulomb equation and the factor of safety using

$$\tau_{fc} = \frac{1}{F}(c' + \sigma' \tan \phi') \qquad (9.3)$$

or, alternatively, the shear stress can be computed from the shear force on the base of each slice from the relationship

$$\tau_{fc} = \frac{S}{\Delta l} \qquad (9.4)$$

where S is the shear force on the base of the slice. These consolidation stresses (σ'_{fc} and τ_{fc}) are used to estimate undrained shear strengths for the second-stage computations.

Second-stage computations. In the second-stage computations, undrained shear strengths and total stress analysis procedures are used for low-permeability zones. Undrained shear strengths are estimated for the second stage using the stresses calculated from the first stage and shear strength envelopes that relate the undrained shear strength to the effective consolidation pressure. The undrained shear strength is expressed as the value of the shear stress on the failure plane at failure, τ_{ff} (Fig. 9.1). The subscript ff distinguishes this value of shear stress from the value of shear stress at consolidation (τ_{fc}). The shear stress on the failure plane at failure is calculated from the principal stress difference, $\sigma_1 - \sigma_3$, at failure and the friction angle, ϕ', using the relationship

$$\tau_{ff} = \frac{\sigma_{1f} - \sigma_{3f}}{2} \cos \phi' \qquad (9.5)$$

The effective stress friction angle is the same one used for the first-stage analysis where effective stress envelopes are used for all soils. If the friction angle varies with effective stress, the value of the friction angle for the applicable range of effective stresses is used.

Undrained shear strength for second stage. The undrained shear strength, τ_{ff}, is plotted versus the effective stress on the failure plane at consolidation, σ'_{fc}. Two shear strength envelopes are plotted (Fig. 9.2): One corresponds to isotropic consolidation ($K_c = \sigma'_{1c}/\sigma'_{3c} = 1$), the other corresponds to anisotropic consolidation with the maximum effective principal stress ratio possible (i.e., $K_c = K_{\text{failure}}$). The first envelope ($K_c = 1$) is obtained from consolidated–undrained triaxial shear tests with isotropic consolidation. The ef-

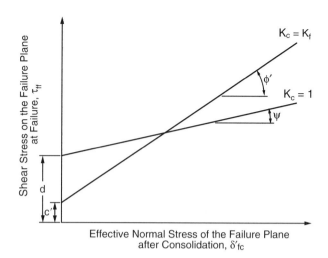

Figure 9.2 Shear strength envelopes used to define the undrained shear strengths for the second stage of a three-stage analysis.

fective stress on the failure plane (σ'_{fc}) for this envelope is the isotropic consolidation pressure, σ'_{3c}. The envelope for $K_c = 1$ is obtained by plotting τ_{ff} calculated from Eq. (9.5) versus the effective consolidation pressure, σ'_{3c}. The second shear strength envelope ($K_c = K_f$) is the same as the effective stress envelope used in the first-stage stability computations. The intercept and slope of the strength envelope for $K_c = K_f$ are the same as the effective cohesion and friction angle, c' and ϕ', respectively. This envelope corresponds to the maximum effective principal stress ratio possible, with the soil at failure during consolidation.

The slope and intercept of the strength envelope for $K_c = 1$ are related to the intercept and slope of the failure envelope that is often referred to as the R or *total stress envelope*. The R envelope is drawn on a Mohr diagram by plotting circles where σ_3 is the minor principal stress at consolidation (σ'_{3c}) and the principal stress difference (diameter of circle) is equal to the principal stress difference at failure, $\sigma_{1f} - \sigma_{3f}$, as illustrated in Fig. 9.3.[1] The intercept and slope of the failure envelopes drawn on such a diagram are designated c_R and ϕ_R. The intercepts and slopes for the τ_{ff} vs. σ'_{fc} envelope and the R envelope are similar but not equal. However, the intercepts and slopes of the two envelopes can be related to one another. If c_R and ϕ_R are the respective intercept and slope angle for the R envelope and the envelope is drawn tangent to the circles on a Mohr diagram (Figure 9.3a), the intercept, d,

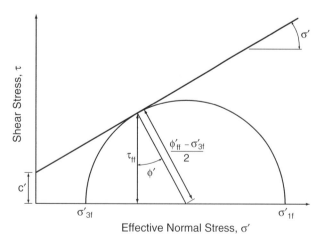

Figure 9.1 Mohr's circle for effective stresses at failure showing the shear stress on the failure plane.

---

[1]*Note:* These are not actually Mohr's circles because one stress (σ'_{3c}) is at consolidation while the other stress ($\sigma_{1f} - \sigma_{3f}$) is at failure.

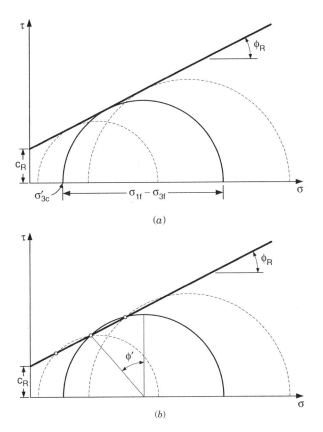

(a)

(b)

Figure 9.3 Shear strength envelopes used to define the undrained shear strengths for the second stage of three-stage analyses: (a) failure envelope tangent to circles; (b) failure envelope through points representing stresses on the failure plane.

and slope, ψ, for the corresponding $K_c = 1$ envelope are related as follows:

$$d_{K_c=1} = c_R \frac{\cos \phi_R \cos \phi'}{1 - \sin \phi_R} \qquad (9.6)$$

$$\psi_{K_c=1} = \arctan \frac{\sin \phi_R \cos \phi'}{1 - \sin \phi_R} \qquad (9.7)$$

The R envelope is sometimes drawn such that it passes through points corresponding to stresses on the failure plane. If the R envelope is drawn in this manner (Figure 9.3b), the slope and intercept of the $K_c = 1$ envelope are computed from

$$d_{K_c=1} = \frac{c_R}{1 + (\sin \phi' - 1) \tan \phi_R / \cos \phi'} \qquad (9.8)$$

$$\psi_{K_c=1} = \arctan \frac{\tan \phi_R}{1 + (\sin \phi' - 1) \tan \phi_R / \cos \phi'} \qquad (9.9)$$

Variation of τ_{ff} with σ'_{fc} and K_c. The undrained shear strength envelopes shown in Figure 9.2, for $K_c = 1$ and $K_c = K_f$, represent the extremes possible for the lowest and highest possible values of K_c. Lowe and Karafiath (1959) showed that for the same value of σ'_{fc}, the undrained strength (τ_{ff}) varies with the value of K_c. They recommended that anisotropically consolidated undrained (ACU) triaxial tests be performed to measure undrained shear strengths using a range of K_c values, to develop data that could be used to evaluate undrained strengths for second-stage analyses. This procedure results in accurate evaluation of undrained strengths for the second-stage analyses but requires extensive and difficult testing. The ACU test specimens must be consolidated slowly to avoid failure during consolidation.

Wong et al. (1983) found that the difficult ACU tests could be avoided by interpolating undrained strengths for values of K_c between $K_c = 1$ and $K_c = K_f$ instead of determining the strengths experimentally. They found that values of τ_{ff} calculated assuming that τ_{ff} varies linearly with K_c were the same as measured values. Using this method of interpolation, only isotropically consolidated undrained (ICU) tests with pore water pressure measurements need be performed. Both of the envelopes shown in Figure 9.2 can be determined from these tests, which are much easier to perform than ACU tests.

Once the $K_c = K_f$ and $K_c = 1$ envelopes shown in Figure 9.2 have been determined, the undrained shear strength for the second-stage computations is obtained based on the effective normal stress on the slip surface, σ' [Eq. (9.2)] and the estimated effective principal stress ratio for consolidation, K_c. The effective principal stress ratio for consolidation is estimated following the recommendations of Lowe and Karafiath (1959). They suggested the assumption that the orientation of the principal stresses at consolidation is the same as the orientation of the principal stresses at failure. This leads to the following equation for the effective principal stress ratio:

$$K_1 = \frac{\sigma' + \tau[(\sin \phi' + 1)/\cos \phi']}{\sigma' + \tau[(\sin \phi' - 1)/\cos \phi']} \qquad (9.10)$$

where K_1 is the effective principal stress ratio for consolidation corresponding to the stage 1 analyses and σ' and τ are the effective normal stress and shear stress on the shear plane at consolidation.

Values of the effective principal stress ratio can be calculated from Eq. (9.10) using the shear stress and effective normal stress calculated from Eqs. (9.2) and (9.3); the friction angle is the effective stress friction angle used in the first-stage computations and in Eq.

(9.5) to calculate τ_{ff}. Values of the undrained shear strength for the effective consolidation pressure, σ', and a consolidation stress ratio $K_c = K_1$ are obtained from the two shear strength envelopes shown in Figure 9.2. Linear interpolation between the values for $K_c = 1$ and $K_c = K_f$ to obtain the value corresponding to K_1 can be expressed by

$$\tau_{ff} = \frac{(K_f - K_1)\tau_{ff(K_c=1)} + (K_1 - 1)\tau_{ff(K_c=K_f)}}{K_f - 1} \quad (9.11)$$

where $\tau_{ff(K_c=1)}$ and $\tau_{ff(K_c=K_f)}$ are the undrained strengths from the two shear strength envelopes shown in Figure 9.2. The undrained shear strengths are determined from these two envelopes using the value of the effective stress, σ'_{fc}, which was calculated from Eq. (9.2).

The effective principal stress ratio at failure shown in Eq. (9.11) can be calculated from the effective stress shear strength parameters. If there is no cohesion, the value for K_f does not depend on the magnitude of the stress and is given by

$$K_f = \tan^2\left(45 + \frac{\phi'}{2}\right) \quad (9.12)$$

If the value of c' is not zero, the value of K_f depends on the effective stress on the slip surface and is given by

$$K_f = \frac{(\sigma' + c'\cos\phi')(1 + \sin\phi')}{(\sigma' - c'\cos\phi')(1 - \sin\phi')} \quad (9.13)$$

where σ' is the effective stress on the slip surface after consolidation (and also at failure because $K_c = K_f$).

If a significant effective cohesion (c') exists, the effective minor principal stress, σ'_3 implicit in Eqs. (9.10) and (9.13) may become negative (i.e., the terms in the denominator of these equations becomes negative). When this occurs, the effective consolidation stress ratios become negative and are nonsensical. The negative stress results from the fact that the soil is assumed to have *cohesion,* and this implies a tensile strength. The corresponding negative values for σ'_3 are, however, not realistic. In cases where negative effective stresses are calculated, the values are rejected and instead of interpolating shear strengths using values of K_c, the shear strength is taken to be the lower of the shear strengths from the $K_c = 1$ and $K_c = K_f$ envelopes. Negative effective stresses can be detected by calculating the effective minor principal stresses after consolidation using Eqs. (9.14) and (9.15):

$$\sigma'_{3c} = \sigma'_{fc} + \tau_{fc}\frac{\sin\phi' - 1}{\cos\phi'} \quad (9.14)$$

$$\sigma'_{3f} = (\sigma'_{fc} - c'\cos\phi')\frac{1 - \sin\phi'}{\cos^2\phi'} \quad (9.15)$$

Equation (9.14) is the effective minor principal stress corresponding to the $K_c = 1$ envelope; Eq. (9.15) corresponds to the $K_c = K_f$ envelope. The values of σ'_{3c} and σ'_{3f} correspond to the stresses in Eqs. (9.10) and (9.13), respectively. If either value is negative (or zero), no interpolation is performed and the lower of the $K_c = 1$ and $K_c = K_f$ strengths is used for the undrained shear strength in the second-stage stability computations. After the undrained shear strength is determined for each slice, a total stress analysis is performed using the undrained shear strengths and the external water loads after drawdown.

The factor of safety calculated in the second-stage analysis assumes that all of the low-permeability materials are undrained during rapid drawdown. Additional, third-stage computations are performed to check if the drained shear strength might be lower than the undrained shear strength, and thus the factor of safety would be lower if these low-permeability materials were drained rather than undrained.

Third-stage computations. For the third stage the undrained shear strengths used for the second-stage computations are compared with the drained strengths for each point along the slip surface. The drained shear strengths are estimated using the total normal stresses from the second-stage analysis and pore water pressures corresponding to the water level after drawdown. The total normal stresses are calculated for each slice using

$$\sigma = \frac{N}{\Delta l} \quad (9.16)$$

where N is the total normal force on the base of the slice calculated for the second stage and Δl is the length of the base of the slice. The effective stress is the total normal stress minus the applicable pore water pressure, u. The drained strength that would exist is then calculated from the Mohr–Coulomb equation,

$$s = c' + (\sigma - u)\tan\phi' \quad (9.17)$$

If at any point on the slip surface the drained shear strength calculated using Eq. (9.17) is lower than the undrained shear strength, an additional slope stability calculation is performed. If, however, the estimated drained strengths are all higher than the undrained shear strengths used for the second-stage computa-tions, no additional computations are performed, and the factor of safety calculated from the second-stage computations is the factor of safety for rapid draw-down.

When a third set of stability computations is per-formed, new shear strengths are assigned to the slices where the drained shear strengths were estimated[2] to be lower than the undrained shear strengths. For these slices effective stress shear strength parameters are as-signed and appropriate pore water pressures are stip-ulated. If the estimated drained strength for any slice is greater than the undrained shear strength, the strength is not changed, and undrained shear strength is used for the third stage. Thus, some portions of the slip surface will have effective stress shear strength parameters assigned and others will still have un-drained shear strengths. Once the appropriate drained or undrained shear strength has been assigned for each slice, the third-stage stability calculations are per-formed. The factor of safety from the third-stage cal-culations represents the factor of safety for rapid drawdown. If the third-stage calculations are not re-quired (i.e., undrained shear strengths are less than drained shear strengths for low-permeability materials everywhere along the slip surface), the factor of safety from the second stage is the factor of safety for rapid drawdown, as mentioned previously.

Example. A simple example problem is described to illustrate the three-stage analysis procedure de-scribed above. To simplify the calculations and allow them to be performed easily by hand, the example con-siders the rapid drawdown stability of an infinite slope. The slope is 3 (horizontal):1 (vertical), as shown in Figure 9.4. At the point in the slope where the stability calculations are performed, the slope is assumed to be submerged beneath 100 ft of water before drawdown. Although a depth of water is assumed, the water depth has no effect on the final results as long as the slope is fully submerged before drawdown. The total (satu-rated) unit weight of the soil is 125 pcf. The effective stress shear strength parameters are $c' = 0$, $\phi' = 40°$. The intercept and slope of the $K_c = 1$ undrained shear strength envelope (d and ψ) are 2000 psf and 20°. The shear strength envelopes are shown in Figure 9.5.

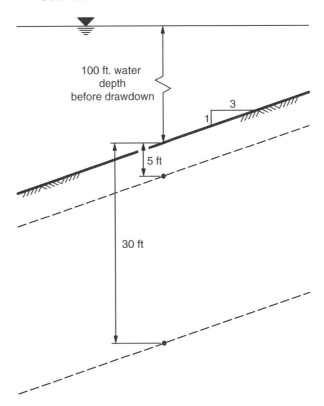

Figure 9.4 Slope used for example calculations of stability due to rapid drawdown using three-stage analysis procedure.

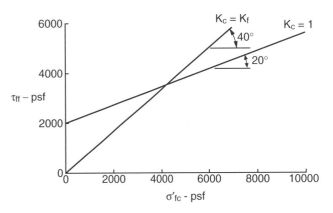

Figure 9.5 Shear strength envelopes for example problem.

Stability calculations are performed for two slip sur-faces. Both of the slip surfaces are planes parallel to the surface of the slope; one is 5 ft deep, the other is 30 ft deep. Total drawdown is assumed to occur (i.e., the water is assumed to be fully removed from the face of the slope). It is further assumed that after drainage has occurred, the pore water pressures are zero. The

---

[2] At the end of the second stage the drained shear strengths can only be estimated because the normal stress that corresponds to drained shear strengths depends to some extent on the drained strengths them-selves. The drained strengths calculated at the end of the second stage are based on total normal stresses that were calculated using un-drained rather than drained shear strengths. Thus, the drained strengths are considered to be "estimated" on the basis of the second stage. However, the degree of approximation involved in this estimate is very small.

calculations for each of the three stages are described below, and key quantities are summarized in Table 9.2.

First-stage analysis. The total normal stress on the slip surface is computed for a submerged infinite slope from the equation

$$\sigma = \gamma z \cos^2\beta + \gamma_w(h_w + z \sin^2\beta) \quad (9.18)$$

Thus, for the slip surface at a depth of 5 ft,

$$\sigma = (125)(5)\cos^2(18.4°) + (62.4)[100 + (5)\sin^2(18.4°)] = 6834 \text{ psf} \quad (9.19)$$

Similarly, for the slip surface at a depth of 30 ft, the total normal stress is 9803 psf. The pore water pressure on the slip plane is computed from

$$u = \gamma_w(z + h_w) \quad (9.20)$$

For the slip surface at a depth of 5 ft, the pore water pressure is

$$u = (62.4)(5 + 100) = 6552 \text{ psf} \quad (9.21)$$

and similarly, the pore water pressure at a depth of 30 ft is $u = 8112$ psf. The effective normal stress is computed by subtracting the pore water pressure form the total stress (i.e., $\sigma' = \sigma - u$). Thus, for the slip surface at a depth of 5 ft the effective stress is

$$\sigma' = 6834 - 6552 = 282 \text{ psf} \quad (9.22)$$

and for the slip surface at a depth of 30 ft, the effective normal stress is 1691 psf (= 9803 − 8112). These represent the effective normal stresses after consolidation, σ'_{fc}. The effective normal stresses can also be calculated directly using submerged unit weights (γ') and the following equation, because there is no flow:

$$\sigma' = \gamma'z \cos^2\beta \quad (9.23)$$

The shear stress on the slip surfaces is calculated from

$$\tau = (\gamma - \gamma_w)z \sin \beta \cos \beta \quad (9.24)$$

For the slip surface at a depth of 5 ft, this gives

$$\tau = (125 - 62.4)(5)\sin(18.4°)\cos(18.4°) = 94 \text{ psf} \quad (9.25)$$

and similarly, for the slip surface at a depth of 30 ft, the shear stress is 562 psf. These stresses represent the shear stresses, τ_{fc}, on the two potential slip surfaces after consolidation. The consolidation stresses are summarized in Table 9.2.

Second-stage analysis. For the second-stage analysis the undrained shear strengths are determined based on σ'_{fc} and τ_{fc}. The undrained shear strengths for the

Table 9.2 Summary of Stability Calculations for Rapid Drawdown of Example Slope

	Quantity	$z = 5$ ft	$z = 30$ ft
Stage 1	Total normal stress on the failure plane, σ_{fc}	6834 psf	9803 psf
	Pore water pressure, u	6552 psf	8112 psf
	Effective normal stress on the failure plane (consolidation pressure), σ'_{fc}	282 psf	1691 psf
	Shear stress on the failure plane after consolidation, τ_{fc}	94 psf	562 psf
Stage 2	Shear strength, τ_{ff} ($K_c = 1$)	2103 psf	2615 psf
	Shear strength, τ_{ff} ($K_c = K_f$)	237 psf	1419 psf
	Effective principal stress ratio for consolidation, K_1	2.0	2.0
	Undrained shear strength (interpolated), τ_{ff}	1585 psf	2283 psf
	Shear stress (after drawdown), τ	187 psf	1123 psf
	Factor of safety (undrained strengths)	8.48	2.03
Stage 3	Total normal stress (after drawdown), σ	563 psf	3376 psf
	Pore water pressure (after drawdown and drainage)	0	0
	Effective normal stress (after drawdown and drainage)	563 psf	3376 psf
	Drained shear strength	472 psf	2833 psf
Final stage	Governing strength after drawdown	472 psf (drained)	2283 psf (undrained)
	Factor of safety after drawdown	2.52 (third stage)	2.03 (second stage)

$K_c = 1$ shear strength envelope are computed using the equation

$$\tau_{ff(K_c=1)} = d + \sigma'_{fc} \tan \psi \qquad (9.26)$$

For the slip surface at a depth of 5 ft, this gives

$$\tau_{ff(K_c=1)} = 2000 + (282)\tan(20°) = 2103 \text{ psf} \qquad (9.27)$$

The corresponding value for the slip surface at a depth of 30 ft is $\tau_{ff(K_c=1)} = 2615$ psf. The shear strengths from the envelope for $K_c = K_f$ are computed from

$$\tau_{ff(K_c=K_f)} = c' + \sigma'_{fc} \tan \phi' \qquad (9.28)$$

For the slip plane at a depth of 5 ft, this gives

$$\tau_{ff(K_c=K_f)} = 0 + (282)\tan(40°) = 237 \text{ psf} \qquad (9.29)$$

Similarly, for the slip surface at 30 ft depth, $\tau_{ff(K_c=K_f)}$ is 1419 psf.

The undrained shear strengths for the second-stage analysis are interpolated using the effective principal stress ratios and the undrained shear strength values determined above. The effective principal stress ratios after consolidation are calculated using the stresses from the first-stage analysis and Eq. (9.10). For the slip surface at a depth of 5 ft,

$$K_1 = \frac{282 + 94\{[\sin(40°) + 1]/\cos(40°)\}}{282 + 94\{[\sin(40°) - 1]/\cos(40°)\}} = 2.0$$

$$(9.30)$$

The effective principal stress ratio for consolidation for the slip surface at a depth of 30 ft is also equal to 2.0.

The effective principal stress ratio at failure is calculated from Eq. (9.12). The effective principal stress ratio at failure is 4.6 and is the same for both depths because $c' = 0$. Undrained shear strengths are determined using Eq. (9.11). For the slip surfaces at depths of 5 and 30 ft, the shear strengths are as follows:

$$\tau_{ff} = \begin{cases} \dfrac{(4.6 - 2.0)2103 + (2.0 - 1)237}{4.6 - 1} = 1585 \text{ psf} \\[4pt] \quad \text{for 5-ft depth} \hspace{4.5cm} (9.31) \\[4pt] \dfrac{(4.6 - 2.0)2615 + (2.0 - 1)1419}{4.6 - 1} = 2283 \text{ psf} \\[4pt] \quad \text{for 30-ft depth} \hspace{4.2cm} (9.32) \end{cases}$$

The next step in the second-stage analysis is to calculate the factor of safety after drawdown using the undrained shear strengths. For the infinite slope and complete drawdown (no water above slope), the shear stress is calculated from

$$\tau = \gamma z \sin \beta \cos \beta \qquad (9.33)$$

For the slip surface at a depth of 5 ft, this gives

$$\tau = (125)(5)\sin(18.4°)\cos(18.4°) = 187 \text{ psf} \qquad (9.34)$$

Similarly, for the slip surface at a depth of 30 ft the shear stress is 1123 psf. The factors of safety for the slip surfaces are calculated by dividing the undrained shear strength by the shear stress. For the slip surfaces at depths of 5 and 30 ft, this gives, respectively,

$$F = \begin{cases} \dfrac{1585}{187} = 8.48 & \text{for 5-ft depth} \quad (9.35) \\[8pt] \dfrac{2283}{1125} = 2.03 & \text{for 30-ft depth} \quad (9.36) \end{cases}$$

These values represent the factors of safety for undrained conditions during drawdown.

Third-stage analysis. The third-stage analysis is begun by estimating the fully drained shear strengths of the soil (assuming that all excess pore water pressures due to drawdown have dissipated). For this example the water level is assumed to be lowered to such a depth that there will be no pore water pressures after drainage is complete. The total normal stress and effective stress are equal since the pore water pressures are zero. The effective stress for drained conditions after drawdown is computed from the equation

$$\sigma' = \gamma z \cos^2 \beta \qquad (9.37)$$

For the slip surface at a depth of 5 ft, this gives

$$\sigma' = (125)(5)\cos^2(18.4°) = 563 \qquad (9.38)$$

The drained shear strength is then

$$\tau_{\text{drained}} = 0 + (563)\tan(40°) = 472 \text{ psf} \qquad (9.39)$$

This value (472 psf) is much less than the value of undrained shear strength (1585 psf) determined earlier. Therefore, the factor of safety would be lower if drainage occurred. The shear stress was calculated above [Eq. (9.34)] to be 187 psf, and the factor of safety for the slip surface at a depth of 5 ft is, therefore,

$$F = \frac{472}{187} = 2.52 \qquad (9.40)$$

For the slip surface at a depth of 30 ft, the effective normal stress after drainage is calculated as

$$\sigma' = (125)(30)\cos^2(18.4°) = 3376 \qquad (9.41)$$

and the corresponding drained shear strength is

$$\tau_{drained} = 0 + (3376)\tan(40°) = 2833 \qquad (9.42)$$

This value of 2833 psf for the drained shear strength is higher than the value of 1125 psf determined earlier for the undrained shear strength. Thus, the undrained shear strength controls, and the factor of safety is equal to the value of 2.02 that was calculated earlier [Eq. (9.36)].

The calculations for this example show that the drained shear strength controls the stability after drawdown for the shallow (5 ft) depth while the undrained shear strength controls the stability for deeper (30 ft) depth. It is commonly found that drained shear strength is smaller at shallower depths, and undrained shear strength is smaller at deeper depths. Generally, this pattern of drained and undrained shear strengths controlling the stability is expected to happen. For slip surfaces that encompass a range of depths, the controlling shear strength may be the drained strength in some parts and the undrained strength in others.

PARTIAL DRAINAGE

Partial drainage during drawdown may result in reduced pore water pressures and improved stability. Theoretically, such improvement in stability could be computed and taken into account by effective stress stability analyses. The computations would be performed like effective stress analyses for long-term stability except that pore water pressures for drawdown would be considered. Although such an approach seems logical, it is beyond the current state of practice and probably beyond the present state of the art. The principal difficulties lie in predicting the pore water pressures induced by drawdown. Approaches based on construction of flow nets, and most of the existing numerical solutions (finite difference, finite element), do not account for changes in pore water pressures induced by shear deformations. Ignoring shear-induced pore water pressures may lead to errors on the unsafe side. For a more complete discussion of procedures for estimating pore water pressures due to sudden drawdown, see Wright and Duncan (1987).

CHAPTER 10

Seismic Slope Stability

Earthquakes expose slopes to dynamic loads that can reduce the soil shear strength and cause instability. During the past 30 years, significant advances have been made in the understanding of earthquake ground motions, nonlinear stress–strain properties of soils, strength losses due to earthquake loading, and dynamic response analyses for earth slopes. This progress has resulted in development of sophisticated procedures for analyzing stability of slopes subjected to earthquakes. At the same time, advances have been made in the use of simpler procedures for screening analyses, to determine if more complex analyses are needed.

ANALYSIS PROCEDURES

Detailed, Comprehensive Analyses

Comprehensive analysis procedures are generally used for any large embankment or any slope or embankment where the consequences of failure are high or significant soil strength losses occur. Although the specific details and steps of these procedures may vary, the general approach that is used is as follows (Seed, 1979; Marcuson et al., 1990):

1. Determine the cross section of the slope and underlying foundation that is to be analyzed.
2. Determine, with the aid of geologists and seismologists working as a team, the anticipated acceleration–time history for the ground beneath the dam. This should account for attenuation of motion away from the causative fault and amplification of motion as waves propagate upward through foundation soils overlying the bedrock.
3. Determine the static and dynamic stress–strain properties of the natural soils and fill materials within and beneath the slope.

4. Estimate the initial static stresses in the slope or embankment prior to the earthquake. This may involve the use of static finite element analyses in which the sequence of construction is simulated, or simpler methods.
5. Perform a dynamic finite element analysis to compute the stresses and strains induced in the embankment by the earthquake acceleration–time history.
6. Estimate the reductions in shear strength and increases in pore water pressure that will result from the earthquake. The most sophisticated dynamic analyses may include computations of reductions in strength as an integral part of the dynamic analysis in step 5.
7. Compute the stability of the slope using conventional limit equilibrium procedures with the reduced shear strengths determined in step 6. This may require analyses using both undrained and drained shear strengths to determine which strengths are most critical.
8. If the analyses indicate that the slope will be stable after the earthquake, compute the permanent displacements. If strength losses due to cyclic loading are small, a *Newmark-type sliding block analysis* may be used for this purpose (Newmark, 1965). However, if strength losses are significant, other methods should be used. For example, Seed (1979) showed that pseudostatic analysis procedures did not adequately reveal the problems with large displacement for the Upper Van Norman Dam, and he used the concept of a *strain potential* to evaluate the displacements. Conceptually, a complete nonlinear finite element analysis should be able to calculate any permanent displacements in a slope or dam; however, such analyses are very complex, involve considerable

uncertainties, and are seldom performed in practice.

The details involved in evaluation of dynamic soil properties and performing the type of dynamic response analyses outlined above are beyond the scope of this book. However, simpler procedures, to determine if detailed analyses are needed, are described in the following sections.

Pseudostatic Analyses

One of the earliest procedures of analysis for seismic stability is the *pseudostatic procedure,* in which the earthquake loading is represented by a static force, equal to the soil weight multiplied by a seismic coefficient, k. The pseudostatic force is used in a conventional limit equilibrium slope stability analysis. The seismic coefficient may be thought of loosely as an acceleration (expressed as a fraction of the acceleration, g, due to gravity) that is produced by the earthquake. However, the pseudostatic force is treated as a static force and acts in only one direction, whereas the earthquake accelerations act for only a short time and change direction, tending at certain instances in time to stabilize rather than destabilize the soil.

The term *pseudostatic* is a misnomer, because the approach is actually a static approach that is more correctly termed *pseudodynamic;* however, the term *pseudostatic* has been used for many years and is common in the geotechnical literature. The vertical components of the earthquake accelerations are usually neglected in the pseudostatic method, and the seismic coefficient usually represents a horizontal force.

Application of a seismic coefficient and pseudostatic force in limit equilibrium slope stability analyses is relatively straightforward from the perspective of the mechanics: The pseudostatic force is assumed to be a known force and is included in the various equilibrium equations. This is illustrated in Figure 10.1 for an infinite slope with the shear strength expressed in terms of total stresses. Similar equations can be derived for effective stresses and for other limit equilibrium procedures, including any of the procedures of slices discussed in Chapter 6.

An issue that arises in pseudostatic analyses is the location of the pseudostatic force. Terzaghi (1950) suggested that the pseudostatic force should act through the center of gravity of each slice or the entire sliding soil mass. This would be true only if the accelerations were constant over the entire soil mass, which they probably are not. Seed (1979) showed that the location assumed for the seismic force can have a small but noticeable effect on the computed factor of safety: For

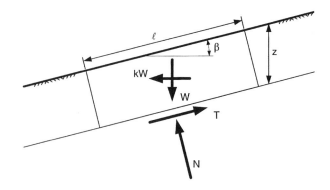

Resolving forces perpendicular to slip plane:
$$N = W\cos\beta - kW\sin\beta \qquad (1)$$

Resolving force parallel to slip plane:
$$T = W\sin\beta + kW\cos\beta \qquad (2)$$

Weight of sliding block:
$$W = \gamma \ell z \cos\beta \qquad (3)$$

Substituting (3) into (1) and (2):
$$N = \gamma \ell z \cos^2\beta - k\gamma \ell z \cos\beta\sin\beta \qquad (4)$$
$$T = \gamma \ell z \cos\beta\sin\beta + k\gamma \ell z \cos^2\beta \qquad (5)$$

For the stresses on the slip plane:
$$\sigma = \frac{N}{\ell} = \gamma z \cos^2\beta - k\gamma z \cos\beta\sin\beta \qquad (6)$$
$$\tau = \gamma z \cos\beta\sin\beta + k\gamma z \cos^2\beta \qquad (7)$$

Finally, for the factor of safety (total stresses):
$$F = \frac{s}{\tau} = \frac{c + \sigma\tan\phi}{\tau} = \frac{c + \left(\gamma z \cos^2\beta - k\gamma z \cos\beta\sin\beta\right)\tan\phi}{\gamma z \cos\beta\sin\beta + k\gamma z \cos^2\beta} \qquad (9)$$

Figure 10.1 Derivation of the equation for the factor of safety of an infinite slope with a seismic force (kW)—total stress analyses.

the Sheffield Dam, changing the location of the pseudostatic force from the centers of gravity to the bottoms of the slices reduced the factor of safety from 1.32 to 1.21 for a seismic coefficient of 0.1.

Dynamic analyses of the response of many dams to earthquakes (Makdisi and Seed, 1978) indicate that peak accelerations increase (i.e., they are amplified) from the bottom to the top of a dam. Thus, the location of the resultant seismic force would be expected to be *above* the center of gravity of the slice. In the case of circular slip surfaces, this would reduce the moment about the center of the circle due to the seismic forces, in comparison to applying the force at the center of gravity of the slice, and the factor of safety would be expected to *increase*. This reasoning is consistent with the results of Seed (1979) for the Sheffield Dam, which showed that the factor of safety *decreased* when the seismic force was located *below* the center of gravity of the slice. Assuming that the pseudostatic force acts through the center of gravity of the slice is probably slightly conservative for most dams. Thus, it appears that Terzaghi's suggestion is reasonable. For most pseudostatic analyses the pseudostatic force is assumed to act through the center of gravity of each slice. If a force equilibrium (only) procedure is used, the location of the pseudostatic force has no effect on the factor of safety computed.

For many years, seismic coefficients were estimated based on empirical guidelines and codes. Typical values for seismic coefficients used ranged from about 0.05 to about 0.25 (Seed, 1979; Hynes-Griffin and Franklin, 1984; Kavazanjian et al., 1997). However, with the development of more sophisticated analyses, particularly displacement analyses such as the sliding block analyses described in the next section, correlations can be made between the seismic coefficient, the expected earthquake accelerations, and the probable displacements. Most seismic coefficients used today are based on experience and results from deformation analyses.

Sliding Block Analyses

Newmark (1965) first suggested a relatively simple deformation analysis based on a rigid sliding block. In this approach the displacement of a mass of soil above a slip surface is modeled as a rigid block of soil sliding on a plane surface (Figure 10.2). When the acceleration of the block exceeds a yield acceleration, a_y, the block begins to slip along the plane. Any acceleration that exceeds the yield acceleration causes the block to slip and imparts a velocity to the block relative to the velocity of the underlying mass. The block continues to move after the acceleration falls below the yield acceleration. Movement continues until the velocity of the block relative to the underlying mass goes to zero, as shown in Figure 10.3. The block will slip again if the acceleration again exceeds the yield acceleration. This stick-slip pattern of motion continues until the accelerations fall below the yield acceleration and the relative velocity drops to zero for the last time. To

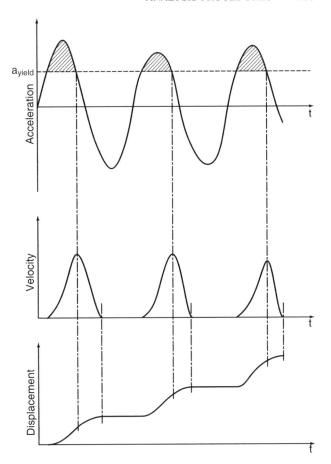

Figure 10.3 Double integration of acceleration–time history to compute permanent displacements.

compute displacements, the accelerations in excess of the yield acceleration are integrated once to compute the velocities and a second time to compute the displacements (Figure 10.3). Given an acceleration–time history and yield acceleration, the integration can be performed numerically. Movements in the upslope direction are neglected (i.e., all displacements are assumed to be "one-way"). Details are beyond the scope of this chapter but can be found in the literature (e.g., Jibson, 1993; Kramer, 1996; Kramer and Smith, 1997).

Limit equilibrium slope stability analyses are used to compute the values of yield acceleration, a_y, used in sliding block analyses. The yield acceleration is usually expressed as a seismic yield coefficient, $k_y = a_y/g$. The seismic yield coefficient is the seismic coefficient that produces a factor of safety of unity in a pseudostatic slope stability analysis.

The value of k_y is determined using conventional slope stability analysis procedures. However, rather than searching for the slip surface that gives the minimum static factor of safety, searches are conducted to

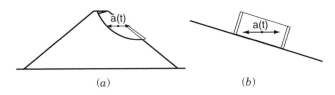

Figure 10.2 (a) Actual slope; (b) sliding block representation used to compute permanent soil displacements in a slope subjected to earthquake shaking.

find the slip surface that gives the minimum value of k_y. The slip surface giving the minimum value of k_y is usually different from the one giving the minimum factor of safety for static conditions.

Recapitulation

- Comprehensive analysis procedures exist and are used to evaluate seismic stability for large embankments and slopes where the consequences of failure are high or significant soil strength losses are anticipated.
- Pseudostatic analysis procedures approximate the earthquake loads as a static force and are less accurate than other procedures, but useful as a screening tool.
- The pseudostatic seismic force is usually assumed to act through the center of gravity of the soil mass, and this assumption seems reasonable.
- Newmark-type sliding block analyses provide a simple way of estimating permanent displacements in a slope caused by an earthquake.
- Improvements have been made in the criteria for selecting seismic coefficients for pseudostatic analyses through studies of embankment displacements using sliding block analyses and case studies.

PSEUDOSTATIC SCREENING ANALYSES

Pseudostatic analyses provide a useful way of screening for potential seismic stability problems, especially when the soils involved are not expected to lose a significant amount of their strength due to the earthquake. Makdisi and Seed (1977) found that for clayey soils, dry or partially saturated cohesionless soils, or very dense saturated cohesionless soils, 80% of the static undrained strength represents an approximate threshold between large and small strains induced by cyclic loading. Substantial permanent strains may be produced when these nonliquefiable soils are subjected to cyclic loads near their full undrained strengths. Essentially elastic behavior was observed when these same soils were subjected to large numbers of cycles (>100) at 80% of their undrained strengths. Accordingly, Makdisi and Seed (1977) recommended the use of 80% of the static undrained strength as the dynamic yield strength for nonliquefiable soils.

Use of pseudostatic analyses as screening analyses can be a simple process. A suitable seismic coefficient is determined based on an appropriate criterion and the factor of safety is computed. The computed factor of safety provides an indication of the possible magnitude

of seismically induced displacements. Criteria for selection of seismic coefficients and for determining what are acceptable factors of safety have been developed by Makdisi and Seed (1978), Hynes-Griffin and Franklin (1984), Bray et al. (1998), and Kavazanjian et al. (1997) by comparing the results of pseudostatic analyses with field experience and the results of deformation analyses.

Several methods for using pseudostatic analyses to determine the need for more detailed studies are summarized in Table 10.1. Each involves these components:

1. *A reference peak acceleration, a_{ref}.* The reference accelerations used are either the peak acceleration in bedrock beneath the slope, or the peak soil acceleration at the top of the slope. Peak bedrock acceleration is easier to use, because determining peak acceleration at the top of the slope requires a dynamic response analysis.
2. *Acceleration multiplier.* The seismic coefficient used in the pseudostatic analysis is equal to a_{ref}/g multiplied by an acceleration multiplier, a/a_{ref} [$k = (a_{ref}/g)(a/a_{ref})$]. Values of acceleration multiplier ranging from 0.17 to 0.75 have been recommended, as shown in Table 10.1.
3. *Shear strength reduction factor.* Most authorities recommend using reduced shear strengths in pseudostatic analyses. As shown in Table 10.1, the strength most often recommended is 80% of the static shear strength, following the findings of Makdisi and Seed (1977). For landfills with geosynthetic liners, Bray et al. (1998) recommended using residual strengths because the geosynthetic interface strength is reached at small deformations that usually are exceeded during landfill construction.
4. *Minimum factor of safety.* All of the screening criteria summarized in Table 10.1 stipulate a minimum acceptable factor of safety. The values are either 1.0 or 1.15.
5. *Tolerable permanent deformation.* Each set of criteria summarized in Table 10.1 carries with it the notion that a certain amount of earthquake-induced deformation is tolerable. The magnitudes of deformation judged to be tolerable vary from 0.15 m in the case of landfill base liners to 1.0 m for dams.

Each of the suggested methods outlined in Table 10.1 is complete within itself, and should be viewed in this way: If a pseudostatic analysis using the specified reference acceleration shown in column (2), the acceleration multiplier shown in column (3), and the strength reduction factor shown in column (4) results

Table 10.1   Suggested Methods for Performing Pseudostatic Screening Analyses

(1) Reference	(2) Reference acceleration, a_{ref}	(3) Acceleration multiplier, a/a_{ref}	(4) Strength reduction factor	(5) Minimum factor of safety	(6) Tolerable displacement
Makdisi and Seed (1978)	0.2 g ($M \approx 6\frac{1}{2}$)	0.5	0.8	1.15	Approx. 1 m
Makdisi and Seed (1978)	0.75 g ($M \approx 8\frac{1}{4}$)	0.2	0.8	1.15	Approx. 1 m
Hynes-Griffin and Franklin (1984)	PHA_{rock}	0.5	0.8	1.0	1 m
Bray et al. (1998)	PHA_{rock}	0.75	Recommend using conservative (e.g., residual) strengths	1.0	0.30 m for landfill covers; 0.15 m for landfill base sliding
Kavazanjian et al. (1997)	PHA_{soil}	0.17 if response analysis is performed	0.8^a	1.0	1 m
Kavazanjian et al. (1997)	PHA_{soil}	0.5 if response analysis is not performed	0.8^a	1.0	1 m

[a]For fully saturated or sensitive clays.

in a factor of safety greater than or equal to the value shown in column (5), this indicates that the permanent displacements induced by the earthquake will not be larger than those shown in column (6).

Although the methods summarized in Table 10.1 differ with regard to details, they employ the same procedure for screening conditions that would lead to development of large permanent seismically induced displacements. It will be noted that the methods that are more stringent with respect to seismic coefficient use tighter criteria for displacements that are considered tolerable.

The criteria by Hynes-Griffin and Franklin (1984) were developed for earth dams and reflect results of extensive analyses not reflected in Makdisi and Seed's (1978) criteria. The criteria by Bray et al. (1998) are applicable to landfills, and like Hynes-Griffin and Franklin's criteria, reflect the results of deformation analyses. The criteria developed by Hynes-Griffin and Franklin and by Bray et al. are based on peak horizontal bedrock acceleration, PHA_{rock}, and do not require site response analyses. However, the criteria are more conservative than those proposed by Kavazanjian et al. (1997), which do require response analysis. The additional effort involved in performing a response

analysis can be weighed against the increased conservatism required if a site response analysis is not performed. Simplified procedures for estimating site response have been discussed by Kavazanjian et al. (1997).

Recapitulation

- Several simple screening criteria have been developed for evaluating seismic stability using pseudostatic analysis procedures.
- The screening criteria differ in the reference seismic acceleration, acceleration multiplier, strength reduction factor, acceptable factor of safety, and tolerable displacement criterion used.

DETERMINING PEAK ACCELERATIONS

Peak accelerations can be determined for highly seismic areas, where extensive data from previous earthquakes has been accumulated, using empirical attenuation relations. For sites where less information is available, peak accelerations can be established using the U.S. Geological Survey Geohazards Internet Web site (*http://eqhazmaps.usgs.gov/*). The Web site provides peak rock acceleration (PGA_{rock}) based on latitude and longitude or zip code. Example results are shown in Table 10.2.

SHEAR STRENGTH FOR PSEUDOSTATIC ANALYSES

The shear strength appropriate for use in a pseudostatic analysis depends on whether the analysis is being performed for short-term (end-of-construction) conditions or for a slope that has been in existence for many years. Pseudostatic analyses may need to be performed for both short- and long-term conditions depending on the particular slope.

Because seismic loading is of short duration, it is reasonable to assume that except for some coarse gravels and cobbles, the soil will not drain appreciably during the period of earthquake shaking. Thus, undrained shear strengths are used for most pseudostatic analyses (with the exception noted later of soils that tend to dilate when sheared and may lose strength after the earthquake as they drain).

Earthquakes Immediately after Construction

Pseudostatic analyses for short-term stability are only appropriate for new slopes. Undrained shear strength can be evaluated using conventional unconsolidated–undrained testing procedures and samples identical to the ones that would be tested to determine the shear strength for static conditions. The analyses are performed using shear strengths expressed in terms of total stresses.

Earthquakes After the Slope Has Reached Consolidated Equilibrium

All slopes that will be subjected to earthquakes should be evaluated for long-term stability using values of undrained shear strength that reflect the eventual long-term conditions, including consolidation or swell after

the slope is constructed. The manner in which the undrained shear strength is determined for this condition depends on whether we are dealing with an existing slope or a slope that is yet to be built.

Existing slopes. If a slope has reached consolidated equilibrium, the shear strength can be determined by taking representative samples of the soil and performing tests using unconsolidated–undrained testing procedures. The stability analysis is then performed much like a short-term stability analysis, using shear strength parameters expressed in terms of total stresses.

New slopes. For new slopes it is necessary to simulate the effects of future consolidation and swell in the laboratory using consolidated–undrained testing procedures (Seed, 1966). The testing and analysis procedures are almost identical to those described in Chapter 9 for rapid drawdown: The soil is first consolidated to a state of effective stress and then sheared with no drainage. The difference between rapid drawdown and seismic loading in a pseudostatic analysis is that the undrained loading for rapid drawdown is due to lowering the water level adjacent to the slope, while the undrained loading for an earthquake is caused by seismic forces.

Once the appropriate shear strength envelopes are determined from the results of consolidated–undrained triaxial shear tests, the slope stability computations are performed using a two-stage analysis procedure nearly identical to the procedures described in Chapter 9 for rapid drawdown. A first-stage analysis is performed for conditions prior to the earthquake (no seismic coefficient) to compute the consolidation stresses, σ'_{fc} and τ_{fc}. These stresses are then used to estimate the undrained shear strength for seismic loading using the same procedures as for rapid drawdown. The undrained shear strength is then used in the second-stage computations (with seismic coefficient) to compute the pseudostatic factor of safety for the slope. For rapid drawdown a third stage of computations is performed to account for the likelihood of drainage during drawdown, but drainage *during* an earthquake is much less likely, and thus a third stage of computations is usually

Table 10.2   Peak Rock Acceleration Results

Zip code	PGA_{rock} with 10% probability of excedence in 50 years (500-year return period)	PGA_{rock} with 5% probability of excedence in 50 years (1000-year return period)	PGA_{rock} with 2% probability of excedence in 50 years (2500-year return period)
24060	0.054 g	0.097 g	0.194 g
78712	0.012 g	0.021 g	0.039 g

not necessary. However, drainage *after* the earthquake could adversely affect stability and should be considered as discussed in the final section of this chapter.

Simplified procedure (*R* envelope and single-stage analysis). Although the two-stage analysis procedure described above is the proper way to perform a pseudostatic analysis for a slope, analyses are sometimes performed using a simple single-stage procedure. In the single-stage procedure the *R* shear strength envelope is used. The *R envelope*, as it is called in U.S. Army Corps of Engineers' terminology, is obtained by plotting results of consolidated–undrained triaxial tests as shown on the Mohr–Coulomb diagram in Figure 10.4. In this diagram circles are plotted as follows: The minor principal stress is the effective stress, σ'_{3c}, at *consolidation;* the diameter of the circle is the principal stress difference *at failure*, $\sigma_{1f} - \sigma_{3f}$. Because the two stresses used to plot the circles (σ'_{3c} and $\sigma_{1f} - \sigma_{3f}$) exist at different times during a test, the circles are not actually Mohr circles—a Mohr's circle represents the state of stress at a point *at an instant in time* (e.g., at consolidation or at failure). Similarly the *R* envelope drawn on such a diagram is not actually a Mohr–Coulomb failure envelope. However, the envelope that is drawn on such a diagram is often quite similar to the corresponding τ_{ff} vs. σ'_{fc} envelope that is plotted from consolidated–undrained tests described in Chapter 9. To illustrate the similarities between the *R* and τ_{ff} vs. σ'_{fc} envelopes, the two envelopes are plotted in Figure 10.5 for four different soils. The four soils and properties are summarized in Table 10.3. It can be seen that the *R* envelope in each case plots below the τ_{ff} vs. σ'_{fc} envelope. There may be other cases where the *R* envelope lies above the τ_{ff} vs. σ'_{fc} envelope.

In the simplified pseudostatic procedure a single set of computations is performed for each trial slip surface using the appropriate seismic coefficient and the intercept and slope (c_R and ϕ_R) of the *R* envelope as Mohr–Coulomb shear strength parameters. In this simplified approach it is important to use the proper pore water pressures: The *R* envelope in this case is being used as an envelope approximating the relationship between the undrained shear strength ($\approx \tau_{ff}$) and effective consolidation pressure ($\approx \sigma'_{fc}$). Thus, pore water pressures equal to those during consolidation (e.g., for steady-state seepage) should be used in the computations.[1]

To illustrate the differences between the simple single-stage procedure using the *R* envelope and the more rigorous two-stage procedure such as the one described earlier, a pseudostatic slope stability analysis was performed for the slope shown in Figure 10.6. A seismic coefficient of 0.15 was used and computations were performed for the slope with both dry (zero pore water pressure) and fully submerged conditions prior to earthquake loading. Computations were performed using each of the four different sets of soil strength properties shown in Figure 10.4 and Table 10.3. Results of the computations are summarized in Table 10.4 and plotted in Figure 10.7. In all cases the factor of safety computed by the simpler procedure is less than the factor of safety computed by the more rigorous two-stage procedure. The factor of safety from the single-stage procedure varies between about 80 and 90% of the value from the two-stage analysis.

Recapitulation

- Unconsolidated–undrained tests performed on undisturbed or laboratory compacted specimens can be used to determine the shear strength for pseudostatic analyses of *existing slopes* or *new slopes at the end of construction.*
- To determine the shear strength for *new slopes after the slope has reached consolidated equilibrium,* shear strengths are determined using consolidated–undrained testing procedures and analyses are performed as two-stage analyses using procedures similar to those used for rapid drawdown.

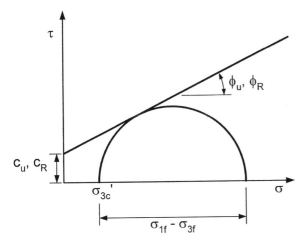

Figure 10.4 *R* (total stress) undrained shear strength envelope plotted from consolidated–undrained triaxial compression tests.

[1] Although the *R* envelope is called a *total stress envelope* in many texts (Terzaghi and Peck, 1967; Peck et al., 1974; Wu, 1976; Sowers, 1979; Dunn et al., 1980; Holtz and Kovacs, 1981; Lee et al., 1983; McCarty, 1993; Liu and Evett, 2001; Abramson et al., 2002; Das, 2002), when the envelope is used in pseudostatic slope stability computations, effective stresses and appropriate pore water pressures must be used.

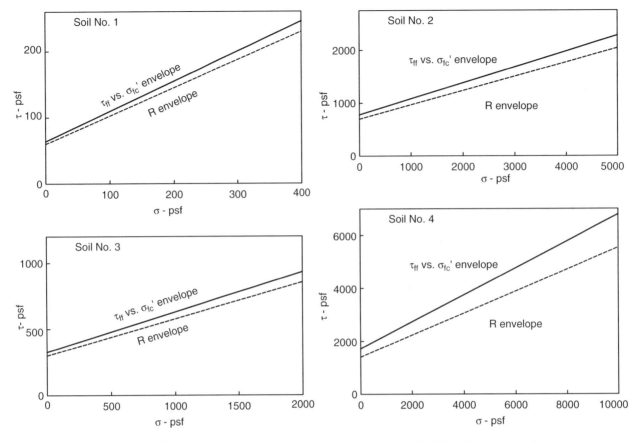

Figure 10.5 Comparison of τ_{ff}-σ'_{fc} shear strength envelope with R (total stress) envelope for undrained shear.

- A simplified single-stage analysis procedure using the R strength envelope appears in many cases to produce conservative estimates for the pseudostatic factor of safety after the slope has reached consolidated equilibrium.
- Reductions in strength of up to 20% caused by cyclic loading during an earthquake are probably offset by the effects of a higher loading rate during an earthquake compared to normal loading rates in static tests.

Effects of Rapid Load Application

Pseudostatic analysis procedures are appropriate only for cases involving soils that do not lose significant strength during an earthquake. Several of the screening guidelines summarized in Table 10.1 allow for moderate strength losses by using a nominal strength reduction factor. However, even if no such factor is used, strength losses of no more than 15 to 20% can probably be safely ignored in selecting shear strength for pseudostatic analyses because of strain rate effects. Most soils that are subjected to undrained loading at

the rates imposed by earthquakes will exhibit strengths that are 20 to 50% higher than the shear strength measured in conventional static loading tests where the time to failure is several minutes or longer. Soils typically show an increase in undrained shear strength of 5 to 25% per tenfold increase in strain rate (decrease in time to failure). Considering earthquake loading with a period of 1 s, the time to increase the load from zero to the peak would be approximately 0.25 s. If a static test is performed with a time to failure of 10 minutes (600 s), and the shear strength of the soil increases 10% per tenfold decrease in the time to failure, the effect of strain rate on the strength during an earthquake would be expected to be about 34% higher:

$$10\% \times \log_{10}\frac{600}{0.25} = 34\%$$

Such increases in strength due to load rate effects will offset a 15 to 20% reduction in strength due to cyclic loading. Thus, there is some basis for not reducing the shear strength used in pseudostatic analyses, provided, of course, that the analysis is only being

Table 10.3   Summary of Soil Properties Used in Comparison of R and τ_{ff} vs. σ'_{fc} Strength Envelopes

Soil no.	Description and reference	Index properties	c' (psf)	ϕ' (deg)	c_R (psf)	ϕ_R (deg)	d^a (psf)	ψ^b (deg)
1	Sandy clay (CL) material from Pilarcitos Dam; envelope for low (0–10 psi) confining pressures. (Wong et al., 1983)	Percent minus No. 200: 60–70 Liquid limit: 45 Plasticity index: 23	0	45	60	23	64	24.4
2	Brown sandy clay from dam site in Rio Blanco, Colorado (Wong et al., 1983)	Percent minus No. 200: 25 Liquid limit: 34 Plasticity index: 12	200	31	700	15	782	16.7
3	Same as soil 1 except envelope fit to 0–100 psi range in confining pressure (Wong et al., 1983)	Percent minus No. 200: 60–70 Liquid limit: 45 Plasticity index: 23	0	34	300	15.5	327	16.8
4	Hirfanli Dam fill material (Lowe and Karafiath, 1960)	Percent minus No. 200: 82 Liquid limit: 32.4 Plastic limit: 19.4	0	35	1400	22.5	1716	26.9

[a]Intercept of τ_{ff} vs. σ'_{fc} envelope—can be calculated knowing c', ϕ', c_R, and ϕ_R.
[b]Slope of τ_{ff} vs. σ'_{fc} envelope—can be calculated knowing c', ϕ', c_R, and ϕ_R.

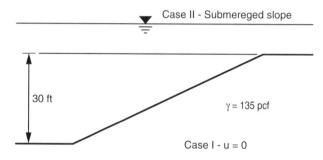

Figure 10.6 Slope used to compare simple, single-stage and rigorous, two-stage pseudostatic analyses.

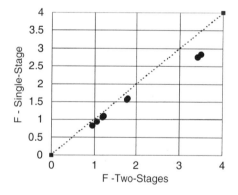

Figure 10.7 Comparison of factors of safety by simplified single-stage pseudostatic and more rigorous two-stage pseudostatic analyses.

used for cases where significant (more than 15 to 20%) strength losses are not anticipated.

POSTEARTHQUAKE STABILITY ANALYSES

Following an earthquake, the stability of a slope may be diminished because cyclic loading has reduced the shear strength of the soil. The reductions in shear strength are generally treated differently depending on whether or not liquefaction occurs. Stability follow-

Table 10.4   Summary of Pseudostatic Safety Factors Computed Using Simple Single-Stage and Rigorous Two-Stage Procedures

	Case I: dry slope		Case II: submerged slope	
Soil	Single-stage analysis	Two-stage analysis	Single-stage analysis	Two-stage analysis
1	0.95	1.06	0.83	0.95
2	1.56	1.77	1.59	1.79
3	1.07	1.19	1.10	1.21
4	2.76	3.42	2.83	3.49

ing an earthquake can be evaluated using a three-step process.

Step 1. Estimate if Liquefaction Will Occur

The first step in evaluating strength losses is to determine if the soil will liquefy. The procedures for doing this are semiempirical, based on results of field tests and case histories, mostly for horizontal ground. According to Youd et al. (2001) four different field tests are suitable for measuring soil resistance to liquefaction: (1) cone penetration tests, (2) Standard Penetration tests, (3) shear-wave velocity measurements, and (4) for gravelly sites, the Becker penetration test. Various correlations have been developed that relate the resistance or stiffness characteristics of the soil measured in these tests to the cyclic shear stresses required to cause liquefaction. The cyclic shear stresses required to cause liquefaction are generally expressed as a normalized ratio of cyclic shear stress to effective vertical consolidation pressure, $\tau_{cyclic}/\sigma'_{vo}$, known as the *cyclic resistance ratio* (CRR). Based on one or more of the field tests described above, an estimate is made of the cyclic resistance ratio using appropriate correlations. The cyclic resistance is then compared with the seismically induced *seismic stress ratio* (CSR) to determine if liquefaction will occur.

Step 2. Estimate Reduced Undrained Shear Strengths

If the soil is expected to liquefy, reduced values of the undrained residual shear strengths are estimated.[2] Seed and Harder (1990) suggested the relationship shown in Figure 10.8 between the undrained residual shear strength and the corrected standard penetration resistance, $(N_1)_{60}$. Poulos, Castro, and their co-workers have proposed an alternative approach based on the concept of *steady-state shear strength* (Poulos et al., 1985). The steady-state strength is estimated based on the void ratio in the field and a relationship between steady-state strength and void ratio determined in the laboratory. A third approach to determining the undrained residual shear strength has been proposed by Stark and Mesri (1992) and Olson and Stark (2002). They have proposed determining an equivalent c/p ratio representing the ratio of the undrained residual shear strength to the effective preearthquake consolidation pressure. Stark and his co-workers have developed empirical correlations that relate values of c/p to Standard

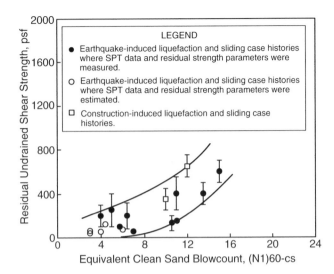

Figure 10.8 Relationship between corrected "clean sand" blowcount (N1)60-cs and undrained residual strength from case studies. (After Seed and Harder, 1990.)

Penetration resistance (Figure 10.9) and cone penetration resistance (Figure 10.10).

Although a soil may not liquefy during an earthquake, it is possible that pore water pressures will increase in the soil and the shear strength may be reduced. Marcuson et al. (1990) suggest that in this case the pore water pressures due to the earthquake can be related to the factor of safety against liquefaction, defined as the cyclic shear stress divided by the cyclic shear stress required to cause liquefaction (based on estimates of the cyclic resistance ratio described earlier). Marcuson et al. present the curves shown in Figure 10.11 for estimating the residual excess pore water pressures. However, care must be exercised in using such curves and defining pore water pressures that will be used in an effective stress representation of shear strength. It is possible that the shear strength corresponding to an effective stress analysis will actually be greater than the original undrained shear strength of the soil because the pore water pressures that are estimated as residual values may not be as large as the pore water pressures when the soil is sheared to failure with no drainage. Accordingly, it is recommended that if pore water pressures are estimated and used in an effective stress analysis, a check be made to ensure that the shear strength does not exceed the undrained shear strength before the earthquake.

As an alternative to the effective stress approach suggested by Marcuson for soils that lose some strength but do not liquefy during an earthquake, reduced values of undrained shear strength can be used.

---

[2] *Undrained residual shear strength,* should be distinguished from *residual shear strength,* used to describe the long-term "drained" shear strength of soils that have previously experienced large static shear strains.

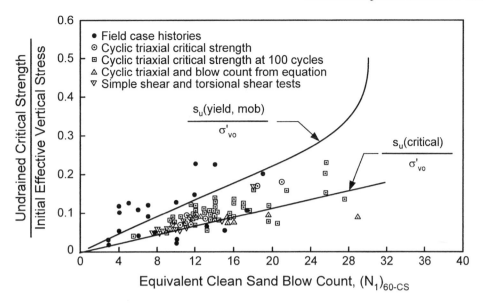

Figure 10.9 Relationship between undrained critical strength ratio and equivalent clean sand blow count. (From Stark and Mesri, 1992.)

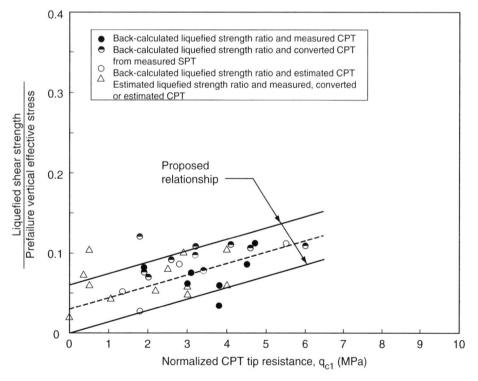

Figure 10.10 Liquefied strength ratio relationship based on normalized CPT tip resistance. (After Olson and Stark, 2002.)

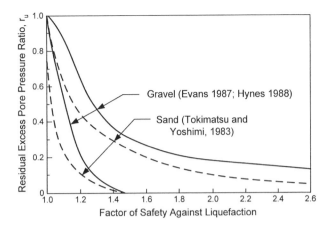

Figure 10.11 Typical residual excess pore water pressure ratios as a function of the factor of safety against liquefaction for sand and gravel. (After Marcuson et al., 1990.)

Reduced undrained shear strengths can be estimated by performing laboratory tests in which specimens are consolidated to stresses comparable to those expected in the field before the earthquake, subjected to loads simulating the earthquake, and finally, sheared to failure in a static load test.

Step 3. Compute Slope Stability

Once the postearthquake shear strengths have been determined, a conventional static slope stability analysis is performed. For some soils and slope geometries, the undrained shear strength after seismic loading may

represent the minimum shear strength, and the shear strength will gradually increase with time after the earthquake. For these soils and slopes, the slope stability computations can be performed using undrained shear strengths that reflect the effects of cyclic loading as discussed in the two preceding sections. However, for other soils, especially those that dilate when sheared, the shear strength may decrease with time after the earthquake as the soil drains and water migrates from zones of high pore water pressure to zones of lower pressure. This was illustrated by Seed (1979) in Figures 10.12 and 10.13 for the Lower San Fernando Dam. The factor of safety computed using undrained strengths immediately after the earthquake (Figure 10.12) was 1.4, while the factor of safety accounting for partial drainage and redistribution of pore water pressure (Figure 10.13) was only 0.8.

In cases where some combination of undrained and drained (or partially drained) shear strengths control the stability, it seems appropriate to perform stability analyses that use the lower of the drained and undrained shear strengths, as is done for rapid drawdown. Procedures similar to the multistage analysis procedures described in Chapter 9 can be used for this purpose. Specifically, the procedures suggested for the analysis of stability following an earthquake involve the following two analysis stages for each trial slip surface:

- *Stage 1*. Stability computations are performed using undrained shear strengths that reflect the effects of cyclic loading for low-permeability ma-

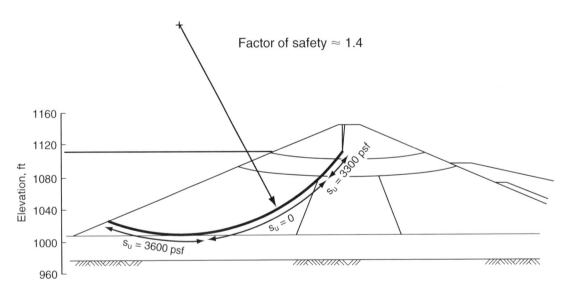

Figure 10.12 Stability of lower San Fernando Dam immediately after earthquake. (After Seed, 1979.)

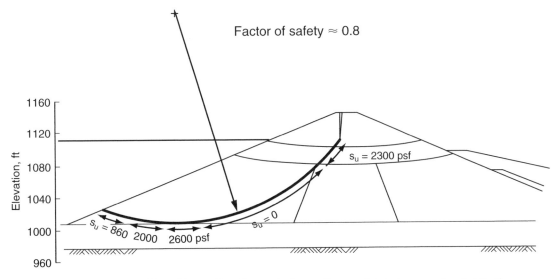

Figure 10.13 Stability of lower San Fernando Dam after partial drainage and redistribution of pore water pressures following the earthquake. (After Seed, 1979.)

terials[3]; effective stresses and drained shear strengths are used for high-permeability soils.

- *Stage 2.* Based on the total normal stress that is computed from the first-stage stability analysis and the pore water pressures that will exist after complete drainage (full excess pore water pressure dissipation), the fully drained shear strength is estimated. This is done slice by slice along the slip surface in all low-permeability soils. If the drained shear strength is less than the undrained shear strength, the drained shear strength is assigned to the slice; otherwise, the undrained shear strength is assumed to be applicable. Stability computa-

tions are then repeated. The computations will involve a mix along the slip surface of total stresses (where undrained strengths are used) and effective stresses (where drained strengths are used). The factor of safety computed from the second-stage analysis is the factor of safety after the earthquake.

Recapitulation

- Static slope stability analyses should be performed using reduced undrained shear strengths that reflect the earthquake load effects to determine that the slope is sufficiently stable after the earthquake.
- For soils that may lose strength as they drain after the earthquake, a two-stage analysis can be performed using the lower of the drained and undrained shear strength along each slop surface.

---

[3] If an effective stress approach, such as the one suggested by Marcuson et al. (1990), where excess pore water pressures are used to represent the postearthquake strengths, effective stress shear strength parameters and excess pore water pressures are used in lieu of undrained shear strengths for the first-stage analysis.

CHAPTER 11

Analyses of Embankments with Partial Consolidation of Weak Foundations

Clay foundations are sometimes too weak to support the entire load of an embankment. Methods of improving stability in these cases include constructing the embankment in stages, allowing time for partial consolidation between stages, constructing the embankment continuously but slowly to allow time for partial consolidation, and using wick drains or sand drains to accelerate the rate of drainage and pore pressure dissipation. As the foundation clay consolidates, its strength increases and stability is improved.

Figure 11.1 shows a stress path (shear stress and effective normal stress on a potential failure plane) for an element of soil in the foundation of an embankment constructed in stages. The most critical periods (the periods when the factor of safety is lowest) are those corresponding to the end of placement of a new portion of the embankment, when the stresses are closest to the failure envelope. The factor of safety increases as consolidation occurs, and the stress point moves away from the failure envelope. After a period of consolidation, it is possible to increase the height of the embankment safely.

The benefits of consolidation during construction are illustrated clearly by studies of two test embankments constructed in Poland (Wolski et al., 1988, 1989). The engineers who performed the studies summarized their findings as follows (Wolski et al., 1989): "[I]t was shown that by constructing the fill in stages, it was possible to safely construct a fill twice as thick as the original shear strength in the ground would have permitted. After another two years of consolidation even this load could be doubled to an 8 m thick fill before failure occurred."

When it is impractical to construct an embankment slowly or to construct it in stages to achieve partial consolidation during construction, wick drains or sand drains can be used to accelerate the rate of consolidation. In any of these cases where consolidation during constuction is essential to the stability of the embankment, consolidation and stability analyses are needed to determine the amount of consolidation, the factor of safety, and the allowable rate of fill placement.

CONSOLIDATION DURING CONSTRUCTION

Although it is frequently assumed that there is no consolidation or dissipation of excess pore water pressure during construction of embankments on weak clays, this may not be a good approximation. The rate of consolidation of clays in the field is frequently fast enough so that a significant amount of dissipation of excess pore water pressure will occur during construction. As shown in Figure 11.2, undrained conditions would correspond to point 2a, whereas partial dissipation during construction would correspond to point 2b or point 2c, which would have higher factors of safety. If some dissipation does occur, stability analyses performed assuming completely undrained conditions at the end of construction (point 2a) will underestimate the factor of safety. This may lead to a perceived need for staged or slow construction, or for wick drains to accelerate the rate of consolidation, when in fact neither would be needed.

An example of such a case is the embankment of La Esperanza Dam in Ecuador. A portion of the dam was built on an ancient incised valley that contained recent alluvium, including as much as 15 m (50 ft) of weak, compressible clay. Stability analyses showed that if the clay was undrained, the embankment would be unstable. Settlement calculations indicated that the

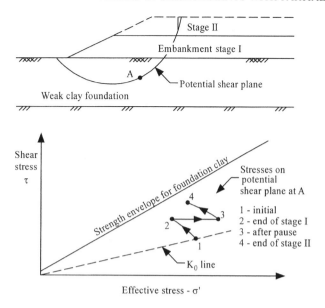

Figure 11.1 Stress changes during staged construction.

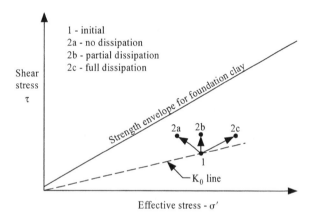

Figure 11.2 Effect of pore pressure dissipation during embankment construction.

settlement of the 45-m (150-ft)-high embankment would be approximately 1.5 m (5 ft). Measurements of the settlement of an instrumented test fill showed that the actual rate of settlement was much faster than the rate calculated using the results of laboratory consolidation tests. The actual rate of consolidation was about 12 times as fast as calculations had indicated they would be. Based on this finding it was concluded that it would be possible to construct the dam without excavating the foundation clay, without restricting the rate of construction of the embankment, and without installing sand drains to accelerate consolidation of the clay. As a result it was possible to reduce the time required for construction of the dam from four years to three. The dam was completed successfully in September 1995.

The lesson to be learned from this experience is that in circumstances where the factor of safety calculated assuming undrained conditions is lower than acceptable, the amount of consolidation and dissipation of excess pore water pressures during construction should be evaluated. It may be found that it is unnecessary to use slow construction, staged construction, or drains to accelerate consolidation, because sufficient dissipation of excess pore pressures will occur without these measures so that the factor of safety will be acceptable throughout and after construction.

ANALYSES OF STABILITY WITH PARTIAL CONSOLIDATION

Idealized variations of embankment height, excess pore water pressure, and factor of safety with time during staged construction are shown in Figure 11.3. The most critical times are those, like points 2 and 4, which correspond to the end of a period of rapid fill placement. These are the conditions for which stability must be evaluated to ensure an adequate factor of safety throughout and after construction. As shown in Fig. 11.3, the factor of safety increases with time after fill placement stops.

The mechanisms through which stability is improved by partial consolidation—dissipation of excess pore pressures and strength increasing as effective stress increases—are founded on well-established principles of soil mechanics. However, specific methods for applying these principles to evaluation of stability with partial consolidation have not been well established, and there is disagreement in the profession regarding whether it is preferable to use effective stress analyses or total stress analyses to evaluate stability for such cases (Ladd, 1991).

Effective Stress Approach

If stability is evaluated using the effective stress approach, the analyses are performed as follows:

1. Conduct drained strength tests, or consolidated–undrained tests with pore pressure measurements, to determine the effective stress shear strength parameters (c' and ϕ') for the foundation clay. Perform either drained or undrained tests to determine strength parameters for the fill. If sufficient experience already exists with the soils, it may be possible to estimate the values of the needed properties.

2. Estimate the variations of excess pore water pressures with depth and laterally in the foundation for a condition of no drainage. These pore water

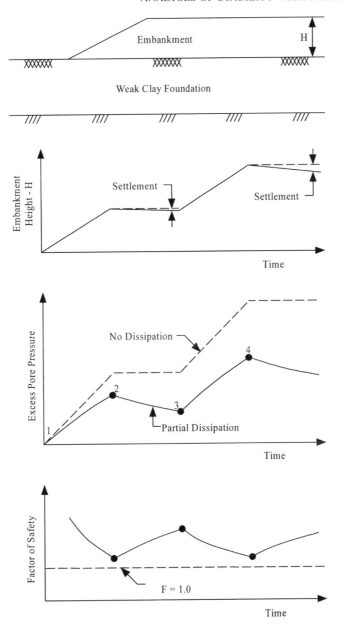

Figure 11.3 Variations of embankment height, excess pore pressure, and factor of safety with time.

pressures would correspond to the no dissipation line in Figure 11.3.

3. Perform consolidation analyses to determine how much excess pore water pressure dissipation will occur during construction. The results of these analyses would correspond to the partial dissipation line in Figure 11.3.

4. Perform stability analyses for stages such as points 2 and 4 in Figure 11.3, when the factor of safety would be lowest.

The primary advantage of the effective stress approach is that stability can be checked during construc-

tion by measuring the pore water pressures and performing additional stability analyses using the measured pore pressures.

Total Stress Approach

If stability is evaluated using the total stress (or undrained strength) approach, the analyses are performed using undrained shear strengths, expressed in terms of total stresses. For saturated soils the undrained shear strength is represented in the analyses by $c = s_u$, with $\phi = 0$. The analyses are performed as follows:

1. Conduct laboratory tests to determine the variation of s_u with effective stress (σ') and *overconsolidation ratio* (OCR) for the foundation clay. Perform either drained or undrained tests to determine strength parameters for the fill. If sufficient experience already exists with the soils, it may be possible to estimate the values of the needed properties.

2. Determine the variation of *preconsolidation pressure, p_p* (also called *maximum past pressure, σ_{vmax}*) with depth in the foundation from the results of laboratory consolidation tests, in situ tests, or past experience.

3. Estimate the excess pore water pressures in the foundation for a condition of no drainage, and perform consolidation analyses to determine how much excess pore water pressure dissipation will occur during construction, as for effective stress analyses. Compute the variation of effective stress (σ') and OCR with depth and laterally for each stage at which stability will be evaluated. Use these to estimate the variation of undrained strength with depth and laterally beneath the embankment.

4. Perform stability analyses for stages such as points 2 and 4 in Figure 11.3, when the factor of safety would be lowest.

The primary advantage cited for the total stress approach is that if failure occurred, it would be undrained. Thus, using undrained strength is more appropriate because it corresponds more closely to the behavior being evaluated.

OBSERVED BEHAVIOR OF AN EMBANKMENT CONSTRUCTED IN STAGES

The studies performed by Wolski et al. (1988, 1989) are uniquely valuable because they are so well documented and because the embankment they studied was eventually loaded to failure. The reports of these studies do not provide information about the effectiveness of undrained pore pressure estimates or consolidation analyses, because the foundation pore pressures were measured rather than calculated. Even so, the studies are very instructive because of the wealth of detail they contain and the clarity with which the results are presented.

The earlier investigation (Wolski et al., 1988) included studies of two embankments constructed in stages, one with wick drains and the other without. The later investigation (Wolski et al., 1989) involved increasing the height of the embankment without wick drains until failure occurred.

The embankments were constructed at a site in Poland where the subsoil contained a layer of peat about 3 m (10 ft) thick, underlain by a layer of weak calcareous soil about 4.7 m (15 ft) thick. The calcareous layer was underlain by sand. Measurements included the consolidation characteristics of the peat and calcareous soil, drained and undrained shear strengths, pore water pressures in the foundation, and horizontal and vertical movements of the embankments.

Undrained strengths (vane shear strengths) in the foundation at three different times during construction are shown in Figure 11.4. It can be seen that the increase in strength with time was very significant, especially near the top of the peat layer and the bottom of the calcareous soil, where consolidation advanced most rapidly. In the center of the calcareous soil, where consolidation was slowest, the undrained strength was found to be relatively low. Although the vane shear strengths shown in Figure 11.4 required correction for use in evaluating stability, they provided a useful means for effective evaluation of strength increase due to consolidation.

Stage 1 of the embankment (1.2 m high) was built in November 1983. Stage 2 (which increased the height to 2.5 m) was added in April 1984. Stage 3 (which brought the height to 3.9 m) was completed in June 1985. In July 1987 the height of the embankment

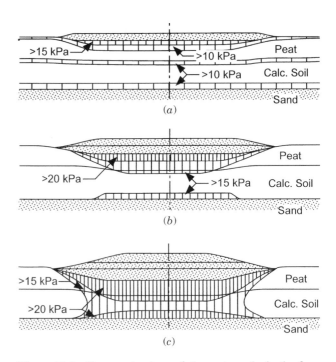

Figure 11.4 Corrected values of shear strengths in the foundation of a test embankment constructed in stages: (*a*) end of stage 1 (4.84); (*b*) end of stage 2 (5.85); (*c*) end of stage 3 (7.87). (From Wolski et al., 1989.)

was increased to 7.95 m (26 ft) in a period of seven days, whereupon the embankment failed. Failure occurred in the middle of the night when there was no construction activity. The measurements made at the end of the previous day had given no sign that failure was imminent.

The shape of the failure zone was estimated based on surface observations and field vane shear tests to locate zones in the foundation where the undrained strength had been reduced by remolding that accompanied the large displacements at failure. The shape of the failure zone inferred from these measurements is shown in Figure 11.5. It can be noted that there is a steep "active" zone beneath the center of the embankment, a nearly horizontal section where failure followed the least consolidated and weakest part of the calcareous soil, and a gently inclined passive zone where the failure extended upward toward the ground surface. These three zones fit well into the active, direct simple shear and passive failure surface orientations described by Ladd (1991).

Factors of safety for the embankment were calculated for several stages during construction. Of greatest interest are those calculated for the conditions at the time of failure, which are shown in Table 11.1. Ideally, the factors of safety calculated would be 1.0 for conditions at failure. Within the range of accuracy of the calculations, this is true for both the total and effective stress analyses.

By the time the embankment reached its final height of 8 m, it was shaped like a truncated pyramid, much narrower at the top than at the base. As a result, the two-dimensional analyses represented the conditions at the maximum section but had to be adjusted to achieve a result that was representative of the average conditions for the entire embankment. These adjustments could be approximated for the undrained case only. Even so, two conclusions are clear from the results: (1) The effective stress and total stress factors of safety

Table 11.1   Factors of Safety for an Embankment Constructed in Stages[a]

	F_t (total stress)	F_e (effective stress)
Two-dimensional factor of safety	0.85	0.89
Estimated increase due to three-dimensional effects	12 to 18%	N.A.
Estimated three-dimensional factor of safety	0.95 to 1.00	N.A.

Source: (Wolski et al. (1989).
[a]Both F_t and F_e were calculated using Janbu's (1973) Generalized Procedure of Slices. F_e was calculated using measured pore pressures.

are very nearly equal at the failure condition. No significance can be attached to the small difference between the calculated values of F_t and F_e. (2) With a reasonable allowance for three-dimensional effects, the factor of safety calculated for the total stress analysis is unity. There is no reason to believe that the same would not be true for the effective stress analysis as well.

DISCUSSION

As noted earlier, methods for analysis of staged construction have not been well established, and there is still disagreement concerning whether effective stress analyses or total stress analyses are preferable. The writers believe that this state of affairs is due to the

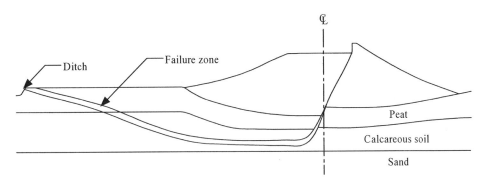

Figure 11.5   Estimated failure zone in a test embankment constructed in stages. (From Wolski et al., 1989.)

fact that the there are not a sufficient number of well-documented case histories of staged construction failures against which to gage the effectiveness of the methods that have been proposed. Some excellent case studies have been performed, notably by Wolski et al. (1988, 1989) and Bromwell and Carrier (1983), but none of these provides information regarding the accuracy of stability evaluations based on consolidation analyses to estimate pore pressures for conditions of partial consolidation.

To the writers' knowledge, only the load test performed by Wolski et al. (1989) was continued to failure. The tailings dam studied by Bromwell and Carrier did not fail, and the one-dimensional consolidation analyses they performed (only vertical flow) did not match measured pore pressures, probably because there was significant horizontal flow during consolidation.

Ladd's exhaustive study (Ladd, 1991) shows how complex, how difficult, and how fraught with uncertainty analyses of staged construction can be. More case studies are needed to advance the state of the art in this area. Until the results of such studies are available, it seems prudent to use both total and effective stress analyses in tandem and to bear in mind the difficult aspects of this class of problems.

Difficulties in Estimating Pore Pressures

A considerable part of the uncertainty in both the effective stress approach and the total stress approach stems from the difficulties in estimating the excess pore pressures due to embankment loading and from the difficulties in estimating their rates of dissipation. The uncertainties in these processes can best be appreciated by considering some of the details of such analyses.

Estimating these pore water pressures requires performing three types of analyses: (1) a stress distribution analysis to calculate the increase in total stress in the clay due to construction of the embankment; (2) an analysis to estimate the values of excess pore water pressure that would result from these changes in total stress with no drainage (these pore water pressure changes should reflect the effects of changes in shear stress as well as changes in mean normal stress); and (3) a consolidation analysis to calculate the remaining excess pore water pressures after a period of dissipation. These remaining excess pore water pressures are added to the initial (before construction) pore pressures to determine the total pore water pressures remaining after dissipation.

Estimating the distribution of pore water pressure that results from undrained loading requires consider-

able effort and is difficult to do accurately. The most straightforward way of estimating these stress changes is by using elastic theory, but elastic theory may result in stresses at some locations that exceed the strength of the clay and would have to be adjusted to values that are consistent with the strength. Alternatively, it would be necessary to perform more sophisticated stress analyses that provide stresses compatible with the strength characteristics of the clay.

The pore pressures that result from the increases in total stress at each point in the foundation depend on (1) the properties of the clay, (2) the overconsolidation ratio (OCR), and (3) the magnitude of the stress increase, particularly how close to failure the clay is loaded. The value of the OCR and the magnitude of the changes in total stress vary from point to point through the foundation.

Skempton (1954) expressed the change in pore water pressure due to changes in total stress in the form

$$\Delta u = B \, \Delta \sigma_3 + \overline{A}(\Delta \sigma_1 - \Delta \sigma_3) \qquad (12.1)$$

where Δu is the change in pore water pressure caused by changes in total stress $\Delta \sigma_1$ and $\Delta \sigma_3$, and B and $\overline{A}$ are Skempton's pore pressure parameters.

If the clay is saturated, the value of B is equal to unity. The value of $\overline{A}$, however, depends on the properties of the clay, the OCR at the location where the pore water pressure is being calculated, and how close the stresses at the point are to the failure envelope. As a result, the value of $\overline{A}$ is difficult to estimate accurately.

Difficulties in Consolidation Analyses

To determine the distribution of pore water pressures after a period of consolidation, it is necessary to perform a consolidation analysis using the undrained condition as the initial condition. Variations in the values of coefficient of consolidation, compressibility, preconsolidation pressure, and change in stress with depth can have a significant effect on the rate of consolidation. In most cases consolidation occurs more rapidly in the field than would be expected based on conventional settlement calculations (Duncan, 1993). To take these factors into account, consolidation analyses should be performed using numerical techniques rather than conventional chart solutions.

Due to lateral flow, pore water pressures may increase in areas of initially low excess pore water pressures (beneath the toe of the embankment) while they are decreasing in other areas (beneath the center of the embankment). To include this effect, it would be necessary to perform two-dimensional consolidation analyses that take horizontal as well as vertical flow into account. Such analyses are possible, but difficult, and are not yet done routinely in practice.

If wick drains or sand drains are used to accelerate the rate of consolidation, suitable analyses are needed to estimate the rate of consolidation with radial flow to the drains. The book by Holtz et al. (1991) is a valuable resource that covers both the theoretical and practical aspects of designing wick drain systems. Hansbo (1981) has developed the most widely used theory for analysis of consolidation with wick drains. The theory includes the effects of smear due to disturbance when the wicks are installed, and the effects of the finite flow capacity of the wicks.

With or without drains to accelerate the rate of dissipation, predicting pore water pressures for staged construction analyses is a difficult task. It is clear from the preceding discussion that making these estimates requires extensive effort and is susceptible to considerable inaccuracy.

Difficulties in Estimating Undrained Shear Strengths

Ladd (1991) has shown that undrained shear strengths of clays depend on several factors:

- The magnitude of the effective consolidation pressure, p'
- The value of the overconsolidation ratio (OCR)
- The ratio of the effective principal stresses during consolidation, $K_c = \sigma_1'/\sigma_3'$
- The amount of reorientation of the principal stresses during loading
- The orientation of the failure plane

Accounting for each of these effects is difficult. Ladd recommends employing simplifying assumptions that result in conservative estimates of undrained shear strength. He assumes that p' is equal to the vertical effective stress, that the K_c ratio is equal to $1/K_0$, and that the amount of stress reorientation and the orientation of the failure plane are uniquely related. The amount of conservatism involved in these simplifications is difficult to estimate.

As discussed in Chapter 5, two methods of laboratory testing can be used to evaluate undrained strengths. Bjerrum (1973) recommended use of what is termed the *recompression procedure,* wherein test specimens are consolidated in the laboratory to their estimated in situ stresses, to overcome some of the effects of disturbance. Ladd and Foott (1974) and Ladd et al. (1977) advocate use of the SHANSEP procedure, wherein test specimens are consolidated to pressures several times higher than the in situ stresses and shear strengths are characterized in terms of ratios of un-

drained strength divided by effective vertical stress during consolidation, s_u/σ_{vc}'. There is fairly general agreement that the recompression procedure is preferable for sensitive and highly structured clays, and that SHANSEP is more suitable for young clays that are not very sensitive and which have no significant bonds or structures that are subject to damage by large strains during consolidation. In addition, if SHANSEP is used, it needs to be established (not just assumed) that s_u/σ_{vc}' is a suitable parameter for characterizing the strength of the clay in question (i.e., that undrained strength divided by consolidation pressure is a constant for the clay).

Intrinsic Difference in Effective Stress and Total Stress Factors of Safety

Effective stress and total stress factors of safety are intrinsically different because they use different measures of shear strength, as illustrated in Figure 3.4. The effective stress factor of safety (F_e) is equal to the shear stress required for equilibrium divided by the shear strength of the soil if the soil fails with no change in the effective stress on the failure plane. The total stress factor of safety (F_t) is equal to the shear stress required for equilibrium divided by the shear strength of the soil if the soil fails with no change in water content. For saturated soils this corresponds to failure with no change in void ratio. In general, the values of F_e and F_t are not the same. For clays that generate positive pore pressures due to changes in shear stress, as shown in Figure 3.4, F_e is greater than F_t. At failure, both F_e and F_t are equal to unity, but for stable conditions, F_e is not equal to F_t.

Instrumentation for Staged Construction

Because the results of analyses of stability during staged construction are so uncertain, it is appropriate to use the observational method (Peck, 1969) to supplement the results of analyses. Two types of instrumentation are especially useful for this purpose.

Piezometers can be used to measure pore water pressures at key points in the foundation, and comparisons of the measured and calculated pore water pressures provide an effective means of determining if the calculated values are high, low, or accurate. Effective stress stability analyses can be performed using the measured pore water pressures to check on stability during construction.

Inclinometers (slope indicators) and settlement plates can be used to measure horizontal movements in the foundation beneath the toe of the fill and settlements under the center of the embankment. Tavenas et al. (1979) have developed criteria that can be used to interpret whether the movements observed are due to consolidation of the foundation clay or whether they indicate impending instability.

Need for Additional Case Histories

As mentioned previously, because there are not enough published case histories of failures of embankments during staged construction, and none that include consolidation analyses, it is difficult to judge the accuracy of the methods of analysis that have been proposed. The studies conducted by Wolski et al. (1988, 1989) are extremely valuable. The instrumentation and testing that they used provide a model of what is desirable in such studies, and much can be learned from review of their results. However, more such studies, including both consolidation and stability analyses, will be needed to determine whether any of the analysis methods that have been proposed are accurate and reliable.

CHAPTER 12

Analyses to Back-Calculate Strengths

When a slope fails by sliding it can provide a useful source of information on the conditions in the slope at the time of the failure as well as an opportunity to validate stability analysis methods. Because the slope has failed, the factor of safety is considered to be unity (1.0) at the time of failure. Using this knowledge and an appropriate method of analysis it is possible to develop a model of the slope at the time that it failed. The model consists of the unit weights and shear strength properties of the soil, groundwater, and pore water pressure conditions and the method of analysis, including failure mechanisms. Such a model can help in understanding the failure better and be used as a basis for analysis of remedial measures. The process of determining the conditions and establishing a suitable model of the slope from a failure is termed *back-analysis* or *back-calculation*.

BACK-CALCULATING AVERAGE SHEAR STRENGTH

The simplest back-analysis is one where an average shear strength is calculated from the known slope geometry and soil unit weights. This is accomplished by assuming a friction angle of zero and calculating a value of cohesion that will produce a factor of safety of 1. This practice of calculating an average strength expressed as a cohesion can, however, lead to erroneous representations of shear strength and potentially unfavorable consequences (Cooper, 1984). For example, consider the natural slope shown in Figure 12.1 and suppose that the slope has failed. We can begin by assuming a value of cohesion and calculating a factor of safety. If we assume a cohesion of 500 psf, the calculated factor of safety is 0.59. The developed cohesion, c_d, can then be calculated as

$$c_d = \frac{c}{F} = \frac{500}{0.59} = 850 \text{ psf} \qquad (12.1)$$

The cohesion developed is the cohesion required for a factor of safety of 1.0. Thus, the back-calculated shear strength is 850 psf. Now, suppose that one remedial measure being considered is to decrease the height of the slope to 30 ft (Figure 12.2). If the slope height is reduced to 30 ft and the cohesion is 850 psf, the new factor of safety is 1.31. Because the shear strength has been calculated from an actual slide, much of the uncertainty normally associated with the measurement of shear strength is eliminated. Thus, a factor of safety of 1.31 may be more than adequate, and based on this analysis we might choose to reduce the slope height to 30 ft as the repair measure.

In the foregoing case we were able to back-calculate an average shear strength expressed as a *cohesion*, c, with $\phi = 0$. Little more can be done if all that we know about the slope is that it failed. However, often there is more information that can be used to obtain a better estimate of the shear strength and other conditions in the slope at the time of failure. Suppose that the slope described above failed many years after the slope was formed. If this is the case, we would analyze the stability using drained shear strengths and effective stresses; we would not consider the friction angle to be zero unless the slope had failed soon after construction. Let's suppose further that from experience with clays like the clay in this slope, we know that the friction angle is about 22° and that there is a small cohesion, c'. Finally, let's suppose that we have found from observations that a piezometric line such as the one shown in Figure 12.3 approximates the seepage conditions in the slope at the time of failure. We can then back-calculate a value for the effective cohesion (c') that will produce a factor of safety of 1.0. The proce-

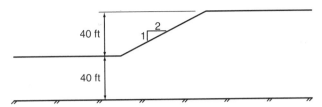

Figure 12.1   Homogeneous natural slope that has failed.

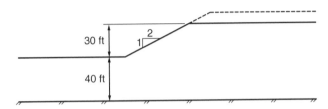

Figure 12.2   Homogeneous slope with reduced height.

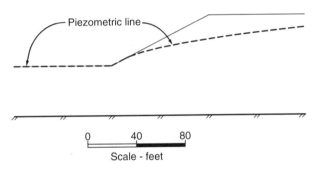

Figure 12.3   Piezometric line for homogeneous slope.

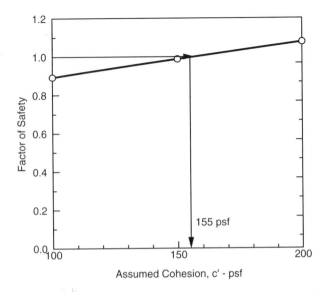

Figure 12.4   Variation in factor of safety with the assumed value for the cohesion (c') for simple homogeneous slope and foundation. $\phi' = 22°$.

dure for back-calculating the cohesion in this case is slightly different from what was done above. Several values of cohesion need to be assumed. With a friction angle of 22° and the piezometric line shown in Figure 12.3, the factor of safety is calculated for each assumed value of cohesion. The results of such calculations are summarized in Figure 12.4. It can be seen that a cohesion of approximately 155 psf produces a factor of safety of 1.0. Using the shear strength parameters ($c' = 155$ psf, $\phi' = 22°$) determined by back-analysis, we can again calculate the stability of the slope with the height reduced from 40 ft to 30 ft. The factor of safety with the height reduced to 30 ft is 1.04. This factor of safety (1.04) is substantially less than the factor of safety (1.31) determined for the slope when the shear strength was back-calculated as a cohesion with $\phi = 0$.

Next, suppose that for the slope described above, another alternative remedial measure is to lower the water level to the elevation of the toe of the slope. If

we apply the first set of shear strengths that were back-calculated ($c = 850$ psf, $\phi = 0$), we will conclude that lowering the water level has no effect on the factor of safety because the friction angle is zero, and thus the shear strength does not depend on either the total or the effective normal stress. However, if we use the effective stress shear strength parameters ($c' = 155$ psf, $\phi' = 22°$) that were determined by the second back-analysis, the factor of safety is increased to 1.38, which would indicate that lowering the groundwater level would be an acceptable remedial measure.

The results of the back-analyses and the analyses of remedial alternatives described above are summarized in Table 12.1. It can be seen that very different conclusions would be reached regarding the effectiveness of remedial measures, depending on how the shear strength is characterized and what information is used

Table 12.1   Summary of Back-Analyses and Analyses of Remedial Measures for Homogeneous Slope

Shear strength parameters from back-analysis	Factor of safety for remedial measure	
	Decrease slope height to 30 ft	Lower water level to toe of slope
$c = 850$ psf, $\phi = 0$	1.31	1.00
$c' = 155$ psf, $\phi' = 22°$	1.04	1.38

for back-analysis. For the natural slope that failed a number of years after formation, back-calculation of an average shear strength expressed as cohesion led to an overestimate of the effectiveness of reducing the slope height and an underestimate of the effectiveness of lowering the water level.

By using back-analysis, it is only possible to back-calculate a single shear strength parameter. In the first case summarized in Table 12.1, ϕ was assumed to be zero and an average shear strength, expressed as a cohesion, was back-calculated. In the second case knowledge that the friction angle was approximately 22° and the approximate location of a piezometric line were used to back-calculate an effective cohesion, c'. In both cases, cohesion was back-calculated while the friction angle (ϕ, ϕ') was either assumed or known from other information. It would also be possible to assume that the cohesion (c, c') was zero and to back-calculate a friction angle; however, only one unknown shear strength parameter can be calculated using back-analysis.

Recapitulation

- Only one strength parameter (c, c' or ϕ, ϕ') can be calculated by back-analysis.
- Back-calculation of an average shear strength expressed as a cohesion, c ($\phi = 0$) can produce misleading results when a slope has failed under long-term drained conditions.

BACK-CALCULATING SHEAR STRENGTH PARAMETERS BASED ON SLIP SURFACE GEOMETRY

Although for any given slope there are an infinite number of pairs of values for cohesion (c, c') and friction angle (ϕ, ϕ') that will produce a factor of safety of 1, each such pair of values will also produce a different location for the critical slip surface. This is illustrated for a simple slope in Figure 12.5. Three sets of shear strength parameters and corresponding critical circles are shown. Each set of shear strength parameters produces a factor of safety of 1, but the critical slip surface is different. For a simple homogeneous slope such as the one shown in Figure 12.5, the depth of the slip surface is related to the dimensionless parameter, $\lambda_{c\phi}$, defined as

$$\lambda_{c\phi} = \frac{\gamma H \tan \phi}{c} \qquad (12.2)$$

where H is the slope height and c and ϕ represent the appropriate total stress or effective stress, shear strength parameters. Values of $\lambda_{c\phi}$ are shown along with the shear strength parameters in Figure 12.5. As $\lambda_{c\phi}$ increases, the depth of the slip surface decreases. When $\lambda_{c\phi}$ is zero, the slip surface is deep, and when $\lambda_{c\phi}$ is infinite (c, $c' = 0$), the slip surface is shallow—essentially a shallow infinite slope failure. Because each pair of shear strength parameters (c–ϕ or c'–ϕ') corresponds to a unique slip surface, the location of the slip surface, along with the knowledge that the slope has failed (i.e., $F = 1$), can be used to back-calculate values for two shear strength parameters (c–ϕ or c'–ϕ').

To illustrate how the location of the slip surface can be used to back-calculate both cohesion and friction, consider the slope illustrated in Figure 12.6. This is a highway embankment constructed in Houston, Texas, of highly plastic clay, known locally as Beaumont Clay. A slide developed in the embankment approximately 17 years after the embankment was built. The estimated location of the slip surface is shown in Figure 12.6. Because the failure occurred many years after construction, drained shear strengths were assumed and slope stability analyses were performed to calculate shear strength parameters in terms of effective stresses. The pore water pressure was assumed to be zero for these particular analyses. The following steps were performed to back-calculate the shear strength parameters and slip surface location:

1. Several pairs of values of cohesion and friction angle (c' and ϕ') were assumed. The pairs of values were chosen such that they represented a range in the dimensionless parameter $\lambda_{c\phi}$, but the values did not necessarily produce a factor of safety of 1.
2. The critical circles and corresponding minimum factors of safety were calculated for each pair of values of the strength parameters.
3. Values of the developed shear strength parameters (c'_d and ϕ'_d) were calculated for each pair of strength parameters from the following equations using the assumed cohesion and friction angle and the computed factor of safety:

$$c'_d = \frac{c'}{F} \qquad (12.3)$$

$$\phi'_d = \arctan \frac{\tan \phi'}{F} \qquad (12.4)$$

The developed cohesion and friction angle represent back-calculated values required to produce a factor of safety of 1.

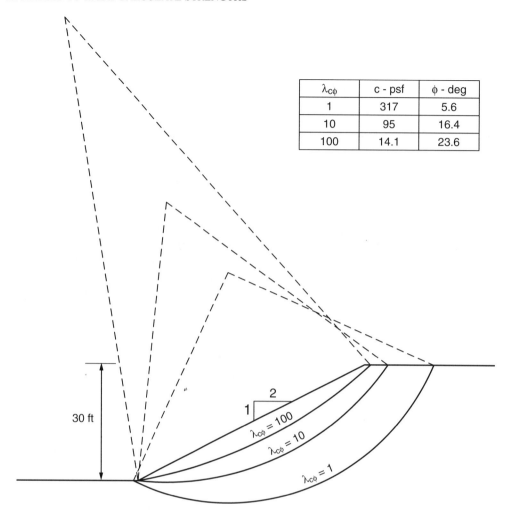

$\lambda_{c\phi}$	c - psf	ϕ - deg
1	317	5.6
10	95	16.4
100	14.1	23.6

Figure 12.5 Critical circles for three different sets of shear strength parameters giving a factor of safety of 1.

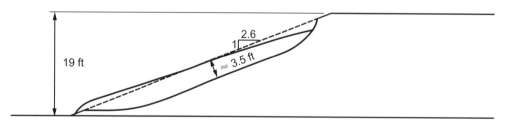

Figure 12.6 Slide in compacted high-PI clay fill.

4. The depth of the critical slip surface for each pair of values of strength parameters was calculated.
5. The back-calculated cohesion and friction angle from step 3 were plotted versus the depth of the slip surface, calculated in step 4 (Figure 12.7).
6. The cohesion and friction angle corresponding to the observed slide depth (3.5) ft were determined from the plotted results.

These steps showed that a cohesion of 5 psf and a friction angle of 19.5° produce a factor of safety of 1 with a slide depth of 3.5 ft. These values seem reasonable for the effective stress shear strength parameters for a highly plastic clay.

Calculations like the ones described above can be simplified by the use of dimensionless stability charts that allow the cohesion and friction angle to be back-

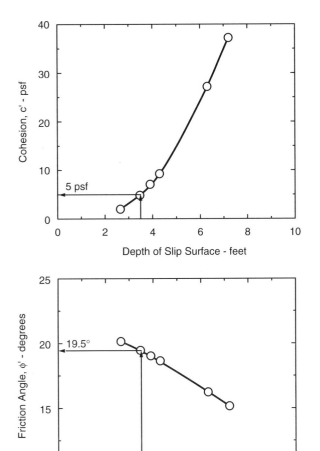

Figure 12.7 Variation in values for cohesion (c') and friction angle (ϕ') that produce a factor of safety of 1 with the depth of the slip surface.

of progressive failure in the slope. Poor agreement may also have been caused by heterogeneity in the slope, which is common in natural slopes. Duncan and Stark also showed that the factor of safety changes only slightly with changes in the position of the slip surface, and thus the position of the slip surface is likely to be influenced to a significant degree by the normal variations in shear strength that occur in a slope.

Back-calculation of cohesion and friction angles by matching the computed critical slip surface with the observed location of the actual slip surface has met with only limited success and should be used cautiously. In many cases greater success is obtained by using other information, such as correlations between Atterberg limits and friction angles, to estimate one of the shear strength parameters and then to back-calculate the other. Several additional examples of back-analyses to determine slope conditions at the time of failure are presented in the next section.

Recapitulation

- Each combination of cohesion and friction angle that produces a factor of safety of 1 produces a unique location for the critical slip surface. Accordingly, the location of the slip surface can be used to calculate values for both cohesion (c, c') and friction angle (ϕ, ϕ').
- Use of the location of the slip surface to back-calculate both cohesion and friction has had mixed success and does not seem to work when there is significant progressive failure or distinct layering and inhomogeneities in the slope.

---

calculated directly. Such charts are based on dimensionless parameters similar to those described for the stability charts used to compute factors of safety, which are described in the Appendix. Abrams and Wright (1972) and Stauffer and Wright (1984) have developed charts for this purpose. Stauffer and Wright used these charts and back-calculated shear strength parameters from a number of slides in embankments constructed of high-PI clays. These analyses were useful in establishing that the effective cohesion values were small for the embankments examined.

Duncan and Stark (1992) also back-calculated shear strength parameters using procedures similar to those described above. They back-calculated values for the Northolt Slip and found that the friction angles that were back-calculated exceed values determined in laboratory tests. They concluded that the procedure was not completely reliable, possibly because of the effects

EXAMPLES OF BACK-ANALYSES OF FAILED SLOPES

The stability of any slope, including the results of any slope stability analysis, depends on numerous variables, including:

1. Unit weight of the soil
2. Loading conditions (i.e., whether the loading is undrained or drained)
3. Shear strength parameters, including whether the soil is anisotropic or the Mohr failure envelope is linear or nonlinear
4. Variability in the undrained shear strength or the shear strength parameters laterally and vertically
5. Seepage conditions and pore water pressures
6. Subsurface stratigraphy, including the presence of thin layers of soil with contrasting hydraulic or shear strength properties

7. Shape of the slip surface
8. Method of analysis, including the assumptions made in the limit equilibrium procedure used

Some amount of uncertainty will exist in each of the foregoing variables, and the outcome of any analysis will reflect this uncertainty. If we seek to determine a shear strength parameter (c, c', ϕ or ϕ') by back-analysis, the value will reflect the uncertainty in all of the other variables that were used in the analysis. The degree of uncertainty in the shear strength parameter will be no less than the degree of uncertainty in all of the other variables that affect the stability analysis. In fact, the back-analysis should actually be conceived as a back-analysis to determine *all* of the variables that are applicable to the failure, rather than only shear strength. To reduce the uncertainty in this determination it is important to utilize all the information that is known or can be estimated by other means prior to performing the back-analysis. The back-analysis will then serve to establish reasonable values for all the variables.

Several examples are presented in this section to illustrate how available information is used in conjunction with back-analyses to establish a complete "model" of the slope at the time of failure. Some of these examples are of actual slopes or patterned after actual slopes and some are hypothetical.

Example 1: Embankment on Saturated Clay Foundation

The first example is of the cohesionless embankment (fill) slope resting on a deep deposit of saturated clay shown in Figure 12.8. The embankment has failed during construction, due to the underlying weak clay foundation. From knowledge of the fill material we can estimate that the friction angle for the embankment is 35° and the fill has a unit weight of 125 pcf. We can

calculate the average undrained shear strength of the foundation by varying the assumed shear strength and calculating the factor of safety. From the results of such calculations it is determined that the average undrained shear strength is approximately 137 psf. Now, instead, suppose that we know from past experience with the soils in the area of the slope that the clay is slightly overconsolidated and that the undrained shear strength increases approximately linearly with depth at the rate of 10 psf per foot of depth. We can calculate a value for the undrained shear strength at the ground surface, assuming that the shear strength increases at the rate of about 10 psf per foot of depth below the surface. Doing so, we find that if the shear strength is approximately 78 psf at the ground surface and increases at the rate of 10 psf per foot of depth, the factor of safety will be 1. The two shear strength representations described above are plotted in Figure 12.9. Both

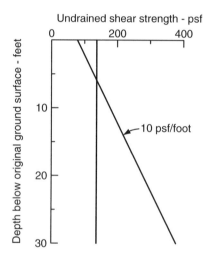

Figure 12.9 Undrained shear strength profiles from back-analysis of embankment on soft clay.

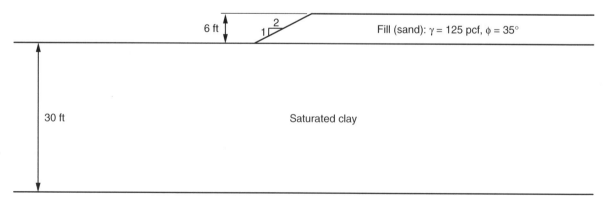

Figure 12.8 Embankment on soft clay foundation.

of these representations give a factor of safety of 1. However, the locations of the corresponding critical slip surfaces are very different as shown in Figure 12.10. In addition, if we use the two shear strength representations to evaluate the effectiveness of reducing the slope height, we will reach different conclusions. Suppose, for example, that we want to increase the factor of safety to 1.5. Decreasing the slope height to 4 ft with the constant shear strength of 137 psf for the foundation is sufficient to achieve a factor of safety of 1.5. However, if the shear strength increases linearly with depth as represented by the second shear strength profile, the factor of safety is only increased to 1.3 by reducing the slope height to 4 ft. A factor of safety of 1.3 may very well be adequate for this embankment if the shear strength has been established from back-analysis, but if a factor of safety of 1.5 is necessary, the slope height must be reduced to something less than 4 ft.

For this example slope, knowledge of the shear strength of the embankment soil and how the shear strength varied with depth was used to establish a representation of shear strength. With the knowledge of the shear strength of the embankment, it was possible to calculate the shear strength of the foundation. Further, with knowledge of how the shear strength increased with depth, it was possible to establish a better representation of strength than was obtained when only an average (constant) shear strength was calculated. Without such information a greater amount of uncertainty would exist in the shear strengths determined by back-analysis.

Example 2: Natural Slope

The second example is of a natural slope located in the western United States. The soil profile consists of approximately 40 ft of weathered shale overlying unweathered shale (Figure 12.11). Substantial movement of the weathered shale was observed. The movement was believed to be taking place by slippage along the bottom of the weathered shale zone. Based on the movements that had already taken place, as well as experience with similar slopes in weathered shale, residual shear strengths were believed to be applicable. From the results of laboratory tests on the shale and correlations presented by Stark and Eid (1994), a residual friction angle (ϕ_r') of 12° was estimated for the shale. In this case the shear strength parameters were believed to be relatively well known and the largest uncertainty was in the pore water pressure conditions in the slope. Therefore, a primary goal of the back-analysis was to estimate the seepage conditions in the slope that would be required to produce a factor of safety of 1. Because of the large lateral extent of the slope movements, infinite slope analysis procedures were used. Assuming a residual friction angle of 12° ($c' = 0$), it was found that a piezometric surface at a depth of approximately 12 ft below the ground surface would produce a factor of safety of 1. The actual water conditions in the slope varied widely over the large area of the slope, but groundwater observations in several borings were consistent with the back-calculated water level.

This slope was stabilized successfully and movement was halted by installation of a number of hori-

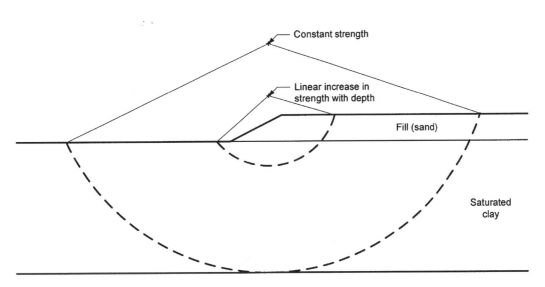

Figure 12.10 Critical circles from back-analysis of embankment on soft clay assuming constant undrained shear strength and linear increase in undrained shear strength with depth for foundation.

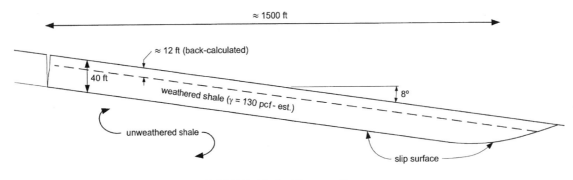

NOT TO SCALE - Vertical Exaggeration

Figure 12.11 Natural slope in weathered shale.

zontal drains that lowered the water level. Lowering the water level 10 ft, from approximately 12 ft below the surface to approximately 22 ft below the surface, produced a factor of safety of 1.2. An increase from 1.0 to 1.2 was judged sufficient to stabilize the slope. In this case the back-analysis was used to confirm the residual shear strength values and determine the pore water pressure conditions at the time of failure. The resulting conditions then provided a basis for assessing the effectiveness of stabilizing the slope with horizontal drains.

Example 3: Earth Dam

The third example is of an earth dam. This example is patterned after the slide that occurred in the Victor Braunig Dam in San Antonio, Texas (Reuss and Schattenberg, 1972). Except for some minor adjustments in the geometry to simplify the problem, conditions are very similar to those of the actual dam. Information on the geometry and shear strength properties was taken from Reuss and Schattenberg. The cross section used for the present analyses is shown in Figure 12.12. The slide that occurred passed through the embankment and nearly horizontally along a clay layer near the top of the foundation. The slide occurred approximately five years after the dam was built. Due to layers and

lenses of sand in the foundation of the dam, it was assumed that steady-state seepage was established in the foundation. Steady-state seepage was also assumed to have developed in the embankment, although the seepage conditions in the embankment did not have a major effect on the computed stability.

Analyses by Reuss and Schattenberg (1972) indicated that low residual shear strengths in the foundation contributed to the failure of the dam. Peak and residual shear strengths were measured for both the embankment soil and the clay in the foundation. Shear strength parameters from these tests are summarized in Table 12.2. A piezometric line representing the pore

Table 12.2   Shear Strength Parameters for Victor Braunig Dam Embankment and Foundation Clay Layer

Description	Peak strengths		Residual strengths	
	c' (psf)	ϕ' (deg)	c'_r (psf)	ϕ'_r (deg)
Embankment	400	22	200	22
Foundation clay	500	18	100	9

Source: Reuss and Schattenberg (1972).

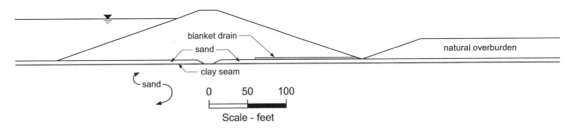

Figure 12.12 Cross section of dam on foundation with weak clay layer.

water pressures in the embankment is shown in Figure 12.13. Pore water pressures in the foundation, particularly beneath the downstream half of the dam where the slide occurred, were assumed to be controlled by sand layers and lenses in the foundation. The sand layers and lenses were connected to the reservoir in the vicinity of the upstream toe of the dam. A simple linear piezometric surface that varied from the level of the reservoir at the upstream toe of the dam to the elevation of the ground surface at the downstream toe of the dam was assumed for the pore water pressures in the foundation (Figure 12.14). Factors of safety were calculated using both peak and residual shear strengths. With peak shear strengths the factor of safety was 1.78, while with residual shear strengths the factor of safety was 0.99. These calculations indicate that residual shear strengths probably developed in the foundation of the dam and contributed to the slide that occurred.

A significant source of uncertainty in the analyses described above was the pore water pressures in the foundation of the dam. Limited measurements of pore water pressures in the foundation were available, and these were in general agreement with the assumed piezometric levels chosen for the analyses. However, to determine if there might have been higher pore water pressures, and thus that possibly peak, rather than residual, shear strengths controlled the stability at the time of failure, additional analyses were performed using higher pore water pressures in the foundation of the dam. Two different piezometric lines chosen for these analyses are illustrated in Figure 12.15. Pore water pressures from these piezometric lines approach the overburden pressure in some areas of the downstream slope, and thus the piezometric lines are considered to represent an extreme condition. Stability calculations were performed using peak shear strengths and the two piezometric lines. Factors of safety for both piezometric lines were about 1.6 (range 1.56 to 1.58). Thus, it seems unlikely that peak shear strengths were applicable to this failure, and it was concluded that the earlier analysis with residual shear strengths was appropriate for the conditions at failure.

The slide in the Victor Braunig was stabilized successfully with a berm at the downstream toe. Analyses with a berm, residual shear strengths, and the piezometric lines shown in Figures 12.13 and 12.14 showed that the factor of safety was approximately 2.0 with the stabilizing berm in place.

Example 4: High-PI Clay Embankment

The failure of a highly plastic clay embankment was described earlier and illustrated in Figure 12.6. Back-analyses to calculate shear strength parameters that matched the location of the observed slip surface were shown to give $c' = 5$ psf, $\phi' = 19.5°$. After these analyses had been performed, consolidated–undrained

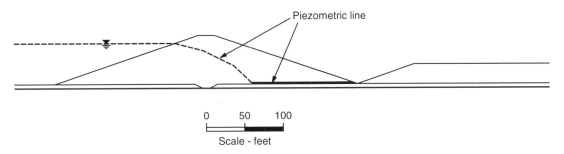

Figure 12.13 Piezometric line assumed for embankment of dam.

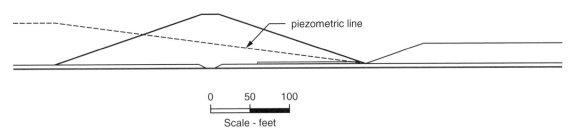

Figure 12.14 Piezometric line assumed for foundation of dam.

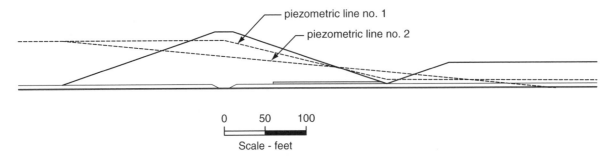

Figure 12.15 Alternative piezometric lines assumed for foundation of dam with peak shear strengths.

triaxial shear tests with pore water pressure measurements were performed to measure the effective stress Mohr failure envelope for the clay. The failure envelope determined from these tests is slightly curved and is shown in Figure 12.16. Also shown in this figure is the failure envelope that was back-calculated earlier from the slide geometry. The failure envelope that was back-calculated is substantially below the failure envelope that was measured in the laboratory tests. One possible explanation for the differences between the measured and back-calculated failure envelopes is that the pore water pressures were assumed to be zero for the back-calculated envelope. To determine if higher pore water pressures might explain the discrepancies, additional stability analyses were performed using the measured failure envelope and higher pore water pressures. For these analyses a piezometric line coincident with the slope face was assumed. This corresponds to uniform horizontal seepage in the entire slope and represents substantial pore water pressures in the slope.

The factor of safety using this piezometric line and the measured shear strength envelope shown in Figure 12.16 was approximately 2.0. Thus, it seems highly unlikely that the discrepancy between the measured and back-calculated shear strengths was due to the assumption of zero pore water pressures used in the back-analyses.

Because of the apparent discrepancies between the back-analyses and the laboratory measurements of shear strength, additional laboratory tests were undertaken. These additional tests on the fill material from the embankment showed that significant softening of the soil occurred when it was subjected to repeated wetting and drying that ultimately led to a lower, "fully-softened" shear strength. The Mohr failure envelope for the fully-softened clay is shown in Figure 12.17. The failure envelopes for peak strength as well as the failure envelope back-calculated earlier are also shown in this figure. The fully softened strength can

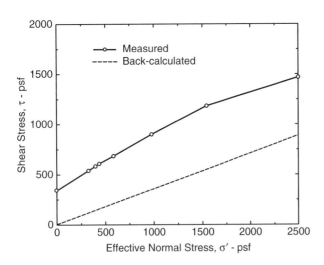

Figure 12.16 Measured failure envelope and failure envelope determined by back-analysis using the critical slip surface location.

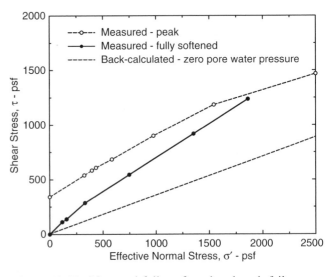

Figure 12.17 Measured fully softened and peak failure envelopes and failure envelope determined by back-analysis using the critical slip surface location.

be seen to be significantly less than the peak strength; however, the fully softened strength is still much higher than the strength that was back-calculated. One possible explanation could again be that the pore water pressures were assumed to be zero in the back-analysis.

To determine if pore water pressures could possibly explain the discrepancies between the fully-softened and back-calculated shear strength envelopes, additional analyses were preformed using the fully-softened strength envelope and assuming various pore water pressure conditions. Pore water pressures represented by a piezometric line coincident with the slope face were found to produce a factor of safety of approximately 1 with the fully softened strength. Based on these analyses, it thus appears that the fully softened shear strengths are applicable and that relatively high pore water pressures were developed in the embankment. The slide in the embankment occurred following a period of wet weather and significant rainfall. It is likely that high pore water pressures developed in the slope at least temporarily. Thus, the back-analyses combined with laboratory testing to measure the fully softened shear strength of the soil were used to establish the probable conditions in the slope when failure occurred. The earlier back-analyses in which the pore water pressures were assumed to be zero and the shear strength parameters were calculated to match the depth of the assumed slip surface were useful in establishing that the shear strengths were relatively low, less than the measured peak shear strengths. However, the initial back-analyses did not fully explain the conditions at failure; only after further laboratory testing and use of the results from the laboratory tests in back-analyses to determine the pore water pressures were the conditions in the slope better understood.

Example 5: Kettleman Hills Landfill Failure

The final example is the Kettleman Hills landfill failure. Mitchell et al. (1990), Seed et al. (1990), Byrne et al. (1992), and Stark and Poeppel (1994) have discussed this failure extensively. One issue that has emerged from the studies by various investigators is whether peak or residual shear strengths were developed in the liner system along the base of the waste fill. Gilbert et al. (1996b) reanalyzed this failure using probabilistic methods. Rather than calculate a single value for the factor of safety based on assumed conditions, they considered the probability of failure. Their analyses accounted for the uncertainties in shear strength due to variability in measured peak and residual shear strengths, as well as the uncertainty associated with the method of analysis, including the effect of interslice force assumptions and potential three-

dimensional effects. For their analyses, Gilbert et al. expressed the strength relative to peak and residual values, by a *mobilized strength factor*, R_s, defined as

$$R_s = \frac{s - s_{av.r}}{s_{av.p} - s_{av.r}}$$ (12.5)

where s is the shear strength, $s_{av.r}$ the mean residual shear strength, and $s_{av.p}$ the mean peak shear strength. Envelopes for both the peak and residual shear strengths were assumed to possess random variability, and a normal distribution was assumed; values from the mean failure envelopes were used to compute R_s. If the residual shear strength is developed, R_s is zero, and if the peak strength is developed, the value of R_s is 1.

Gilbert et al. (1996b) performed analyses for the cross section shown in Figure 12.18 and calculated the probability of failure as a function of the value for the factor R_s. Their results are summarized in Figure 12.19. The most probable value of R_s is 0.44, indicating that the shear strength developed at failure was approximately halfway between the peak ($R_s = 1$) and residual ($R_s = 0$) values. Previous studies by Byrne et al. (1992) and Stark and Poeppel (1994) had concluded that peak shear strengths were developed along the base of the landfill, while another study by Gilbert et al. (1996a) concluded that residual shear strengths were developed. The subsequent analyses by Gilbert et al. (1996b) indicate that the probabilities of peak and residual shear strengths being developed are approximately equal. However, the analyses also indicate that probably neither peak nor residual shear strengths, but rather some intermediate values of shear strengths, were developed. The analyses indicated that a progressive failure probably took place, and this is supported by subsequent finite element analyses by Filz et al. (2001).

Back-analyses of the Kettleman Hills landfill using probabilistic methods were helpful in understanding the failure that occurred and what shear strengths were developed at the time of failure. However, for analysis and design of remedial measures in this case, it is probably more appropriate to use residual shear strengths rather than the higher strengths that were developed at the onset of failure, because the slope experienced relatively large deformations.

Summary

In each of the examples presented above, some information about the shear strength parameters or pore water pressure conditions was available to guide the back-analyses. This information, along with the knowledge that the factor of safety was 1, was used to arrive

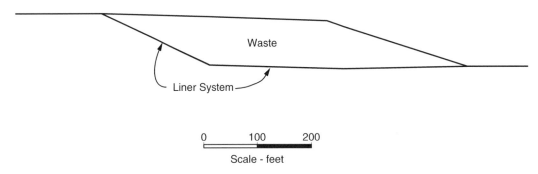

Figure 12.18 Prefailure cross section of Kettleman Hills landfill.

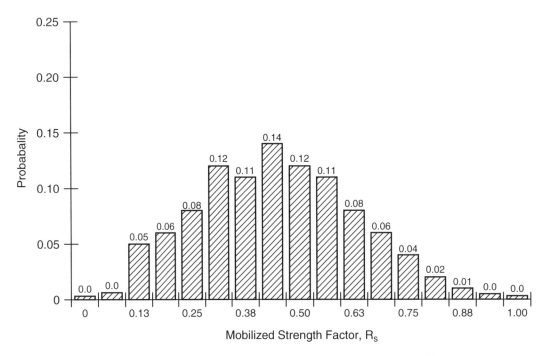

Figure 12.19 Probabilities of failure for various relative amounts of mobilized peak and residual shear strengths in Kettleman Hills landfill liner.

at a complete set of conditions that were believed to be representative of those in the slope at the time of failure. In three of the five examples, the pore water pressures at failure were uncertain, and some information was available about the shear strength parameters. Back-analyses were used to establish the pore water pressures as well as to confirm the values of the shear strength parameters.

Duncan (1999) presented results of back-analyses for three case histories where procedures similar to those described above were used to establish the conditions in the slope at the time of failure. These three examples are summarized in Table 12.3 along with the five examples presented in the preceding section. For each example the important conditions that were established by the back-analyses are indicated. In each of these cases there was at least some uncertainty in several variables, and the back-analyses served to establish a reasonable set of conditions at the time of failure. Also, by performing the analyses with the same limit equilibrium procedures used subsequently to analyze the remedial measures, the validity of the computational procedure was established along with the soil and slope properties. This validation gave increased confidence in the analyses that were performed to evaluate remedial measures.

Table 12.3   Summary of Examples of Back-Analyses and Results

Example	Conditions defined by back-analysis
Example 1: embankment on saturated clay foundation	Back-analyses established the undrained shear strength at the ground surface using estimated rates of strength increase with depth.
Example 2: natural slope	Back-analyses confirmed residual shear strengths and established a piezometric level in agreement with groundwater observations.
Example 3: earth dam	Back-analyses showed that residual shear strengths were developed and established piezometric levels used later for evaluation of remedial measures.
Example 4: high-PI clay embankment	Initial back-analyses showed that negligible cohesion intercept existed. Further analyses supported by laboratory data showed that shear strengths were reduced from peak to fully softened and that relatively high pore water pressures must have existed at the time of failure.
Example 5: Kettleman Hills landfill failure	Probabilistic analyses were used to establish that the shear strength developed at failure was somewhere between peak and residual values and suggested that progressive failure occurred.
San Luis Dam (Duncan, 1999)	Back-analyses established that residual shear strengths were developed in highly plastic clay slopewash.
Olmsted landslide (Duncan, 1999; Filz et al., 1992)	Back-analyses established anisotropic shear strength values for McNairy I formation and piezometric levels in slope.
La Esperanza Dam (Duncan, 1999)	Back-analyses established piezometric levels and confirmed estimated shear strengths of brecciated shale.

PRACTICAL PROBLEMS AND LIMITATION OF BACK-ANALYSES

Back-analyses can provide a useful insight into the conditions in a slope at the time of failure; however, several limitations and factors can complicate such analyses. These are discussed below.

Progressive Failure

A fundamental assumption in all limit equilibrium slope stability analyses is that the shear strength is mobilized simultaneously along the entire slip surface. If a single set of shear strength parameters (i.e., a single value for c and ϕ) is assumed, while in reality the values vary as a result of progressive failure, the back-calculated values represent only an "average" of the shear strength parameters that were mobilized on the failure surface; the average may not represent the actual shear strength parameters at any point on the failure surface. This is very likely to be the case where progressive failure occurs. As Duncan and Stark (1992) showed for the analysis of the Northolt Slip, the shear strength parameters back-calculated from the slide geometry did not agree with either peak or residual shear strength values. Gilbert et al. (1996b) showed similar results for the Kettleman Hills landfill failure. In the case of Kettleman Hills, probabilistic methods were useful in understanding the shear strengths that

were developed, as were the finite element analyses presented by Filz at al. (2001).

If progressive failure occurs, the back-calculated shear strengths are likely to be inappropriate for re-design. For most slopes where progressive failure occurs, the movements are likely to be large enough that once the slide movement has ceased, the strengths are reduced to residual values, and residual values should be used for redesign even though higher values may be determined by back-analysis. The use of average values calculated from the initial slide geometry may be unconservative.

Decreasing Strengths with Time

In most cases back-analyses are performed for either short-term conditions with undrained shear strengths or for long-term conditions with effective stress shear strength parameters and steady-state seepage or groundwater levels. The corresponding back-analyses assuming the appropriate one of these conditions, however, may not result in the most critical value of shear strength. For example, consider an excavated slope in clay that fails during construction. Ordinarily, undrained conditions would be assumed and used to back-calculate shear strengths for such a slope. The undrained shear strength calculated from such analyses should reflect the shear strength when the failure occurred. If the slope was excavated very recently, or if it fails during construction, the soil is probably undergoing swelling (expansion) and the shear strength and stability will continue to decrease after failure occurs. For redesign, a strength significantly lower than the one determined by back-analysis may be appropriate.

Even when slopes fail a number of years after construction, the strength that is calculated only represents the strength at the time of failure, and the strength may still be decreasing further. For back-analysis of slopes that fail a number of years after they are built, the pore water pressures are often assumed based on an estimated groundwater level, and if seepage occurs, the seepage is assumed to have reached a steady-state condition. This is probably the case with early investigations of a number of slopes in London Clay. Skempton (1964) first suggested that many slopes that failed in London Clay failed at different times after construction because progressive failure was occurring and the strength (effective stress shear strength parameters) was decreasing with time. Although Skempton did not clearly indicate what pore water pressure conditions were assumed for his analyses, it appears that steady-state seepage was assumed and that groundwater levels were assumed to have reached steady-state equilibrium levels. Vaughan and Walbancke (1973) later measured pore water pressures in slopes in London Clay and showed that the pore water pressures increased gradually over many years. This led to the realization that a significant portion of the time-related delay in failure may have been due to changes in pore water pressure; progressive failure over time may actually have played only a small role in the failures.

Complex Shear Strength Patterns

Back-calculation of shear strength in most cases consists of back-calculating one quantity (c or ϕ) or at most two quantities (c and ϕ) to represent the shear strength of the soil. In reality the shear strength is generally more complex and knowledge of the shear strength is helpful in establishing what should be back-calculated. For example, for the Example 1 slope shown in the preceding section, the undrained shear strength was known to increase with depth, and when this was taken into account, very different results were obtained compared to what was found when the shear strength was assumed to be constant.

In addition to the shear strength varying with depth, there are other forms of shear strength variations that may affect and complicate the back-calculation of shear strength. One such case is where the shear strength varies with the orientation of the failure plane (i.e., where the shear strength is anisotropic); another case is where the shear strength varies nonlinearly with

the normal stress (i.e., where the Mohr failure envelope is curved). In both these cases, back-calculation of a single c or ϕ value may lead to significant errors that vary with the location of the slip surface. If the slip surface that was used for the back-analysis is very different in its orientation or depth from the critical slip surface for the redesigned slope, the back-calculated shear strengths may not be applicable and may not reflect the proper shear strength for other than the original slip surface.

In back-calculating shear strengths it is important to have the proper model for the shear strength. Laboratory data or estimates of shear strength based on correlations between shear strength and index properties can be useful in estimating shear strength parameters. It is also essential to know whether the shear strengths should be represented by undrained shear strength parameters and total stresses or by drained shear strengths and effective stresses. Similarly, it is important to know if:

1. The soil is likely to be anisotropic and anisotropy will play a significant role in the location of the failure surface.
2. The shear strength envelope is curved such that c and ϕ are stress dependent.
3. The undrained shear strength ($s_u = c$, $\phi = 0$) is constant or varies significantly with depth.

As noted previously, it is also important to judge whether the appropriate shear strength to be calculated is the undrained shear strength or the drained shear strength, and if the shear strength may decrease after failure.

Recapitulation

- Back-calculated shear strengths must be used cautiously if progressive failure has occurred.
- Laboratory data and experience can provide useful information to guide the back-calculation of shear strengths. Even when laboratory data are not available, it is possible to make reasonable estimates of the friction angle, ϕ', based on index properties.
- If the shear strengths are decreasing significantly at the time of failure due either to changes in pore water pressure or softening of the soil structure, the back-calculated shear strengths may not be appropriate for use in designing remedial measures.
- It is important to assume the appropriate model to back-calculate the shear strength parameters. Curved Mohr failure envelopes and anisotropy may influence the validity of back-calculated shear strengths.

OTHER UNCERTAINTIES

Slope stability analyses can involve numerous uncertainties, and some of the uncertainties can be difficult to quantify. One of the benefits of back-analyses is that many of the same errors exist for both the back-analysis and the redesign. Thus, there are compensating effects and the net result of the errors is diminished or removed entirely. This needs to be kept in mind when the results from back-analysis are compared with data obtained by other means. For example, results of laboratory tests may not compare favorably with results of back-analysis if there are significant three-dimensional effects that were not considered in the slope stability analyses. However, if the slope is to be redesigned using two-dimensional analyses and, again, three-dimensional effects are neglected, the back-calculated values may be the more appropriate values to use. Caution must also be exercised, however, because neglect of three-dimensional effects will cause back-calculated shear strengths to be too high, and if comparable three-dimensional effects do not exist for the slope redesign, the results may be on the unsafe side.

CHAPTER 13

Factors of Safety and Reliability

Factors of safety provide a quantitative indication of slope stability. A value of $F = 1.0$ indicates that a slope is on the boundary between stability and instability; the factors tending to make the slope stable are in precise balance with those tending to make the slope unstable. A calculated value of F less than 1.0 indicates that a slope would be unstable under the conditions contemplated, and a value of F greater than 1.0 indicates that a slope would be stable.

If we could compute factors of safety with absolute precision, a value of $F = 1.1$ or even 1.01 would be acceptable. However, because the quantities involved in computing factors of safety are always uncertain to some degree, computed values of F are never absolutely precise. We need larger factors of safety to be sure (or sure enough) that a slope will be stable. How large the factor of safety should be is determined by experience, by the degree of uncertainty that we think is involved in calculating F, and by the consequences that would ensue if the slope failed.

The reliability of a slope (R) is an alternative measure of stability that considers explicitly the uncertainties involved in stability analyses. The reliability of a slope is the computed probability that a slope will not fail and is 1.0 minus the probability of failure:

$$R = 1 - P_f \qquad (13.1)$$

where P_f is the probability of failure and R is the reliability or probability of no failure. A method for computing P_f is described later in the chapter. Factors of safety are more widely used than R or P_f to characterize slope stability. Although R and P_f are equally logical measures of stability, there is less experience with their use, and therefore less guidance regarding acceptable values.

Another consideration regarding use of reliability and probability of failure is that it is sometimes easier to explain the concepts of reliability or probability of failure to laypeople. However, some find it disturbing that a slope has a probability of failure that is not zero, and may not be comfortable hearing that there is some chance that a slope might fail. Factors of safety and reliability complement each other, and each has its own advantages and disadvantages. Knowing the values of both is more useful than knowing either one by itself.

DEFINITIONS OF FACTOR OF SAFETY

The most widely used and most generally useful definition of factor of safety for slope stability is

$$F = \frac{\text{shear strength of the soil}}{\text{shear stress required for equilibrium}} \qquad (13.2)$$

Uncertainty about shear strength is often the largest uncertainty involved in slope stability analyses, and it is therefore logical that the factor of safety—called by George Sowers the *factor of ignorance*—should be related directly to shear strength. One way of judging whether a value of F provides a sufficient margin of safety is by considering the question: What is the lowest conceivable value of shear strength? A value of $F = 1.5$ for a slope indicates that the slope should be stable even if the shear strength was 33% lower than anticipated (if all the other factors were the same as anticipated). When shear strength is represented in terms of c and ϕ, or c' and ϕ', the same value of F is applied to both of these components of shear strength.

It can be said that this definition of factor of safety is based on the assumption that F is the same for every point along the slip surface. This calls into question whether such analyses are reasonable, because it can be shown, for example by finite element analyses, that the factor of safety for every slice is *not* the same, and it therefore appears that an underlying assumption of limit equilibrium analysis is not true. However, despite the fact that the local factor of safety may be more or less than the value of F calculated by conventional limit equilibrium methods, the average value calculated by these methods is a valid and very useful measure of stability. The factor of safety determined from conventional limit equilibrium analyses is the answer to the question: By what factor could the shear strength of the soil be reduced before the slope would fail? This is a significant question, and the value of F calculated as described above is the most generally useful measure of stability that has been devised.

Alternative Definitions of F

Other definitions of F have sometimes been used for slope stability. For analyses using circular slip surfaces, the factor of safety is sometimes defined as the ratio of resisting moment divided by overturning moment. Because the resisting moment is proportional to shear strength, and the shear stress required for equilibrium of a mass bounded by a circular slip surface is proportional to the overturning moment, the factor of safety defined as the ratio of resisting to overturning moment is the same as the factor of safety defined by Eq. (13.1).

In times past, different factors of safety were sometimes applied to cohesion and friction. However, this is seldom done any more. The strength parameters c and ϕ, or c' and ϕ', are empirical coefficients in equations that relate shear strength to normal stress or to effective normal stress. There is no clear reason to factor them differently, and the greater complexity that results if this is done seems not to be justified by additional insight or improved basis for judging the adequacy of stability.

Reinforcing and anchoring elements within a slope impose stabilizing forces that, like the soil strength, should be factored to include a margin of safety to reflect the fact that there is uncertainty in their magnitudes. The issues causing uncertainties in reinforcing and anchoring forces are not the same as the issues leading to uncertainties in soil strength, and it is therefore logical to apply different factors of safety to reinforcement and soil strength. This can be achieved by prefactoring reinforcement and anchor forces and including them in stability analyses as known forces that are not factored further in the course of the analysis.

FACTOR OF SAFETY CRITERIA

Importance of Uncertainties and Consequences of Failure

The value of the factor of safety used in any given case should be commensurate with the uncertainties involved in its calculation and the consequences that would ensue from failure. The greater the degree of uncertainty about the shear strength and other conditions, and the greater the consequences of failure, the larger should be the required factor of safety. Table 13.1 shows values of F based on this concept.

Corps of Engineers' Criteria for Factors of Safety

The values of factor of safety listed in Table 13.2 are from the U.S. Army Corps of Engineers' slope stability manual. They are intended for application to slopes of embankment dams, other embankments, excavations, and natural slopes where conditions are well understood and where the properties of the soils have been studied thoroughly. They represent conventional, prudent practice for these types of slopes and conditions, where the consequences of failure may be significant, as they nearly always are for dams.

Recommended values of factor of safety, like those in Table 13.2, are based on experience, which is logical. It is not logical, however, to apply the same values of factor of safety to conditions that involve widely varying degrees of uncertainty. It is therefore significant that the factors of safety in Table 13.2 are intended for Corps of Engineers' projects, where methods of exploration, testing, and analysis are consistent from one project to another and the degree of uncertainty regarding these factors does not vary widely. For other situations, where practices and circumstances differ, the values of F in Table 13.2 may not be appropriate.

RELIABILITY AND PROBABILITY OF FAILURE

Reliability calculations provide a means of evaluating the combined effects of uncertainties and a means of distinguishing between conditions where uncertainties are particularly high or low. Despite the fact that it has potential value, reliability theory has not been used much in routine geotechnical practice because it involves terms and concepts that are not familiar to many geotechnical engineers, and because it is commonly perceived that using reliability theory would require more data, time, and effort than are available in most circumstances.

Table 13.1   Recommended Minimum Values of Factor of Safety

Cost and consequences of slope failure	Uncertainty of analysis conditions	
	Small[a]	Large[b]
Cost of repair comparable to incremental cost to construct more conservatively designed slope	1.25	1.5
Cost of repair much greater than incremental cost to construct more conservatively designed slope	1.5	2.0 or greater

[a]The uncertainty regarding analysis conditions is smallest when the geologic setting is well understood, the soil conditions are uniform, and thorough investigations provide a consistent, complete, and logical picture of conditions at the site.

[b]The uncertainty regarding analysis conditions is largest when the geologic setting is complex and poorly understood, soil conditions vary sharply from one location to another, and investigations do not provide a consistent and reliable picture of conditions at the site.

Table 13.2   Factor of Safety Criteria from U.S. Army Corps of Engineers' Slope Stability Manual

Types of slopes	Required factors of safety[a]		
	For end of construction[b]	For long-term steady seepage	For rapid drawdown[c]
Slopes of dams, levees, and dikes, and other embankment and excavation slopes[c]	1.3	1.5	1.0–1.2

[a]For slopes where either sliding or large deformations have occurred, and back analyses have been performed to establish design shear strengths, lower factors of safety may be used. In such cases probabilistic analyses may be useful in supporting the use of lower factors of safety for design. Lower factors of safety may also be justified when the consequences of failure are small.

[b]Temporary excavated slopes are sometimes designed only for short-term stability, with knowledge that long-term stability would be inadequate. Special care, and possibly higher factors of safety, should be used in such cases.

[c]$F = 1.0$ applies to drawdown from maximum surcharge pool, for conditions where these water levels are unlikely to persist for long enough to establish steady seepage. $F = 1.2$ applies to maximum storage pool level, likely to persist for long periods prior to drawdown. For slopes in pumped storage projects, where rapid drawdown is a normal operating condition, higher factors of safety (e.g., 1.3 to 1.4) should be used.

Harr (1987) defines the engineering definition of *reliability* as follows: "Reliability is the probability of an object (item or system) performing its required function adequately for a specified period of time under stated conditions." As it applies in the present context, the *reliability of a slope* can be defined as follows: The reliability of a slope is the probability that the slope will remain stable under specified design conditions. The design conditions include, for example, the end-of-construction condition, the long-term steady seepage condition, rapid drawdown, and earthquake of a specified magnitude.

The design life of a slope and the time over which it is expected to remain stable are usually not stated explicitly but are generally thought of as a long time, probably beyond the lifetime of anyone alive today. The element of time may be considered more explicitly when design conditions involve earthquakes with a specified return period, or other loads whose occurrence can be stated in probabilistic terms.

Christian et al. (1994), Tang et al. (1999), Duncan (2000), and others have described examples of the use of reliability for slope stability. Reliability analysis can be applied in simple ways, without more data, time, or effort than are commonly available. Working with the same quantity and types of data, and the same types of engineering judgments that are used in conventional analyses, it is possible to make approximate but useful evaluations of probability of failure and reliability.

The results of simple reliability analyses are neither more accurate nor less accurate than factors of safety calculated using the same types of data, judgments, and approximations. Although neither deterministic nor reliability analyses are precise, they both have value, and each enhances the value of the other. The simple types of reliability analyses described in this chapter require only modest extra effort compared to that required to calculate factors of safety, but they can add considerable value to the results of slope stability analyses.

STANDARD DEVIATIONS AND COEFFICIENTS OF VARIATION

If several tests are performed to measure a soil property, it will usually be found that there is scatter in the values measured. For example, consider the undrained strengths of San Francisco Bay mud measured at a site on Hamilton Air Force Base in Marin County, California, that are shown in Table 13.3. There is no discernible systematic variation in the measured values of shear strength between 10 and 20 ft depth at the site. The differences among the values in Table 13.3 are due to natural variations in the strength of the Bay mud in situ, and varying amounts of disturbance of the test specimens. Standard deviation is a quantitative measure of the scatter of a variable. The greater the scatter, the larger the standard deviation.

Statistical Estimates

If a sufficient number of measurements have been made, the standard deviation can be computed using the formula

$$\sigma = \sqrt{\frac{1}{N-1}\sum_1^N (x - x_{av})^2} \quad (13.3)$$

where σ is the standard deviation, N the number of measurements, x the measured variable, and x_{av} the average value of x. Standard deviation has the same units as the measured variable.

The average of the 20 measured values of s_u in Table 13.3 is 0.22 tsf (tons/ft^2). The standard deviation, computed using Eq. (13.3), is

Table 13.3   Undrained Shear Strength Values for San Francisco Bay Mud at Hamilton Air Force Base in Marin County, California[a]

Depth (ft)	Test	s_u (tons/ft^2)
10.5	UU	0.25
	UC	0.22
11.5	UU	0.23
	UC	0.25
14.0	UU	0.20
	UC	0.22
14.5	UU	0.15
	UC	0.18
16.0	UU	0.19
	UC	0.20
	UU	0.23
	UC	0.25
16.5	UU	0.15
	UC	0.18
17.0	UU	0.23
	UC	0.26
17.5	UU	0.24
	UC	0.25
19.5	UU	0.24
	UC	0.21

[a]Values measured in unconfined compression (UC) and unconsolidated–undrained (UU) triaxial compression tests.

$$\sigma_{s_u} = \sqrt{\frac{1}{19}\sum_1^{20}(s_u - s_{u,av})^2} = 0.033 \text{ tsf} \quad (13.4)$$

where s_u is the undrained shear strength and $s_{u,av}$ is the average undrained shear strength = 0.22 tsf. The *coefficient of variation* is the standard deviation divided by the expected value of a variable, which for practical purposes can be taken as the average:

$$\text{COV} = \frac{\sigma}{\text{average value}} \quad (13.5)$$

where COV is the coefficient of variation, usually expressed in percent. Thus the coefficient of variation of the measured strengths in Table 13.3 is

$$\text{COV}_{s_u} = \frac{0.033}{0.22} = 15\% \quad (13.6)$$

where COV_{s_u} is the coefficient of variation of the undrained strength data in Table 13.3.

The coefficient of variation is a very convenient measure of scatter in data, or uncertainty in the value of the variable, because it is dimensionless. If all of the strength values in Table 13.3 were twice as large as those shown, the standard deviation of the values would be twice as large, but the coefficient of variation would be the same. The tests summarized in Table 13.3 were performed on high-quality test specimens using carefully controlled procedures, and the Bay mud at the Hamilton site is very uniform. The value of $COV_{s_u} = 15\%$ for these data is about as small as could ever be expected. Harr (1987) suggests that a representative value of $COV_{s_u} = 40\%$.

Estimates Based on Published Values

Frequently in geotechnical engineering, the values of soil properties are estimated based on correlations or on meager data plus judgment, and it is not possible to calculate values of standard deviation or coefficient of variation as shown above. Because standard deviations or coefficients of variation are needed for reliability analyses, it is essential that their values can be estimated using experience and judgment. Values of COV for various soil properties and in situ tests are shown in Table 13.4. These values may be of some use in estimating COVs for reliability analysis, but the values cover wide ranges, and it is not possible to use this type of information to make refined estimates of COV for specific cases.

The 3σ Rule

This rule of thumb, described by Dai and Wang (1992), uses the fact that 99.73% of all values of a normally distributed parameter fall within three standard deviations of the average. Therefore, if HCV is the highest conceivable value of the parameter and LCV is the lowest conceivable value of the parameter, these are approximately three standard deviations above and below the average value.

The 3σ rule can be used to estimate a value of standard deviation by first estimating the highest and lowest conceivable values of the parameter, and then dividing the difference between them by 6:

$$\sigma = \frac{HCV - LCV}{6} \quad (13.7)$$

where HCV is the highest conceivable value of the parameter and LCV is the lowest conceivable value of the parameter.

Consider, for example, how the 3σ rule can be used to estimate a coefficient of variation for a friction angle for sand that is estimated based on a correlation with standard penetration test blow count: For a value of $N_{60} = 20$, the *most likely value* (MLV) of ϕ' might be estimated to be 35°. However, no correlation is precise, and the value of ϕ' for a particular sand with an SPT blow count of 20 might be higher or lower than 35°. Suppose that the HCV was estimated to be 45°, and the LCV was estimated to be 25°. Then, using Eq. (13.7), the COV would be estimated to be

$$\sigma'_\phi = \frac{45° - 25°}{6} = 3.3° \quad (13.8)$$

and the coefficient of variation = 3.3°/35° = 9%.

Studies have shown that there is a tendency to estimate a range of values between HCV and LCV that is too small. One such study, described by Folayan et al. (1970), involved asking a number of geotechnical engineers to estimate the possible range of values of $C_c/(1 + e)$ for San Francisco Bay mud, with which

Table 13.4 Coefficients of Variation for Geotechnical Properties and In Situ Tests

Property or in situ test	COV (%)	References
Unit weight (γ)	3–7	Harr (1987), Kulhawy (1992)
Buoyant unit weight (γb)	0–10	Lacasse and Nadim (1997), Duncan (2000)
Effective stress friction angle (ϕ')	2–13	Harr (1987), Kulhawy (1992), Duncan (2000)
Undrained shear strength (S_u)	13–40	Kulhawy (1992), Harr (1987), Lacasse and Nadim (1997)
Undrained strength ratio (s_u/σ'_v)	5–15	Lacasse and Nadim (1997), Duncan (2000)
Standard penetration test blow count (N)	15–45	Harr (1987), Kulhawy (1992)
Electric cone penetration test (q_c)	5–15	Kulhawy (1992)
Mechanical cone penetration test (q_c)	15–37	Harr (1987), Kulhawy (1992)
Dilatometer test tip resistance (q_{DMT})	5–15	Kulhawy (1992)
Vane shear test undrained strength (S_v)	10–20	Kulhawy (1992)

they all had experience. The data collected in this exercise are summarized below:

Average value of $\dfrac{C_c}{1 + e}$ estimate by experienced

engineers = 0.29

Average value of $\dfrac{C_c}{1 + e}$ from 45 laboratory tests

= 0.34

Average COV of $\dfrac{C_c}{1 + e}$ estimate by experienced

engineers = 8%

Average COV of $\dfrac{C_c}{1 + e}$ from 45 laboratory tests

= 18%

The experienced engineers were able to estimate the value of $C_c/(1 + e)$ for Bay mud within about 15%, but they underestimated the COV of $C_c/(1 + e)$ by about 55%.

Christian and Baecher (2001) showed that people (experienced engineers included) tend to be overconfident about their ability to estimate values, and therefore estimate possible ranges of values that are narrower than the actual range. If the range between the highest conceivable value (HCV) and the lowest conceivable value (LCV) is too small, values of coefficient of variation estimated using the 3σ rule will also be too small, introducing an unconservative bias in reliability analysis.

Based on statistical analysis, Christian and Baecher (2001) showed that the expected range of values in a sample containing 20 values is 3.7 times the standard deviation, and the expected range of values in a sample of 30 is 4.1 times the standard deviation. This information can be used to improve the accuracy of estimated values of standard deviation by modifying the 3σ rule. If the experience of the person making the estimate encompasses sample sizes in the range of 20 to 30 values, a better estimate of standard deviation would be made by dividing the range between HCV and LCV by 4 rather than 6:

$$\sigma = \frac{HCV - LCV}{4} \qquad (13.9)$$

If Eq. (13.9) is used to estimate the coefficient of variation of ϕ', the value is

$$\sigma = \frac{45° - 25°}{4} = 5° \qquad (13.10)$$

and the coefficient of variation is $5°/35° = 14\%$.

With the 3σ rule it is possible to estimate values of standard deviation using the same amounts and types of data that are used for conventional deterministic geotechnical analyses. The 3σ rule can be applied when only limited data are available and when no data are available. It can also be used to judge the reasonableness of values of the coefficient of the variation from published sources, considering that the lowest conceivable value would be two or three standard deviations below the mean, and the highest conceivable value would be two or three standard deviations above the mean. If these values seem unreasonable, some adjustment of values is called for.

The 3σ rule uses the simple normal distribution as a basis for estimating that a range of three standard deviations covers virtually the entire population. However, the same is true of other distributions (Harr, 1987), and the 3σ rule is not tied rigidly to any particular probability distribution.

Graphical 3σ Rule

The concept behind the 3σ rule of Dai and Wang (1992) can be extended to a graphical procedure that is applicable to many situations in geotechnical engineering, where the parameter of interest, such as undrained shear strength, varies with depth. An examples is shown in Figure 13.1.

The steps involved in applying the graphical 3σ rule are as follows:

1. Draw a straight line or a curve through the data that represent the most likely average variation of the parameter with depth.
2. Draw straight lines or curves that represent the highest and lowest conceivable bounds on the data. These should be wide enough to include all valid data and an allowance for the fact that the natural tendency is to estimate such bounds too narrowly, as discussed previously. Note that some points in Figure 13.1 are outside the estimated highest and lowest conceivable lines, indicating that these data points are believed to be erroneous.
3. Draw straight lines or curves that represent the average plus one standard deviation and the average minus one standard deviation. These are one-third of the distance (or one-half of the distance) from the average line to the highest and lowest conceivable bounds.

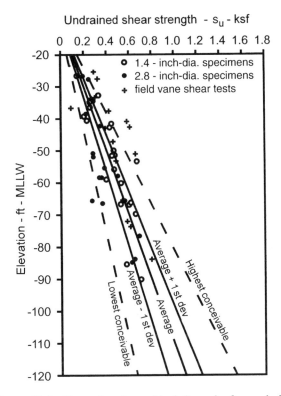

Figure 13.1 Example of graphical 3σ rule for undrained strength profile.

The average-plus-1σ and average-minus-1σ curves or lines are used in the Taylor series method described below in the same way as are parameters that can be represented by single values.

The graphical 3σ rule is also useful for characterizing strength envelopes for soils. In this case the quantity (shear strength) varies with normal stress rather than with depth, but the procedure is the same. Strength envelopes are drawn that represent the average and the highest and lowest conceivable bounds on the data, as shown in Figure 13.2. Then average-plus-1σ and average-minus-1σ envelopes are drawn one-third of the distance (or one-half of the distance) from the average envelope to the highest and lowest conceivable bounds.

Using the graphical 3σ rule to establish average-plus-1σ and average-minus-1σ strength envelopes is preferable to using separate standard deviations for the strength parameters c and ϕ. Strength parameters (c and ϕ) are useful empirical coefficients that characterize the variation of shear strength with normal stress, but they are not of fundamental significance or interest by themselves. The important parameter is shear strength, and the graphical 3σ rule provides a straightforward means for characterizing the uncertainty in shear strength.

COEFFICIENT OF VARIATION OF FACTOR OF SAFETY

Reliability and probability of failure can be determined easily once the factor of safety and the coefficient of variation of the factor of safety (COV_F) have been determined. The value of factor of safety is determined in the usual way, using a computer program, slope stability charts, or spreadsheet calculations. The value of COV_F can be evaluated using the Taylor series method, which involves these steps:

1. Estimate the standard deviations of the quantities involved in analyzing the stability of the slope: for example, the shear strengths of the soils, the unit weights of the soils, the piezometric levels, the water level outside the slope, and the loads on the slope.
2. Use the Taylor series numerical method (Wolff, 1994; U.S. Army Corps of Engineers, 1998) to estimate the standard deviation and the coefficient of variation of the factor of safety, using these formulas:

$$\sigma_F = \sqrt{\left(\frac{\Delta F_1}{2}\right)^2 + \left(\frac{\Delta F_2}{2}\right)^2 + \cdots + \left(\frac{\Delta F_N}{2}\right)^2}$$

(13.11)

$$\mathrm{COV}_F = \frac{\sigma_F}{F_{\mathrm{MLV}}}$$

(13.12)

where $\Delta F_1 = (F_1^+ - F_1^-)$. F_1^+ is the factor of safety calculated with the value of the first parameter increased by one standard deviation from its most likely value, and F_1^- is the factor of safety calculated with the value of the first parameter decreased by one standard deviation.

In calculating F_1^+ and F_1^-, the values of all of the other variables are kept at their most likely values. The other values of ΔF_2, ΔF_3, . . . Δ, F_N are calculated by varying the values of the other variables by plus and minus one standard deviation from their most likely values. F_{MLV} in Eq. (13.12) is the most likely value of factor of safety, computed using most likely values for all the parameters.

Substituting the values of ΔF into Eq. (13.11), the value of the standard deviation of the factor of safety (σ_F) is computed, and the coefficient of variation of the factor of safety (COV_F) is computed using Eq. (13.12). With both F_{MLV} and COV_F known, the probability of failure (P_f) can be determined using Table

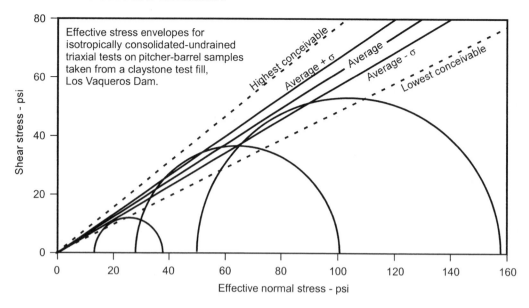

Figure 13.2 Example of graphical 3σ rule for shear strength envelope.

13.5, Figure 13.3, or the reliability index, as explained below.

Table 13.5 and Figure 13.3 assume that the factor of safety is lognormally distributed, which seems reasonable because calculating the factor of safety involves many multiplication and division operations. The central limit theorem indicates that the result of adding and subtracting many random variables approaches a normal distribution as the number of operations increases. Since multiplying and dividing amounts to adding and subtracting logarithms, it follows that the factor of safety distribution can be approximated by a lognormal distribution. Thus, although there is no proof that factors of safety are lognormally distributed, it is at least a reasonable approximation. The assumption of a lognormal distribution for factor of safety does not imply that the values of the individual variables must be distributed lognormally. It is not necessary to make any particular assumption concerning the distributions of the variables to use this method.

RELIABILITY INDEX

The *reliability index* (β) is an alternative measure of safety, or reliability, which is uniquely related to the probability of failure. The value of β indicates the number of standard deviations between $F = 1.0$ (failure) and F_{MLV}, as shown in Figure 13.4. The usefulness of β lies in the fact that probability of failure and reliability are uniquely related to β, as shown in Figure

13.5. The *lognormal reliability index*, β_{LN}, can be determined from the values of F_{MLV} and COV_F using Eq. (13.13):

$$\beta_{LN} = \frac{\ln (F_{MLV}/\sqrt{1 + COV_F^2})}{\sqrt{\ln(1 + COV_F^2)}} \quad (13.13)$$

where β_{LN} is the lognormal reliability index, F_{MLV} the most likely value of factor of safety, and COV_F the coefficient of variation of factor of safety.

The relationship between β and P_f shown in Figure 13.5 is called the *standard cumulative normal distribution function*, which can be found in many textbooks on probability and reliability. Values of P_f corresponding to a given value of β can be calculated using the NORMSDIST function in Excel. The argument of this function is the reliability index, β_{LN}. In Excel, under "Insert Function," "Statistical," choose "NORMSDIST" and type the value of β_{LN}. The result is the reliability, R. For example, for $\beta_{LN} = 2.32$, the result is 0.9898, which corresponds to $P_f = 0.0102$. Table 13.5 and Figures 13.3 and 13.5 were developed using this Excel function.

PROBABILITY OF FAILURE

Once the most likely value of factor of safety (F_{MLV}) and the coefficient of variation of factor of safety (COV_F) have been evaluated, the probability of failure (P_f) can be determined in any of the following ways:

Table 13.5   Probabilities of Failure (%) Based on Lognormal Distribution of F

| | COV$_F$ = coefficient of variation of factor of safety | | | | | | | | |
F_{MLV}[a]	10%	12%	14%	16%	20%	25%	30%	40%	50%
1.05	33.02	36.38	38.95	41.01	44.14	47.01	49.23	52.63	55.29
1.10	18.26	23.05	26.95	30.15	35.11	39.59	42.94	47.82	51.37
1.15	8.83	13.37	17.53	21.20	27.20	32.83	37.10	43.24	47.62
1.20	3.77	7.15	10.77	14.29	20.57	26.85	31.76	38.95	44.05
1.25	1.44	3.54	6.28	9.27	15.20	21.68	26.98	34.95	40.66
1.30	0.49	1.64	3.49	5.81	11.01	17.30	22.75	31.26	37.48
1.35	0.15	0.71	1.86	3.53	7.83	13.66	19.06	27.88	34.49
1.40	0.04	0.29	0.95	2.08	5.48	10.69	15.88	24.80	31.70
1.50	0.00	0.04	0.23	0.67	2.57	6.38	10.85	19.49	26.69
1.60	0.00	0.01	0.05	0.20	1.15	3.71	7.29	15.21	22.40
1.70	0.00	0.00	0.01	0.06	0.49	2.11	4.84	11.81	18.75
1.80	0.00	0.00	0.00	0.01	0.21	1.18	3.18	9.13	15.67
1.90	0.00	0.00	0.00	0.00	0.08	0.65	2.07	7.03	13.08
2.00	0.00	0.00	0.00	0.00	0.03	0.36	1.34	5.41	10.91
2.20	0.00	0.00	0.00	0.00	0.01	0.10	0.56	3.19	7.59
2.40	0.00	0.00	0.00	0.00	0.00	0.03	0.23	1.88	5.29
2.60	0.00	0.00	0.00	0.00	0.00	0.01	0.09	1.11	3.70
2.80	0.00	0.00	0.00	0.00	0.00	0.00	0.04	0.66	2.60
3.00	0.00	0.00	0.00	0.00	0.00	0.00	0.02	0.39	1.83

[a]F_{MLV}, factor of safety computed using most likely values of parameters.

1. Using Table 13.5
2. Using Figure 13.3
3. Using Figure 13.5, with β_{LN} computed using Eq. (13.13)
4. Using the Excel function NORMSDIST, with β_{LN} computed using Eq. (13.13)

Interpretation of Probability of Failure

The event whose probability is described as the *probability of failure* is not necessarily a catastrophic failure. In the case of shallow sloughing of a slope, for example, failure very likely would not be catastrophic. If the slope could be repaired easily and there were no serious secondary consequences, shallow sloughing would not be catastrophic. However, a slope failure that would be very expensive to repair, or that would have the potential for delaying an important project, or that would involve threat to life, would be catastrophic. Although the term *probability of failure* would be used in both of these cases, it is important to recognize the different nature of the consequences.

In recognition of this important distinction between catastrophic failure and less significant performance problems, the Corps of Engineers uses the term *prob-ability of unsatisfactory performance* (U.S. Army Corps of Engineers, 1998). Whatever terminology is used, it is important to keep in mind the real consequences of the event analyzed and not to be blinded by the word *failure* where the term *probability of failure* is used.

Probability of Failure Criteria

There is no universally appropriate value of probability of failure. Experience indicates that slopes designed in accord with conventional practice often have a probability of failure in the neighborhood of 1%, but like factor of safety, the appropriate value of P_f should depend on the consequences of failure.

One important advantage of probability of failure is the possibility of judging an acceptable level of risk based on the potential cost of failure. Suppose, for example, that two alternative designs for the slopes on a project are analyzed, with these results:

- *Case A.* Steep slopes, construction and land costs = $100,000, $P_f = 0.1$.
- *Case B.* Flat slopes, construction and land costs = $400,000, $P_f = 0.01$.

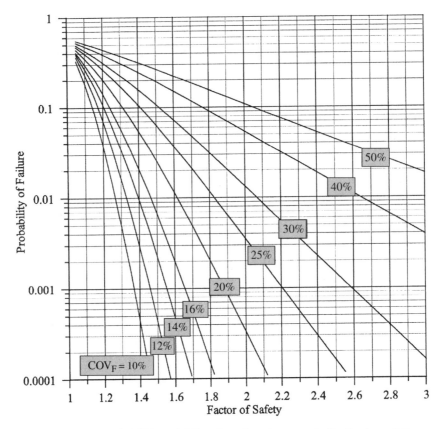

Figure 13.3   Probabilities of failure based on lognormal distribution of *F*.

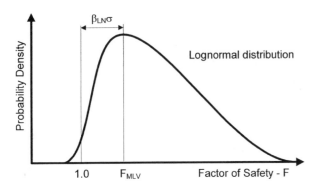

Figure 13.4   Relationship of β_{LN} to probability distribution.

Suppose further that the consequences of failure are estimated to be the same in either case, $5,000,000, considering primary and secondary consequences of failure. In case A, the total cost of construction, land, and probable cost of failure is $100,000 + (0.1)($5,000,000) = $600,000. In case B, the total cost is $400,000 + (0.01)($5,000,000) = $450,000. Considering the probable cost of failure, as well as construction and land costs, case B is less costly overall.

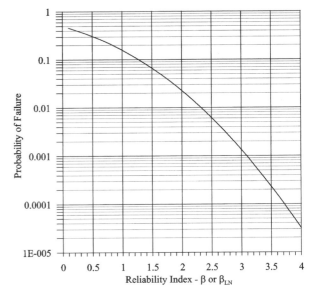

Figure 13.5   Variation of P_f with β.

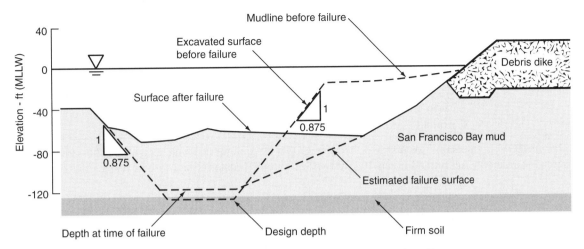

Figure 13.6   Underwater slope failure in San Francisco Bay.

Even without cost analysis, P_f may provide a better basis for judging what is an acceptable risk than does factor of safety. Many people may find that comparing one chance in 10 with one chance in 100 provides a more understandable basis for decision than does comparing a factor of safety of 1.3 with a factor of safety of 1.5.

Example.   In August 1970, a trench about 100 ft deep was excavated underwater in San Francisco Bay. The trench was to be filled with sand to stabilize the adjacent area and reduce seismic deformations at a new lighter aboard ship (LASH) terminal. The trench slopes were made steeper than was the normal practice in order to reduce the volume of excavation and fill. As shown in Figure 13.6, the slopes were excavated at an inclination of 0.875 horizontal to 1.0 vertical.

On August 20, after a section of the trench about 500 ft long had been excavated, the dredge operator found that the clamshell bucket could not be lowered to the depth from which mud had been excavated only hours before. Using the side-scanning sonar with which the dredge was equipped, four cross sections

made within two hours showed that a failure had occurred that involved a 250-ft-long section of the trench. The cross section is shown in Figure 13.6. Later, a second failure occurred, involving an additional 200 ft of length along the trench. The rest of the 2000-ft-long trench remained stable for about four months, at which time the trench was backfilled with sand. Additional details regarding the failure can be found in Duncan and Buchignani (1973).

Figure 13.1 shows the variation of undrained strength of the Bay mud at the site, and the average, average $+ \sigma$ and average $- \sigma$ lines established by Duncan (2000) for a reliability analysis of the slope. The average buoyant unit weight of the Bay mud was 38 pcf and the standard deviation was 3.3 pcf, based on measurements made on undisturbed samples.

Factors of safety calculated using average values of strength and unit weight (F_{MLV}) and using average $+\sigma$ and average $-\sigma$ values are shown in Table 13.6. The ΔF value for variation in Bay mud strength is 0.31, and the ΔF value for unit weight variation is 0.20. The ΔF due to strength variation is always significant, but

Table 13.6   Reliability Analysis for 0.875 Horizontal on 1.0 Vertical Underwater Slope in San Francisco Bay Mud

Variable	Values	F	ΔF
Undrained shear strength	Average line in Figure 13.1 ⎫	$F_{\mathrm{MLV}} = 1.17$	—
Buoyant unit weight	$\gamma_{b(\mathrm{av})} = 38$ pcf ⎭		
Undrained shear strength	Average $+ \sigma$ line in Figure 13.1	$F^+ = 1.33$	0.31
	Average $- \sigma$ line in Figure 13.1	$F^- = 1.02$	
Buoyant unit weight	Average $+ \sigma = 41.3$ pcf	$F^+ = 1.08$	0.20
	Average $- \sigma = 34.7$ pcf	$F^- = 1.28$	

Table 13.7   Summary of Analyses of LASH Terminal Trench Slope

Case	Slope (H on V)	F_{MLV}	COV_F[a] (%)	P_f (%)	Trench volume[b] (yd^3)
As constructed	0.875 on 1.0	1.17	16%	18%	860,000
Less-steep A	1.25 on 1.0	1.3	16%	6%	1,000,000
Less-steep B	1.6 on 1.0	1.5	16%	1%	1,130,000

[a]The value of COV_F is the same for all cases because COVs of strength and unit weight are the same for all cases.
[b]For the as-constructed case, an additional 100,000 yd^3 of slumped material had to be excavated after the failure.

it is unusual for the ΔF due to unit weight variation to be as large as it is in this case. Its magnitude in this case is due to the fact that the buoyant unit weight is so low, only 38 pcf. Therefore, the variation by ± 3.3 pcf has a significant effect.

The standard deviation and coefficient of variation of the factor of safety are calculated using Eqs. (13.11) and (13.12):

$$\sigma_F = \sqrt{\left(\frac{0.31}{2}\right)^2 + \left(\frac{0.20}{2}\right)^2} = 0.18 \quad (13.14)$$

$$COV_F = \frac{0.18}{1.17} = 16\% \quad (13.15)$$

The probability of failure corresponding to these values of F_{MLV} and COV_F can be determined using any of the four methods discussed previously. A value of $P_f = 18\%$ was determined using the Excel function NORMSDIST. Such a large probability of failure is not in keeping with conventional practice. Although it appeared in 1970, when the slope was designed, that the conditions were known well enough to justify using a very low factor of safety of 1.17, the failure showed otherwise. Based on this experience, it is readily apparent that such a low factor of safety and such a high probability of failure exceed the bounds of normal practice. The probability of failure was computed after the failure (Duncan, 2000) and was not available to guide the design in 1970. In retrospect, it seems likely that knowing that the computed probability of failure was 18% might have changed the decision to make the trench slopes so steep.

The cost of excavating the mud that slid into the trench, plus the cost of extra sand backfill, was approximately the same as the savings resulting from the use of steeper slopes. Given the fact that the expected savings were not realized, that the failure caused great alarm among all concerned, and that the confidence of the owner was diminished as a result of the failure, it is now clear that using 0.875 (horizontal) on 1 (vertical) slopes was not a good idea.

Further analyses have been made to determine what the probability of failure would have been if the inboard slope had been excavated less steep. Two additional cases have been analyzed, as summarized in Table 13.7. The analyses were performed using the chart developed by Hunter and Schuster (1968) for shear strength increasing linearly with depth, which is given in the Appendix. The factors of safety for the less-steep alternatives A and B are in keeping with the factor of safety criteria of the Corps of Engineers, summarized in Table 13.2.

A parametric study such as the one summarized in Table 13.7 provides a basis for decision making and for enhanced communication among the members of the design team and with the client. The study could be extended through estimates of the costs of construction for the three cases and estimates of the potential cost of failure. This would provide a basis for the design team and clients to decide how much risk should be accepted. This type of evaluation was not made in 1970 because only factors of safety were computed to guide the design.

Recapitulation

- Uncertainty about shear strength is usually the largest uncertainty involved in slope stability analyses.
- The most widely used and most generally useful definition of factor of safety for slope stability is

$$F = \frac{\text{shear strength of the soil}}{\text{shear stress required for equilibrium}}$$

- The value of the factor of safety used in any given case should be commensurate with the uncertainties involved in its calculation and the consequences of failure.
- Reliability calculations provide a means of evaluating the combined effects of uncertainties and a means of distinguishing between conditions where uncertainties are particularly high or low.

- Standard deviation is a quantitative measure of the scatter of a variable. The greater the scatter, the larger the standard deviation. The coefficient of variation is the standard deviation divided by the expected value of the variable.
- The 3σ rule can be used to estimate a value of standard deviation by first estimating the highest and lowest conceivable values of the parameter and then dividing the difference between them by 6. If the experience of the person making the estimate encompasses sample sizes in the range of 20 to 30 values, a better estimate of standard deviation can be made by dividing by 4 rather than 6.

- Reliability and probability of failure can readily be determined once the factor of safety (F_{MLV}) and the coefficient of variation of the factor of safety (COV_F) have been determined, using the Taylor series numerical method.
- The event whose probability is described as the probability of failure is not necessarily a catastrophic failure. It is important to recognize the nature of the consequences of the event and not to be blinded by the word *failure*.
- The principal advantage of probability of failure, in contrast with factor of safety, is the possibility of judging acceptable level of risk based on the estimated cost and consequences of failure.

CHAPTER 14

Important Details of Stability Analyses

The reliability of slope stability computations depends on the validity of the soil properties, the slope and subsurface geometry, and the pore water pressures used in the analyses. The reliability of the results is also dependent on several other aspects of the computation procedures. These include the following:

- Method of searching for the critical slip surface and verifying that the most critical slip surface has been located
- Detection and elimination of tensile forces between slices at the top of the slip surface
- Detection and elimination of unreasonably large compressive or tensile forces between slices at the toe of the slip surface
- Evaluation of possible three-dimensional effects

These and several other aspects of slope stability computations are discussed in this chapter.

LOCATION OF CRITICAL SLIP SURFACES

For simple slopes it is possible to estimate the location of the critical slip surface relatively well. For example, for a homogeneous slope composed of dry cohesionless soil with a constant friction angle (linear failure envelope), the critical slip surface is a plane coincident with the face of the slope; the factor of safety is given by the equation for an infinite slope: $F = \tan \phi' / \tan \beta$. For most cases the critical slip surface must be determined by trial and error. Even for a homogeneous slope composed of cohesionless soil, the critical slip surface must be located by trial and error if the Mohr failure envelope is curved or if there is a nonuniform seepage pattern (hydraulic gradients vary) near the face of the slope.

Infinite Slope

The easiest critical slip surface to locate is the one for an infinite slope. For an infinite slope the slip surface is a plane parallel to the face of the slope. Location of the critical slip surface involves finding the depth of the shear plane that produces the minimum factor of safety. For cohesionless soils ($c = 0$ or $c' = 0$) the critical plane is in the layer with the lowest friction angle, unless seepage conditions dictate otherwise. For purely cohesive soils ($\phi = 0$) the critical slip surface will usually be at the depth where the ratio of shear strength to depth, c/z, is a minimum; the critical slip surface will pass to the bottom of any layer that has a constant shear strength. If stability is checked for slip surfaces located at the top and bottom of each soil layer, the critical slip surface will almost always be found.

Circular Slip Surfaces

Locating a critical circle is more complicated than finding the depth of the critical slip surface for an infinite slope. Locating a critical circle requires a systematic search in which the center point of the circle and its radius are varied. Such searches are usually performed using a computer program in which the search is automated. Details of the particular search methods vary from one computer program to another. It is important to understand the particular method used, and to be able to control the search, to ensure that the search is thorough.

Most of the schemes used in computer programs to locate critical circles require that an initial estimate be made of the starting location of the circles used in the search. The initial estimate is usually based on an estimate of the location of the center point and the radius of the critical circle. Depending on the search scheme employed either a specific initial trial circle for starting the search is designated or the extent of a grid of center points over which the search will be conducted is specified. Usually, the radius is estimated and designated by specifying one of the following:

- The depth of a line to which circles are tangent
- A point through which circles pass
- The radius of circles

Depending on the specific search scheme used, the initial information may define either the radius for a single initial trial circle or a range in radii to be investigated. The following guidelines can be used to estimate the location of a critical circle for initiating a search.

Center point location. A likely center point for a critical circle can be estimated from what is known about the critical circles for the cases of simple, homogeneous slopes in purely frictional soil (c, $c' = 0$) and purely cohesive ($\phi = 0$) soil. For a cohesionless slope the critical slip surface is a plane coincident with the face of the slope. If a search is performed with circles, the critical "circle" is very shallow and has a very large radius. The critical circle degenerates to essentially a plane parallel to the slope surface. The center of the critical circle is located along a line passing through the midpoint of the slope, perpendicular to the face of the slope (Figure 14.1a). In contrast, for a purely cohesive slope, the critical circle passes as deep as possible. In this case the center of the critical circle lies along a vertical line passing through the midpoint of the slope as shown in Figure 14.1b.[1] In both cases ($c = 0$ and $\phi = 0$) the centers of the critical circles lie on a line passing through the midpoint of the slope, and the line is inclined at an angle, ϕ_d, from vertical (Figure 14.2), where ϕ_d is the "mobilized" friction angle (i.e., $\tan \phi_d = \tan \phi / F$). Based on this knowledge, an estimate of the center of the critical circle for other slopes can be made by drawing a line from the midpoint of the slope inclined at an angle, ϕ_d, from vertical (see O–P in Figure 14.2). A sufficient estimate of the factor of safety can usually be made for the purpose of determining the developed friction angle, ϕ_d. A good starting point for searching for the critical circle is a point along the line (O–P) at a distance equal to one or two times the slope height above the crest of the slope (see point C in Figure 14.2). Also, a starting center anywhere in the region between a vertical line and a line drawn perpendicular to the midpoint of the

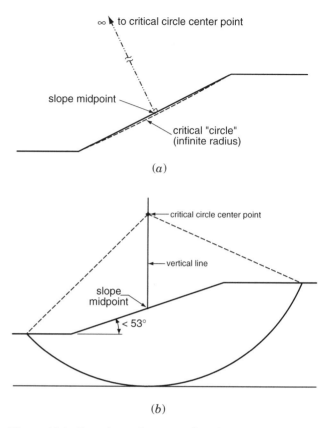

Figure 14.1 Locations of center points for critical circles in (a) purely cohesionless (c, $c' = 0$) and (b) purely cohesive (ϕ, $\phi' = 0$) slopes.

slope (the shaded area in Figure 14.2) would be a reasonable starting center.

Radius (depth of circles). When a search is conducted to locate a critical slip surface, it is important to explore circles that pass through all of the different materials in the cross section. If the shear strength of a given stratum is characterized by c = constant, ϕ = 0, the critical circle will usually, but not always, pass to the bottom of the layer. Circles should be analyzed that pass to the bottom of each stratum that has a constant shear strength. Also, it is usually more effective to start with deep circles and search upward rather than searching downward. By starting the search with circles that are deep, the trial circles should intersect most layers in the cross section, and the circles will be more likely to migrate into the layers that produce the lower factors of safety. Another useful strategy is to search for the critical circle that passes through the toe of the slope and to examine other types of circles (radii and tangent depths), using the critical toe circle as the beginning circle.

Increments for center point coordinates and radius. Most computer programs that employ search schemes using circles vary the coordinates of the cen-

---

[1] This is true only for slopes flatter than 53°, but covers many of the cases of practical interest.

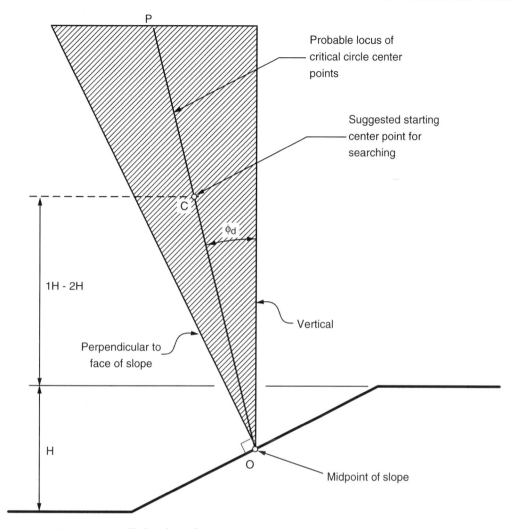

Figure 14.2   Estimation of starting center point to search for critical circles.

ter point systematically and the radius in increments until a minimum factor of safety is found. It is important that the increments used to vary the center point coordinates and the radius be small enough so that important features in the cross section are detected by the search. This requires that the increments be no greater than some fraction of the thickness of the thinnest layer in the cross section. A distance no larger than one-half to one-fourth of the later thickness is effective. Search increments from one-tenth (0.1) to one-hundredth (0.01) of the slope height are suitable if there are no thin layers in the cross section. With the computational speed of current computers a search increment of 1% of the slope height (0.01 H) can be used with little concern for the computation time required.

Multiple minimums.   For even simple slopes more than one circle may be found that produces a local minimum for the factor of safety. That is, if the circle is moved slightly in any direction away from the local

minimum, the factor of safety will increase. As an illustration, consider the three simple, purely cohesive ($\phi = 0$) slopes shown in Figure 14.3. The slopes shown in this figure are identical except for the thickness of the foundation. The slope shown in Figure 14.3a has a 30-ft-thick foundation. The two circles shown in this figure both represent local minimums, with factors of safety of 1.124 for the shallow circle and 1.135 for the deep circle. The shallower circle through the toe of the slope is slightly more critical and thus represents the overall minimum factor of safety. The slope shown in Figure 14.3b has a slightly deeper foundation, 46.5 ft deep. This slope also has two circles that produce locally minimum factors of safety, but both circles have the same factor of safety (1.124); there are two critical circles with identical minimum factors of safety. The third slope, shown in Figure 14.3c, has the deepest foundation: 60 ft deep. There are again two local minimums for this slope,

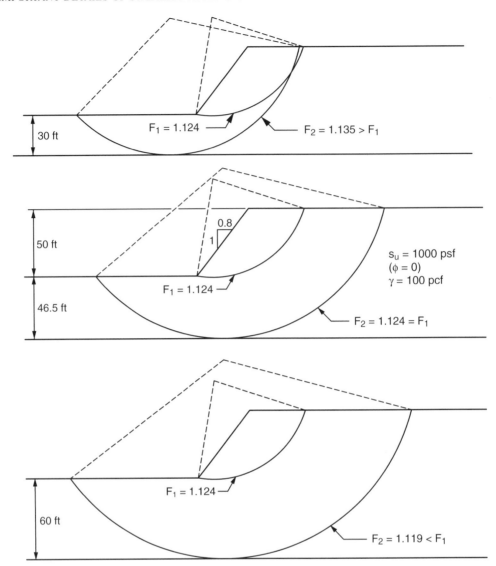

Figure 14.3   Local critical circles with similar or the same factor of safety.

with the shallow circle producing a factor of safety of 1.124 and the deeper circle having a factor of safety of 1.119. In this case the deeper circle represents the more critical of the two local minimums. This example shows that with relatively small changes in the depth of the foundation there may be quite different locations for the critical circle, and in some cases there may even be two different critical circles that have the same minimum factor of safety.

Another example of a slope with two local minimums for the factor of safety is shown in Figure 14.4. For the cohesionless embankment there is a local minimum factor of safety corresponding to a shallow, infinite slope mechanism. The factor of safety for this shallow slip surface is 1.44. Another local minimum

occurs for a deeper circle passing to the bottom of the clay foundation. The factor of safety for the deeper circle is 1.40 and is the overall minimum factor of safety for this slope. In cases like the ones illustrated in Figure 14.3 and 14.4, it is important to conduct multiple searches to explore both deep and shallow regions of the slope profile.

When searches for the critical circle are done manually, it is straightforward to explore all regions of the cross section. However, when searches are done using computer programs in which the search is conducted automatically, care is required to ensure that all feasible regions of the slope and foundation have been explored and that the critical circle has been found. Most computer programs with automatic search routines per-

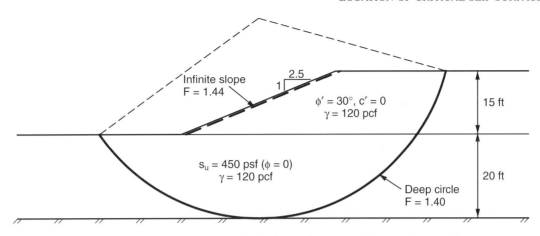

Figure 14.4 Slope with two local critical slip surfaces and minimum factors of safety.

mit the starting point for the search to be designated as input data. Therefore, it is possible to perform a number of independent searches by using different starting locations. It is good practice to perform several searches using different starting locations to ensure that all regions of the cross section are explored and that the most critical slip surface is found.

One scheme for locating a critical circle is to vary the radius for each trial center point until a critical radius, corresponding to the minimum factor of safety for the center point, is found. When the radii are varied for a given center point, multiple local minimums may be found. For example, Figure 14.5 shows the variation in factor of safety with the radius (depth) of a circle whose center point is the center of the most critical circle overall. It can be seen that there are several local minimums for the factor of safety for the designated center point. It is important in such cases to vary the radius between wide limits and in relatively small increments to capture this behavior and to ensure that the minimum factor of safety is found. An appropriate scheme for doing this is to select the minimum and maximum radius (or tangent line elevation) of interest, along with a suitable increment for varying the radius. The increment used to vary the radius should not exceed one-half to one-fourth of the thickness of the *thinnest* layer.

Although varying the radius to find a critical radius for each trial center point is relatively inefficient computationally—many more circles may be analyzed than perhaps are necessary—the scheme can provide useful information. This scheme is particularly useful when contours are to be drawn to show how the factor of safety varies with the location of the center point. The most meaningful contours that can be drawn for a series of selected center points are those for factors of safety corresponding to the critical radius (minimum factor of safety) for each point.

Noncircular Shear Surfaces

Critical noncircular slip surfaces are much more difficult to locate because they involve many more variables than circles. Circles are described by three values: the x and y coordinates of the center, and the radius. Noncircular slip surfaces are described by two values (x and y) for each point on the slip surface.[2] Depending on the number of points required to characterize the noncircular slip surface, the number of variables in the search can be quite large.

Two basically different approaches have been used to vary the positions of noncircular slip surfaces and find the slip surface with the minimum factor of safety. One approach employs a systematic shifting of the slip surface, with an appropriate minimization or optimization scheme. Implementations of this method vary from sophisticated schemes employing linear programming methods to much simpler, direct numerical techniques (e.g., Baker, 1980; Celestino and Duncan, 1981; Arai and Tagyo, 1985; Nguyen, 1985; Li and White, 1987; Chen and Shao, 1988).

The second approach for finding a critical noncircular slip surface employs a random process to select trial slip surfaces. This scheme generally proceeds until a specified number of slip surfaces have been explored (e.g., Boutrop and Lovell, 1980; Siegel et al., 1981). A random number generator is used to establish the location of the random slip surfaces.

---

[2]The exception to this is at the ends of the slip surface, where the points are constrained additionally to be either on the surface of the slope or at the bottom of a crack profile.

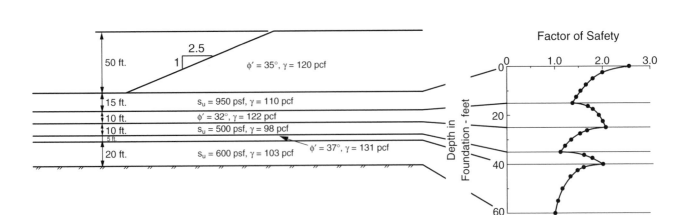

Figure 14.5 Variation of the factor of safety with depth (radius) of circles for an embankment on a stratified soil foundation.

Regardless of the scheme used to find a critical noncircular slip surface, the potential location of the critical slip surface must usually be estimated either as a starting point for the search or to establish limits on the search. A good estimate for the starting location of the noncircular slip surface is the location of the critical circular slip surface. Locating the critical circle and then using several (4 to 10) points along the critical circle to define an initial trial noncircular slip surface works well in many cases. The locations of the initial points are varied in accordance with whatever search scheme is being used to find a more critical noncircular slip surface. Additional points may be added to the noncircular slip surface as the search progresses. This scheme works well in most cases, particularly where there are not very thin, weaker zones of soil. Alternatively, if distinctly weaker zones exist, a slip surface passing through the weaker zones may be selected as a starting point.

When a particularly thin, weak layer is likely to control stability, it is usually better to start with a noncircular slip surface that has a portion of the slip surface following along the weak layer. Such a surface may enter or exit the slope through stronger overlying material. When it does, the inclination of the slip surface at the crest and at the toe of the slope should be chosen to conform to the inclinations of active and passive shear planes, respectively. Active and passive zones are discussed later in this chapter. Near the crest of the slope the slip surface will usually be chosen to enter

at an angle of 45 to 65° from the horizontal. Near the toe of the slope the slip surface will usually be chosen to exit at an angle of 25 to 45° from the horizontal.

As with circles, it is important with noncircular slip surfaces to choose several starting locations for the slip surface and perform searches beginning with each. The initial slip surfaces should pass through the various zones of soil that are expected to influence stability. It is also important that sufficiently small increments be used to vary the position of points on the slip surface so that the effects of thin layers are evaluated accurately. Increments no greater than one-half the thickness of the thinnest significant stratum and as small as 1% of the slope height (0.01 horizontal) are appropriate. In most cases it should be anticipated that more than one local minimum exists.

Importance of Cross-Section Details

In embankment dams with complex cross sections, there may be numerous zones of different materials that result in at least several local minimum factors of safety. It is important that each potential slip surface and mechanism of sliding be explored. It is also important to model the cross section in as much detail as possible, because these details may have an effect on the stability. For example, the slope protection (riprap) and associated filters on the upstream face of an earth dam can have a large effect on the stability and safety against shallow sliding due to drawdown. Omission of the shallow layers of slope protection in the geometry used for an analysis of sudden drawdown may result in the factor of safety being significantly underestimated, and unnecessary conservatism in the design.

To illustrate the importance of cross-section details, consider the embankment dam shown in Figure 14.6. The embankment is to be subjected to a rapid drawdown of 19 ft, and slope stability computations were performed using the U.S. Army Corps of Engineers (1970) procedure for rapid drawdown. Computations were performed with and without the riprap and filter zones included in the cross section used for analysis. When the filter and riprap were excluded, the material was assumed to be the same as the embankment material. Factors of safety are summarized in Figure 14.6. The factor of safety with the riprap and filter zones included in the analyses was 1.09. The factor of safety dropped to 0.84 when the riprap was excluded—a decrease in factor of safety of approximately 23%. Considering that the U.S. Army Corps of Engineers (1970) recommended factor of safety for this case is 1.0, the effect of omitting the riprap and filters from the analysis was to change the factor of safety from acceptable to unacceptable.

Recapitulation

- An initial estimate of the critical circle can be made from what is known about the critical circles for homogeneous slopes (Figure 14.2).
- Searches should be conducted beginning at several starting points to fully explore the soil profile and detect multiple minimums.
- Searches are usually more successful when the search is started deep rather than shallow.
- Increments of distance used to move the center and change the radius of circles in a search should not exceed one-half the thickness of the thinnest stratum of interest. Increments are typically chosen to be 0.01 to 0.1 times the slope height. The speed of current computers allows use of small increments with no practical penalty.
- Good starting locations for a search with noncircular slip surfaces can be estimated by starting from the critical circle, or by examining the slope cross section to identify layers of weakness.
- The smallest increments used to move the points when searching with noncircular slip surfaces should not exceed one-half the thickness of the thinnest stratum of interest and are typically chosen to be 0.01 to 0.1 times the slope height.
- All cross section details that may influence the position of the critical slip surface and factor of safety should be included in the geometry used for the slope stability analysis.

EXAMINATION OF NONCRITICAL SLIP SURFACES

In some cases the slip surface with the minimum factor of safety may not be the slip surface of greatest interest. For example, the minimum factor of safety for the embankment shown in Figure 14.7 is 1.15. The minimum factor of safety corresponds to an infinite slope failure in the cohesionless fill. Also shown in Figure 14.7 is a deeper circle that has a locally minimum factor of safety. This deeper circle has a factor of safety of 1.21, which is higher than the factor of safety for the shallower, infinite slope slip surface. However, if sliding occurred along the deep circle, it would have a far more severe consequence. Failure along the shallow surface in the embankment would consist simply of raveling of material down the slope and might, at the most, represent a maintenance problem. In contrast, failure along the deeper surface might require reconstruction of the embankment. Consequently, the deeper surface and the associated factor of safety of 1.21 would be considered unacceptable, while a factor of safety of 1.15 for the shallower, most critical slip surface might be acceptable.

More than one mechanism of failure and associated factor of safety must sometimes be examined for design. It is good practice to locate as many locally critical slip surfaces as possible. In some cases it is also useful to examine slip surfaces that are important but do not represent local minimum factors of safety. A good example of cases where noncritical slip surfaces need to be studied is older mine tailings disposal dams. Many older tailing dams consist predominately of cohesionless materials where the critical slip surface is a shallow "skin" slide. The consequences of shallow sliding may be minor, but deeper sliding must be avoided. To address this concern, stability computations can be performed using slip surfaces that extend to various depths. The variation of the factor of safety with the assumed depth of slide is then examined and a judgment is made regarding the depth of sliding that can be tolerated. Minimum factor of safety requirements are established for slip surfaces that extend deeper than the tolerable limit. To investigate deeper slip surfaces that do not necessarily represent minimum factors of safety, several techniques can be used to define the slip surfaces that will be analyzed. Three techniques for doing this are illustrated in Figure 14.8. They consist of:

1. Forcing trial slip surfaces to be tangent to a line that is parallel to and below the surface of the slope (Figure 14.8a)
2. Forcing trial slip surfaces to pass through a designated point that is located at some depth below the surface (Figure 14.8b)

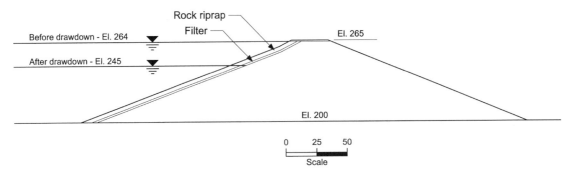

Figure 14.6   Effect of included cross-section detail on computed factor of safety for rapid drawdown.

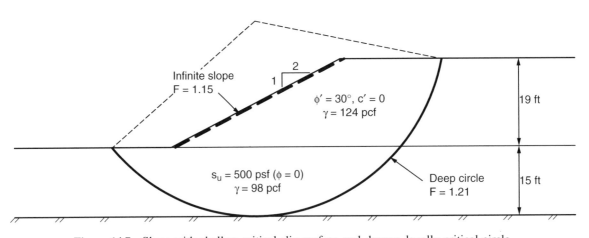

Figure 14.7   Slope with shallow critical slip surface and deeper, locally critical circle.

3. Requiring that the soil mass above the slip surface have some minimum weight (Figure 14.8c)

Many computer programs provide some capability for limiting the extent of a search for a critical slip surface in one or more of the ways described above.

Earth dams represent another case where there are slip surfaces besides the slip surface producing the minimum factor of safety that are of interest. Stability analyses must be performed for both the upstream and downstream slopes of dams. One slope (upstream or downstream) typically has a lower factor of safety than the other. In searching for a critical circle on one face of the dam it is possible that the search will "hop" to the other face and abandon searching further for the critical slip surface beneath the slope face of interest.

If this occurs, it is necessary to employ some means to restrict the search to a specific slope face. Some computer programs automatically restrict the search to the slope face where the search was started; other programs may require that additional constraints be used.

Recapitulation

- The slip surface with the minimum factor of safety is not always the one of greatest interest.
- It is often desirable to examine slip surfaces and factors of safety that are not minimums, especially when the consequences of other mechanisms of failure are important.

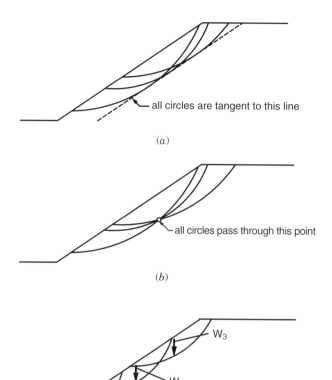

Figure 14.8 Artificial constraints imposed to examine non-critical slip surfaces of interest: (*a*) all circles tangent to a line; (*b*) all circles pass through a fixed point; (*c*) all circles have minimum weight.

TENSION IN THE ACTIVE ZONE

When there are cohesive soils in the upper portion of a slope, slope stability calculations usually will reveal tension at the interfaces between slices as well as on the bottom of the slices. The existence of tension may be of concern for two reasons:

1. Most soils do not have significant tensile strength and thus cannot withstand tension. Calculated tensile stresses are unrealistic and inappropriate.
2. When significant tension develops it can cause numerical problems in the slope stability calculations.

For both reasons it is desirable to eliminate the tensile forces from the analyses.

Rankine Active Earth Pressures

The tension at the top of the slope that occurs in slope stability analyses is analogous to the tensile stresses that are computed from Rankine active earth pressure theory. In fact, the mechanisms that produce tension in slope stability analyses and active earth pressure calculations are the same. For total stresses the Rankine active earth pressure beneath a horizontal ground surface, σ_h, is given by

$$\sigma_h = \gamma z \tan^2\left(45 - \frac{\phi}{2}\right) - 2c \tan\left(45 - \frac{\phi}{2}\right) \quad (14.1)$$

where z is the depth below the ground surface (Figure 14.9). At the ground surface, where $z = 0$, the expression for the horizontal stress is

$$\sigma_h = -2c \tan\left(45 - \frac{\phi}{2}\right) \quad (14.2)$$

which indicates that stresses are negative (tensile) if c is greater than zero. Negative stresses extend to a depth, z_t, given by

$$z_t = \frac{2c}{\gamma \tan(45 - \phi/2)} \quad (14.3)$$

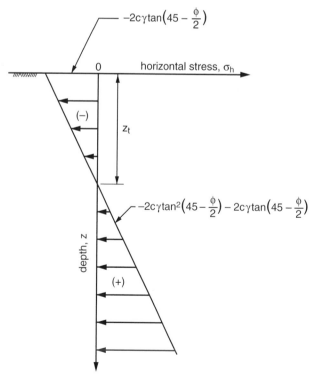

Figure 14.9 Active Rankine horizontal earth pressures beneath a horizontal ground surface.

The active earth pressures increase linearly from the negative value given by Eq. (14.2) to zero at the depth, z_t. Below the depth z_t the active earth pressures are positive and continue to increase linearly with depth.

Rankine's theory for active earth pressures and the limit equilibrium slope stability analysis procedures described in Chapter 6 are similar for soil near the crest of a slope. Both employ equations of static equilibrium to compute the stresses on vertical planes; in the case of slope stability analyses the interslice forces represent the stresses on vertical planes.

There are also some differences between Rankine earth pressures and limit equilibrium slope analysis procedures. For Rankine earth pressures the forces on vertical planes are parallel to the ground surface; while for limit equilibrium slope stability analyses, various assumptions are made for the direction of forces, depending on the particular procedure of slices used. Also, for Rankine earth pressures the shear strength is assumed to be fully developed ($F = 1$), while for slope stability the shear strength is generally not fully developed (i.e., $F > 1$).

Despite the differences between slope stability and earth pressure calculations, the fundamental cause of tension is the same: Tension is due to some or all of the shear strength of the soil being mobilized, when the soil is assumed to have cohesion. As shown in Figure 14.10, if a Mohr–Coulomb failure envelope with cohesion is assumed, tensile strength is implied. The Mohr circle shown in Figure 14.10 has a major principal stress that is positive while the minor principal stress is negative. Although cohesion may be appropriate for describing the position of a failure envelope for positive normal stresses (σ, $\sigma' > 0$), the same envelope is generally not appropriate for negative (tensile) normal stresses. To achieve more realistic re-

sults, it is necessary to ignore the implied tensile strength.

Tension can appear in the results of limit equilibrium slope stability computations in three different ways:

1. The interslice forces become negative.
2. The total or effective normal forces on the bases of slices become negative.
3. The line of thrust moves outside the slice.

Theoretically, at the point where the compressive and tensile stresses are equal on the boundary between slices the line of thrust must lie an infinite distance away from the slope (Figure 14.11). At the point where the compressive and tensile stresses produce equal resultant forces, there is a finite moment but no resultant force to produce the moment. Thus, an infinite moment arm (line of thrust) is required.

Tension may appear in one or more of the three forms listed above. For the Simplified Bishop procedure, interslice forces are not computed, and thus tension appears only in the form of negative normal forces on the bottom of slices. For force equilibrium procedures such as the Modified Swedish and Simplified Janbu procedures, the interslice forces are computed, but not their locations. Thus, in these procedures tension may appear in the form of negative interslice forces and negative normal forces on the base of slices. Finally, for procedures such as Spencer's that consider complete equilibrium, tension may appear in any of the three forms listed above.

Eliminating Tension

Two different techniques can be used to eliminate the effects of tension from the results of slope stability calculations:

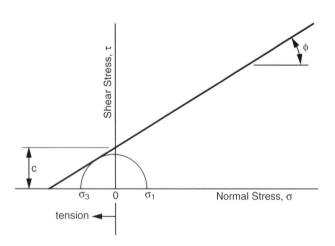

Figure 14.10 Mohr's circle with tensile stresses for a cohesive soil.

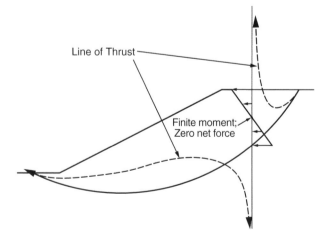

Figure 14.11 Lateral compressive and tensile stresses on an interslice boundary that produce a line of thrust at infinity.

1. A tension crack can be used in the slope stability calculations.
2. The Mohr failure envelope can be adjusted so that there is no shear strength when the normal stress becomes negative.

Tension crack.   A tension crack is introduced into slope stability calculations by terminating the slip surface at the edge of a slice at an appropriate depth below the ground surface (Figure 14.12). The depth of tension can be estimated from Eq. (14.3) using the developed shear strength parameters c_d and ϕ_d. The estimated depth for the vertical crack is given by

$$d_{\text{crack}} = \frac{2c_d}{\gamma z \, \tan(45 - \phi_d/2)} \qquad (14.4)$$

Although c_d and ϕ_d depend on the factor of safety, values can usually be estimated with sufficient accuracy for calculating the depth of crack before the factor of safety is calculated. If necessary, the values for c_d and ϕ_d can be adjusted and the calculations repeated with a revised crack depth once an initial factor of safety is calculated with the estimated crack depth.

In general, when a vertical crack is introduced, the crack should not extend significantly beyond the depth of tension. If the crack depth is overestimated, compressive forces will be eliminated and the factor of safety will be overestimated. In many cases a tension crack has only a minor effect on the computed factor of safety. However, one reason for introducing a tension crack is to eliminate numerical stability problems and inappropriate negative stresses. Thus, even though a tension crack may not have a significant effect on the computed factor of safety, it is a good practice to introduce a crack when there are cohesive soils present along the upper portion of the slip surface.

In most cases only an approximate estimate of the depth of tension crack is needed and can be made based on considerations of active earth pressures as described above. However, if the effect of tension is unclear, a series of stability computations can be performed in which a vertical crack is introduced and the depth is varied. Typically, these analyses will show that the factor of safety first decreases as the crack depth is increased and tension is eliminated and then increases as the crack depth becomes greater and compressive stresses are eliminated as well (Figure 14.13). Such analyses are useful where there are questions regarding the effects of tension and the appropriate crack depth.

Zero tensile strength envelope.   Instead of introducing a tension crack, the shear strength envelope can be adjusted so that the shear strength is zero when there is tension. This can be accomplished using a nonlinear Mohr failure envelope like either of the ones shown by the heavy lines in Figure 14.14. Trial-and-error procedures are then required to determine the appropriate shear strength because the shear strength parameters (c, c' and ϕ, ϕ') depend on the normal stress (σ, σ'). If the Mohr failure envelope has abrupt changes in slope, numerical instability and convergence problems can occur in these trial-and-error procedures. That is, the shear strength may oscillate between zero and some finite value of shear strength on successive iterations. Thus, although use of a nonlinear failure envelope may be a good approach from a fundamental viewpoint, appropriate envelopes are difficult to determine, and their use can cause additional numerical problems. Use of a tension crack as described earlier

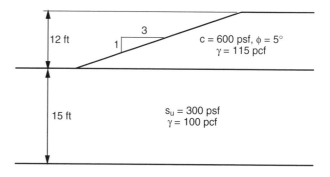

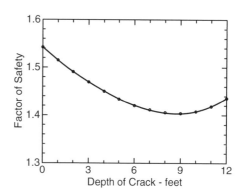

Figure 14.13 Variation in factor of safety with the assumed depth of crack.

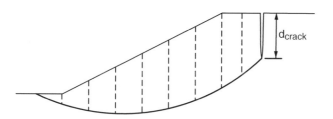

Figure 14.12 Slope and slip surface with a tension crack to eliminate tensile stresses.

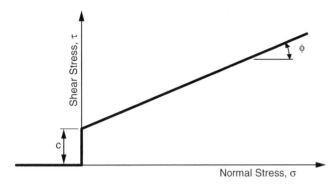

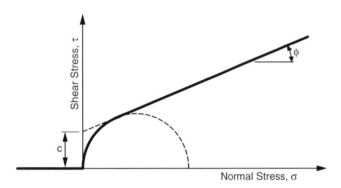

Figure 14.14 Nonlinear Mohr failure envelopes (heavy lines) used to prevent tension in cohesive soils.

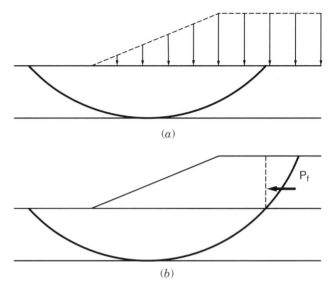

Figure 14.15 Load representations for embankments where the shear strength is neglected: (*a*) vertical surcharge; (*b*) lateral thrust produced by weak fill.

is usually a more practical alternative for eliminating tension.

Replacing Cracked Embankments by Surface Loads

In some cases, particularly where relatively strong cohesive embankments rest on much weaker foundations, the crack depth calculated from Eq. (14.4) may exceed the height of the embankment. If a crack depth equal to the height of the embankment is assumed, the embankment strength plays no role in the computed factor of safety. The factor of safety calculated with a tension crack extending through the full height of the embankment is identical to the factor of safety that is calculated assuming a vertical surcharge, represented by a distribution of vertical stress such as the one shown in Figure 14.15*a*.

When an embankment is treated as a vertical surcharge, the shear strength of the embankment has no effect on the computed factor of safety. This, however, is not equivalent to the case where an embankment is assumed to have no, or very little, shear strength. If an embankment or fill has negligible strength, there will be a significant horizontal thrust that is not represented by a vertical surcharge (see P_f in Figure 14.15*b*). It is

appropriate in the latter instances to model the fill as a low strength material by assigning small or zero values for the shear strength parameters (*c* and ϕ). If the embankment is assigned a small or zero shear strength, the appropriate lateral thrust will be reflected in the limit equilibrium calculations.

Recapitulation

- Tension can cause numerical stability problems in the solution for the factor of safety.
- Tension can imply tensile strength that does not exist and will result in a calculated factor of safety that is too high.
- Introducing a tension crack can eliminate the adverse effects of tension.
- The depth for a tension crack can be estimated from simple equations derived from earth pressure theory.
- Neglecting the strength of an embankment by introducing a tension crack or treating the embankment as a vertical surcharge is appropriate when the embankment is very strong, but is not correct when the embankment is very weak.

INAPPROPRIATE FORCES IN THE PASSIVE ZONE

In the preceding section problems that can develop near the crest of the slope were discussed. Other problems can develop near the toe of the slope. As with the problems near the crest of the slope, earth pressure theories are helpful in understanding the stresses and problems that can develop near the toe of a slope and slip surface. The region near the toe of a slip surface corresponds to a zone of passive earth pressures. Problems in the form of very large compressive or even tensile stresses can develop near the toe of the slope.

Cause of Problems

Problems develop near the toe of a slope when the direction of the resultant force, R, on the base of the last slice is very close to the direction of the interslice force, Z. In this case the resultant force on the base of the slice and the interslice force become either very large or negative. This is illustrated for a typical slice in Figure 14.16. It is assumed in this case that the soil is cohesionless. The resultant force (R) due to the normal stress and the mobilized shear strength acts at the angle, ϕ_d, from a line perpendicular to the base of the slice. The slice shown in Figure 14.16 has a base slope angle (α) of $-55°$ and the mobilized friction angle is $25°$. The resultant interslice force inclination is $10°$. The interslice force and the resultant force on the base of the slice both have the same line of action. There is no force perpendicular to these two forces (R and Z) to balance the weight of the slice. Consequently, math-

ematically the interslice force and the force on the base of the slice become infinite. This condition is reflected by the following term that appears in the denominator of the equations for the interslice force resultant, Q, presented in Chapter 6 for Spencer's procedure:

$$\cos(\alpha - \theta) + \frac{\sin(\alpha - \theta)\tan\phi'}{F} \quad (14.5)$$

This quantity appears in the denominator of one of the terms for the interslice force resultant (Q). Substituting values, $\alpha = -55°$, $\tan\phi'/F = \tan\phi_d = \tan 25°$, and $\theta = 10°$ into (14.5) produces a value of zero. As a result, the equations used to compute the factor of safety by Spencer's procedure contain zero in the denominator for one or more slices, and the forces Q and N become infinite. Whitman and Bailey (1967) noted that a similar problem occurs in the Simplified Bishop procedure when the following term, designated as m_α, becomes small:

$$m_\alpha = \cos\alpha + \frac{\sin\alpha\tan\phi'}{F} \quad (14.6)$$

This term is identical to the term in Eq. (14.5) from Spencer's procedure when the interslice force inclination (θ) is set to zero, as is done in the Simplified Bishop procedure. The problem that is illustrated in Figure 14.16 and is shown by Eqs. (14.5) and (14.6) exists with all limit equilibrium procedures of slices except the Ordinary Method of Slices, which ignores the forces on vertical slice boundaries.

Eliminating the Problem

When the conditions described in the preceding paragraph occur, any of the following may happen:

1. The trial-and-error solution for the factor of safety may not converge—the solution may "blow up."
2. Forces may become extremely large and produce very high shear strengths in frictional materials.
3. Forces may become negative (tensile).

Mathematically, negative stresses in frictional materials will produce negative shear strengths! If this occurs, the factor of safety may be much smaller than reasonable. In this case, if an automatic search is being performed to locate a critical slip surface, the search may suddenly pursue an unrealistic minimum as a solution. The solution may be correct from a mathematical perspective, being a valid root to the simultaneous equations and representing a minimum value for F; however, such a solution is clearly not realistic phys-

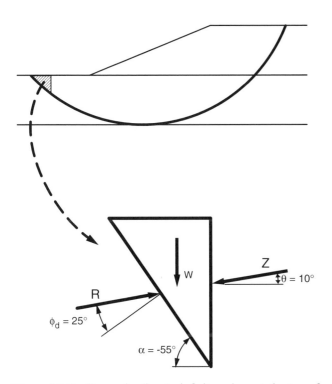

Figure 14.16 Forces leading to infinite values at the toe of the slip surface.

ically and should be rejected. Whitman and Bailey (1967) suggested that when the Simplified Bishop procedure is being used and the value of m_α expressed by Eq. (14.6) becomes less than 0.2, alternative solutions should be explored.

The problem of negative stresses near the toe of the slip surface can be eliminated in at least four different ways. These are described below.

Change the slip surface inclination. Large or negative forces at the toe of the slip surface occur because the inclination of the interslice force and slip surface are different from those corresponding to critical conditions. In other words, the inclinations of the interslice force and slip surface are significantly different from the inclinations corresponding to the minimum passive earth pressure. The relationship between the slip surface inclination and the interslice force inclination for minimum passive earth pressures has been presented by Jumikis (1962) and is illustrated in Figure 14.17. If the assumption pertaining to the inclinations of the interslice forces in a solution are reasonable, as they usually are, the most realistic remedy for inappropriate forces in the passive zone is to change the inclination of the slip surface near the toe of the slope. The inclination of the slip surface that should be used for a given interslice force inclination, and developed shear strength (ϕ_d) can be estimated from Figure 14.17.

Changing the slip surface inclination is the most appropriate way to eliminate inappropriate stresses in the passive zone. However, if calculations are being performed using circular slip surfaces, this remedy cannot be used because the overall geometry of the circle determines the orientation of the toe of the slip surface. In such cases one of the three remaining alternatives described below may be adopted.

Compute strength independently. Ladd (personal communication) has suggested one alternative remedy for tension in the passive zone. He suggested that the shear strength in the passive zone can be estimated independent of the slope stability calculations based on the stresses calculated using Rankine passive earth pressure theory. If the overlying ground surface is horizontal and the soil is cohesionless, the vertical and horizontal stresses, σ_h and σ_v, respectively, are principal stresses that can be calculated as follows:

$$\sigma_v = \sigma_3 = \gamma z \tag{14.7}$$

$$\sigma_h = \sigma_1 = \gamma z \tan^2\left(45 + \frac{\phi'}{2}\right) \tag{14.8}$$

The corresponding shear strength on the slip surface is then determined from the following:

$$\tau_{ff} = \frac{\sigma_1 - \sigma_3}{2}\cos\phi' = \frac{1}{2}\gamma z\left[\tan^2\left(45 + \frac{\phi'}{2}\right) - 1\right] \tag{14.9}$$

where τ_{ff} is the shear stress on the failure plane at failure.

The shear strength, $s = \tau_{ff}$, can be calculated from Eq. (14.9) and entered into the slope stability calculations as a cohesion ($s = c = \tau_{ff}$) that increases linearly with depth and with $\phi = 0$. This approach is simple and seems to work reasonably well. If the zone where the problem occurs does not have a major impact on the stability of the slope, Ladd's simple assumption is adequate.

Use the Ordinary Method of Slices. Very large or negative normal stresses at the toe of the slope do not occur in the Ordinary Method of Slices. The normal force on the base of the slice varies from a value equal to the weight of the slice (when the base is horizontal) to zero (as the base of the slice approaches vertical). Thus, if problems occur in the passive zone with other limit equilibrium procedures, the Ordinary Method of Slices can be used.

Change the side force inclination. The final remedy for eliminating tension near the toe of the slip surface is to change the inclination of the interslice forces in the area where the problem occurs. Figure 14.17 can be used to determine an appropriate interslice force inclination that is consistent with the slip surface inclination in the passive shear zone. Depending on the developed friction angle (ϕ_d) and slip surface inclination (α), an appropriate interslice force inclination (θ) can be determined from Figure 14.17. The inclination can be used directly as the assumed interslice force inclination in force equilibrium slope stability calculations. However, this can only be done with force equilibrium procedures. The interslice force inclination cannot be changed directly in other limit equilibrium procedures, although the inclination can be changed indirectly in the Morgenstern and Price and Chen and Morgenstern procedures through the assumed values for $f(x)$ and $g(x)$.

Discussion. Slope stability calculations in which critical noncircular slip surfaces have been located show that the critical slip surface exits the toe of the slope at angles very similar to what would be expected based on passive earth pressure theory (i.e., the inclinations agree well with those shown in Figure 14.17). Thus, it is generally sufficient to initiate a search with a reasonable starting angle (45° or less, depending on the shear strength properties of the soil), and the critical slip surface is then usually found without problems. It is also helpful to place constraints on the

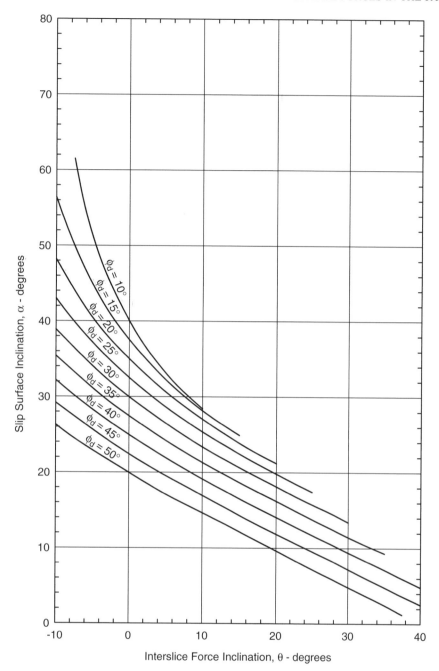

Figure 14.17 Combinations of slip surface inclination and interslice (earth pressure) force inclination for minimum passive earth pressures.

inclination of the slip surface in an automatic search to avoid very steep slip surfaces being considered where unreasonable solutions may exist due to negative stresses at the toe of the slope. Many computer programs that search for critical noncircular slip surfaces contain provisions for limiting the steepness of the slip surfaces where they exit the slope.

Use of the appropriate passive inclination for the slip surface appears to be the most realistic solution to problems at the toe of the slope and probably conforms most closely with conditions in the field. The only instances where one of the other techniques described needs to be employed is when circular slip surfaces are being used. In these cases the second and third options—calculation of the strength independently, and using the Ordinary Method of Slices—usually work well.

Recapitulation

- Very large positive or negative forces may develop on the sides and bottom of slices when the slip surface is steep near the toe of the slope, and the forces act in directions that are significantly different from the directions corresponding to minimum passive earth pressures.
- The problem of inappropriate stresses at the toe of the slip surface is best remedied by changing the inclination of the slip surface so that it corresponds more closely to the inclination of the shear plane for passive earth pressures.
- The problem of inappropriate stresses at the toe of a slope may also be remedied by (1) estimating shear strength directly based on Rankine passive earth pressure theory, (2) using the Ordinary Method of Slices, or (3) changing the interslice force inclination.

OTHER DETAILS

Additional details that affect the results of slope stability calculations include the number of slices into which the soil mass is divided, and the tolerances used to define convergence in the iterative procedures for calculating factors of safety.

Iteration Tolerances and Convergence

All procedures except the infinite slope and the Ordinary Method of Slices procedures require an iterative (trial-and-error) process to calculate the factor of safety for a given slip surface. The iterative process requires one or more convergence criteria to define when iterations can be stopped and a sufficiently accurate solution for the factor of safety has been found. Convergence criteria can be specified in terms of maximum allowable changes in the calculated values of factor of safety on successive iterations, or limits on the magnitude of allowable force and moment imbalance, or on a combination of these criteria. The tolerances for convergence may be embedded in the software or they may be included as part of the input data. In either case, awareness of the criteria and the possible consequences is important.

When a search is being conducted to locate a critical slip surface the convergence criteria for the factor of safety must be smaller than the changes in factor of safety between adjacent slip surfaces analyzed in the search process. If the convergence criteria produce factors of safety that are less precise than the changes that occur when the slip surface is moved small distances, the search can result in false minimums for the factor of safety. For typical slopes and search schemes, it is appropriate for convergence criteria to produce factors of safety that are precise to 0.0001.

Convergence criteria for force and moment imbalances need to be chosen carefully, depending on the problem being solved. Criteria that require force and moment equilibrium to be satisfied to within 100 lb and 100 ft-lb usually work well. However, for very shallow slides these limits should be reduced. Slip surfaces for very shallow slides that approximate infinite slope failures sometimes involve less than 100 lb of total weight and a precision of 100 lb force imbalance is not adequate; a 100-lb force imbalance could cause twofold variations in the computed factors of safety. In contrast to very shallow slides where very small imbalance tolerances are required, massive slides may require larger than usual equilibrium tolerances. For example, Gucwa and Kehle (1978) describe slope stability analyses for a massive landslide in the Bearpaw Mountains of Montana. The lateral extent of the slide from the crest to the toe of the slide mass was in excess of 10 km, and the depth of the slip surface may have approached 1 km. For analyses of the slide it was necessary to use as acceptable force imbalances values that were several orders of magnitude greater than the usual 100 lb.

An additional level of iteration is required to calculate the factor of safety when the Mohr failure envelope is curved rather than linear. In such cases the shear strength parameters (c, c' and ϕ, ϕ') vary with the normal stress, but the normal stress can only be determined once the shear strength parameters are known. Only the infinite slope and Ordinary Method of Slices procedures permit the normal stress to be calculated independently of the shear strength parameters. Any of the other limit equilibrium procedures requires trial-and-error solutions when curved (nonlinear) Mohr failure envelopes are used. In the trial-and-error procedures, the shear strength is estimated, the factor of safety and normal stresses are calculated, new strengths are calculated, and the process is repeated until the assumed and calculated shear strengths are sufficiently alike. Some tolerance must either be built into or entered as input data into any computer program that performs such calculations. If the tolerances are too large, they can result in inaccurate solutions or false local minima for the factor of safety.

Number of Slices

All of the procedures of slices involve subdividing the soil above the slip surface into slices. The number of slices used depends on several factors, including the complexity of the soil profile, whether the calculations are performed by hand or by a computer program, and accuracy requirements.

Required slices. Most slopes have several points where it is either convenient or necessary to place slice boundaries. For example, boundaries are usually placed wherever there is a break in the slope profile (points *b* and *f* in Figure 14.18), wherever a stratum outcrops on the slope surface (point *c* in Figure 14.18), or wherever the slip surface crosses a boundary between two different strata (points *a*, *d* and *e* in Figure 14.18). By placing slice boundaries at such points, the base of each slice will be in only one material, and the soil strata will vary continuously and smoothly across each slice. Slice boundaries may also be placed where there is a change in the slope of a piezometric surface, where the slip surface crosses either the water table or piezometric line, and where there is an abrupt change in distributed load on the surface of the slope.

For circular slip surfaces, once the required slice boundaries are established, additional slices are added to achieve a more precise approximation of the curved slip surface by straight-line segments (the bases of the slices). Additional slices may also be appropriate to account better for variations in shear strength or pore water pressure along a slip surface. Also, the number of slices used depends on whether the calculations are being performed by hand or by computer.

Hand calculations. When calculations are done by hand, the achievable level of accuracy is on the order of 1%. The number of slices needed to be consistent with this accuracy is usually no more than 8 to 12. Once the required number of slices is set, a few additional slices may be added to make the slices more uniform in size. For complex geometries with a variety of piezometric surfaces and distributed loads, as many as 30 to 50 slices are sometimes required to capture

all of the details in the cross section. This number (30 to 50), however, is exceptional. It would be unusual to use hand calculations for such a problem except to verify a computer solution.

Computer calculations. When calculations are performed using a computer, rather than by hand, many more slices are used. Very little time is required to perform the calculations with a computer, even when 50 or more slices are used.

For circles, the effect of the number of slices on the factor of safety is best related to the angle, θ_s, subtended by radii drawn to each side of the base of each slice (Figure 14.19). It has been found that the accuracy of a solution is related more to the angle, θ_s, than to the number of slices (Wright, 1969). To illustrate the effect of the subtended angle and number of slices, slope stability computations were performed for the two slopes shown in Figure 14.20. For the slope shown in the lower part of Figure 14.20, two different representations of undrained shear strength were used for the foundation, and separate computations were performed for each. The subtended angle used to generate slices was varied from 1 to 40°. Results for the three different slope shear strength combinations are summarized in Table 14.1. Several conclusions can be drawn from these results:

- Factors of safety are essentially the same (maximum difference 0.1%) using 1° and 3° as the maximum subtended angle for subdivision into slices.
- The factor of safety tends to increase very slightly (less than 1%) as the subtended angle is increased from 3° to 15°.
- For very large subtended angles (more than 15°) the factor of safety may either increase or decrease, depending on the particular slope geometry and variation in soil properties.

It is very likely that the factor of safety could vary differently with the number of slices, depending on the particular algorithms employed to compute the slide weights, length of slip surface, and the center of gravity and moment arms for slices. However, it seems unlikely that there will be any significant differences or errors if the slices are subdivided using a subtended angle of 3° or less.

Based on the discussion above, the most appropriate way to set the number of slices is to set an upper limit for the subtended angle, θ_s. A value of 3° works well for this purpose, and 3° typically results in from 30 to 60 slices, assuming total subtended angles for the slip surface of 90 to 180°. For deep slip surfaces with large total subtended angles, 50 or more slices may be re-

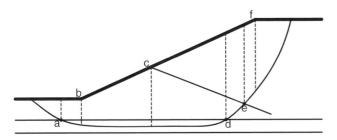

Figure 14.18 Locations where slice boundaries are required.

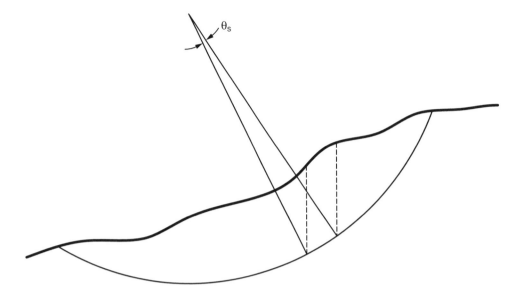

Figure 14.19   Subtended angle, θ_s, used to characterize the size of slices.

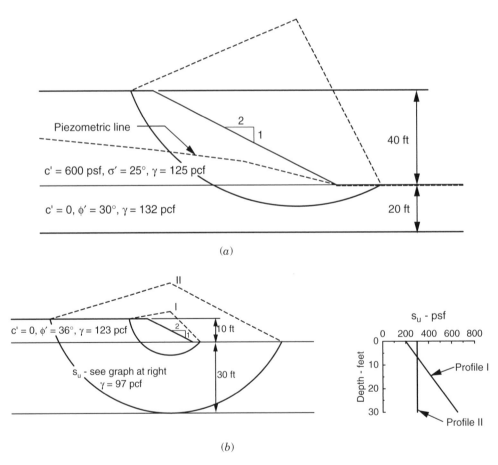

Figure 14.20   Slopes used to illustrate the effect of subdivision into slices: (*a*) example slope 1; (*b*) example slope 2.

Table 14.1   Variations in Factors of Safety with the Subtended Angle Used for Subdivision of the Soil Mass into Slices[a]

Subtended angle (deg)	Factor of safety		
	Example slope 1	Example slope 2–strength profile I	Example slope 2–strength profile II
1	1.528	1.276	1.328
3	1.529	1.276	1.328
	(0.1)	(0.0)	(0.0)
5	1.530	1.277	1.329
	(0.1)	(0.1)	(0.1)
10	1.535	1.279	1.332
	(0.5)	(0.2)	(0.3)
20	1.542	1.291	1.331
	(0.9)	(1.2)	(0.2)
30	1.542	1.323	1.314
	(0.9)	(3.7)	(−1.1)
40	1.542	1.323	1.295
	(0.9)	(3.7)	(−2.5)

[a]Numbers in parentheses represent the percent increase (+) or decrease (−) in factor of safety compared to the factor of safety when the soil mass is subdivided into slices using a subtended angle of 1°.

quired for a subtended angle of 3°. On the other hand, for very shallow slip surfaces that approximate the infinite slope mechanism, very few slices are required to achieve the same degree of accuracy. The number of slices will also depend on the complexity of the slope geometry and the number of interslice boundaries required.

In some cases where circles are very shallow, a maximum subtended angle of 3° may produce only one slice. This causes the base of the slice (slip surface) to coincide with the surface of the slope, and the slice has zero area. When this occurs, an arbitrary number of slices is used. The number of slices is not important—as few as two or three slices will suffice. In these cases it is more appropriate to subdivide the slip surface into slices using a prescribed arc length rather than a prescribed subtended angle. The arc length can be chosen so that any slip surface that is of sufficient length to be of interest will be divided into 5 to 10 slices. For example, if slip surfaces with lengths of 10 ft or more are judged to be of interest, a maximum arc length of 2 ft could be used for subdividing into slices.

The number of slices used to represent noncircular slip surfaces has an effect on the computed factor of safety, but the criterion for the number of slices cannot be represented by a subtended angle. Instead, a mini-

mum number of slices is usually chosen; usually, 30 or more slices works well. Similarly to circles, the best results are obtained by subdividing the soil mass such that the lengths of the base of slices, Δl, are approximately equal, rather than using slices with equal widths, Δx. Use of a constant base length along the slip surface produces more slices at the more steeply inclined ends of the slip surface and fewer where the slip surface is flatter (Figure 14.21).

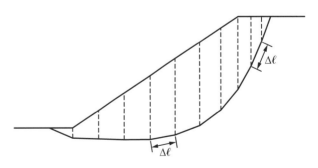

Figure 14.21 Subdivision for slices with a noncircular slip surface so that the bases of slices are of approximately constant length.

Recapitulation

- Convergence criteria that are too coarse can result in false minima and an incorrect location for the critical slip surface.
- Convergence criteria for force and moment imbalance should be scaled to the size of the slope. Tolerances of 100 lb/ft (0.1 kN/m) and 100 ft-lb/ft (0.1 kNm/m) are suitable for most slopes.
- The number of slices does not have a large effect on the computed factor of safety, provided that details of the slope and subsurface stratigraphy are represented.
- For hand calculations only a small number of slices (6 to 12) is required to be consistent with the accuracy achievable by hand calculations.
- For computer solutions with circular slip surfaces, the number of slices is usually chosen by selecting a maximum subtended angle, θ_s, of 3° per slice.
- For computer solutions with noncircular slip surfaces, 30 or more slices are used. Slices are subdivided to produce approximately equal lengths for the base of the slices along the slip surface.

VERIFICATION OF CALCULATIONS

Any set of slope stability calculations should be checked by some independent means. Numerous examples of how analyses can be checked have already been presented in Chapter 7. Also, Duncan (1992) summarized several ways in which the results of slope stability calculations can be checked, including:

1. Experience (what has happened in the past and what is reasonable)
2. By performing extra analyses to confirm the method used by comparison with known results
3. By performing extra analyses to be sure that changes in input causes changes in results that are reasonable
4. By comparing key results with computations performed using another computer program, slope stability charts, spreadsheet, or detailed hand calculations

Many slope stability calculations are performed with a computer program that uses a complete equilibrium procedure of slices. Most of these complete equilibrium procedures (Spencer, Morgenstern and Price, Chen and Morgenstern, etc.) are too complex for calculation by hand. In this case suitable manual checks

of the calculations can be made using force equilibrium procedures and assuming that the interslice forces are inclined at the angle(s) obtained from the computer solution. Suitable spreadsheets for this purpose and example calculations are presented in Chapter 7.

Published benchmark problems also provide a useful way for checking the validity of computer codes, and although benchmarks do not verify the solution for the particular problem of interest, they can lend confidence that the computer software is working properly and that the user understands the input data. Several compilations of benchmark problems have been developed and published for this purpose (e.g., Donald and Giam, 1989; Edris and Wright, 1987).

In addition to benchmark problems there are several ways that simple problems can be developed to verify that computer codes are working properly. Most of these simple problems are based on the fact that it is possible to model the same problem in more than one way. Examples of several simple test problems are listed below and illustrated in Figures 14.22 and 14.23:

1. Computation of the factor of safety for a submerged slope under drained conditions using:
   a. Total unit weights, external loads representing the water pressures, and internal pore water pressures
   b. Submerged unit weights with no pore water pressure or external water loads
   This is illustrated in Figure 14.22a. Both approaches should give the same factor of safety.[3]
2. Computation of the factor of safety with the same slope facing to the left and to the right (Figure 14.22b). The factor of safety should not depend on the direction that the slope faces.
3. Computation of the factor of safety for a partially or fully submerged slope, treating the water as:
   a. An externally applied pressure on the surface of the slope
   b. A "soil" with no strength ($c = 0$, $\phi = 0$) and having the unit weight of water.
   This is illustrated in Figure 14.22c.
4. Computation of the factor of safety for a slope with very long internal reinforcement (geogrid, tieback) applying the reinforcement loads as:
   a. External loads on the slope
   b. As internal loads at the slip surface

---

[3] See also the discussion in Chapter 6 of water pressures and how they are handled. Large differences between the two ways of representing water pressures may occur if force equilibrium procedures are used. See also Example 2 in Chapter 7.

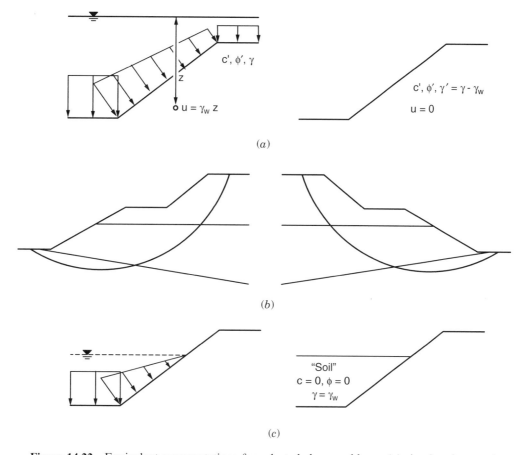

Figure 14.22 Equivalent representations for selected slope problems: (*a*) simple submerged slope, no flow; (*b*) left- and right-facing slope; (*c*) partially submerged slope.

This is illustrated in Figure 14.23*a*. Although intuitively the location where the force is applied might be expected to have an effect, the location does not have a large effect on the computed factor of safety, provided that the slip surface does not pass behind the reinforcement and the force does not vary along the length of the reinforcement.

5. Computation of the bearing capacity of a uniformly loaded strip footing on a horizontal, infinitely deep, purely cohesive foundation (Figure 14.23*b*). For circular slip surfaces and a bearing pressure equal to 5.53 times the shear strength (*c*) of the soil, a factor of safety of unity should be calculated.

6. Computation of the seismic stability of a slope using:
   a. A seismic coefficient, *k*
   b. No seismic coefficient, but the slope is steepened by rotating the entire slope geometry through an angle, θ, where the tangent of θ is

the seismic coefficient (i.e., $k = \tan\theta$) *and* the unit weight is increased by multiplying the actual unit weight by $\sqrt{1 + k^2}$

This is illustrated in Figure 14.23*c*. In both solutions the magnitude and direction of the forces will be the same and should produce the same factor of safety.

Additional test problems can also be created using slope stability charts like the ones presented in the Appendix.

THREE-DIMENSIONAL EFFECTS

All the analysis procedures discussed in Chapter 6 assume that the slope is infinitely long in the direction perpendicular to the plane of interest; failure is assumed to occur simultaneously along the entire length of the slope. A two-dimensional (plane strain) cross section is examined, and equilibrium is considered in

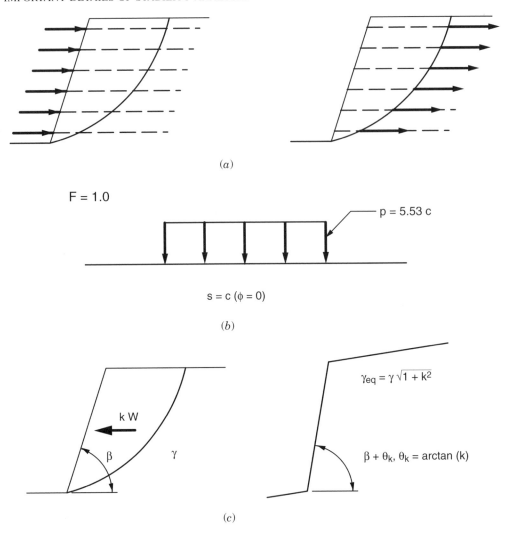

Figure 14.23 Equivalent representations for selected slope problems: (*a*) reinforced slope; (*b*) bearing capacity on saturated clay; (*c*) pseudostatic analyses.

just two directions. On the other hand, most slope failures are finite, and most failure surfaces are three-dimensional, often being bowl-shaped.

Many three-dimensional failures occur because soil properties and pore water pressures vary along the length of the slope (i.e., subsurface conditions and soil properties are not uniform). Generally, there are not enough data to describe the variation in properties along the length of a slope in sufficient detail to perform more rigorous analyses that account for spatial variations.

Three-dimensional slope failures may also occur due to the three-dimensional geometry of the slope. Several analysis methods have been developed to address the effects of three-dimensional geometry. Comparisons can be and have been made between results of two-

and three-dimensional analyses. In comparing results of three-dimensional slope stability analysis with those from two-dimensional analyses, it is important that the cross section used for the two-dimensional analyses be stipulated. Two-dimensional analyses are generally performed either for the maximum cross section or for the cross section that gives the lowest factor of safety. The maximum cross section is the cross section where the slope is highest or where the maximum amount of soil is involved in potential sliding. The maximum cross section for a dam is usually near the center of the valley (Figure 14.24*a*); the maximum cross section for a waste fill may be perpendicular to an exposed face of the fill, approximately midway between two opposing, lateral side slopes (Figure 14.24*b*). However, the maximum cross section does not always produce

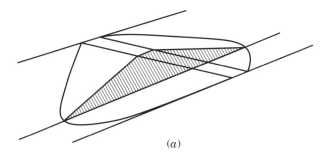

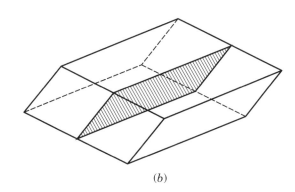

Figure 14.24 Typical "maximum" cross sections used for the two-dimensional analysis of three-dimensional geometries: (*a*) earth dam; (*b*) waste fill.

the minimum factor of safety. This was clearly demonstrated by Seed et al. (1990) for the analyses of a slide that occurred at the Kettleman Hills landfill. Factors of safety calculated for the maximum sections were 1.10 to 1.35, while factors of safety for smaller sections near the side slopes were as low as 0.85. Conclusions drawn regarding differences between two- and three-dimensional analyses will thus depend on whether the two-dimensional analyses are performed for the maximum cross section or for the cross section with the minimum factor of safety. Very different conclusions may be drawn regarding differences depending on which two-dimensional cross section is used for the analyses.

The two-dimensional factor of safety for the most critical (lowest factor of safety) section will always be lower than the three-dimensional factor of safety, provided that comparable limit equilibrium approaches are used (i.e., similar assumptions and rigor are used to satisfy static equilibrium).

Most of the general-purpose three-dimensional slope stability analysis procedures are based on a method of

columns. The methods of columns are the three-dimensional equivalent of the two-dimensional procedures of slices. In the method of columns the soil mass is subdivided into a number of vertical columns, each with an approximately square cross section in plan view. A considerable number of assumptions must be made to achieve a statically determinate solution with the method of columns. Several procedures employ simplifying approaches comparable to the Ordinary Method of Slices, rather than fully satisfying the six equations[4] of static equilibrium. The effects of these assumptions may be as large as the three-dimensional effects themselves. Thus, a considerable amount of uncertainty exists in the results from many of the three-dimensional procedures, and the procedures should be used cautiously, especially when they are being used as a basis for acceptance when two-dimensional analyses might indicate unacceptably low factors of safety.

Hutchinson and Sharma (1985) and Leshchinsky and Baker (1986) pointed out that for cohesionless soils two- and three-dimensional analyses should give the same factor of safety because the critical slip surface is a plane coincident with the surface of the slope. Unless there is significant cohesion or the geometry and soil strengths are such that the failure surface is deep, the difference between two- and three-dimensional analyses is likely to be small.

Azzouz et al. (1981) and Leshchinsky and Huang (1992) both noted that when shear strengths are back-calculated using two-dimensional analyses for three-dimensional conditions, the shear strengths will be too high. This error is compensated, however, if the shear strengths are subsequently used in two-dimensional analyses of similar conditions.

Most slope stability analyses neglect three-dimensional effects without significant consequence. However, there are two practical instances where two-dimensional analyses might not be adequate, and three-dimensional analyses would be warranted:

1. When strengths are back-calculated from three-dimensional failures and the bias in the back-calculated strengths would not be compensated in subsequent analyses
2. When, due to the slope geometry, there may be a significant benefit of the three-dimensional effects that will improve stability

In these cases, three-dimensional analyses may be warranted.

---

[4]Three equations for force equilibrium and three equations for equilibrium of moments about three axes.

CHAPTER 15

Presenting Results of Stability Evaluations

Clear and comprehensive presentations of results of slope stability evaluations are important for several reasons:

1. Results need to be checked. Results may be checked and reviewed by several persons, including both engineers in the organization doing the stability analyses and engineers in other organizations that have review and regulatory responsibilities.
2. Results need to be clear to the client or other person for whom the evaluations were made.
3. Responsibility for engineering is sometimes transferred to another engineer within the organization or to an entirely different organization. The person assuming responsibility needs a clear understanding of what has been done and the basis for decisions that have been made.
4. Engineers often need to "revisit" their work many years later or resume work after delays cause it to be put aside. Clear documentation of the work that has already been done is important.

The level and form of documentation may vary depending on the progress of the job. In general, some form of written documentation should be prepared and maintained at all stages of the job so that if the job is stopped or delayed prematurely, work can be resumed later and prior work can be understood. The primary emphasis of this chapter is on the presentation of results of a slope stability evaluation once it has been completed. However, many of the components of this final documentation should be prepared as the work progresses.

SITE CHARACTERIZATION AND REPRESENTATION

Any presentation should designate the location of the site and slope being evaluated. As a minimum the pre-

sentation should contain sufficient information so that the site can be located and visited by the person reviewing the work. Presentation of the site location may vary from a simple written description of the site location to elaborate site maps, plan sheets, and photographs showing the location in detail. Where such plans and drawings are presented they may also include topographic contours and identification of potentially important features. For example, the site plan may identify nearby reservoirs, rivers, streams, areas of extensive fill work or excavation, structures, and transportation and other infrastructure elements (roads, pipelines, electricity and telecommunication transmission lines, etc.).

An appropriate description of the geologic setting is essential. Geologic details often play a major role in slope stability and thus the geologic information is very important. The extent to which such information is included as a specific part of a stability evaluation as compared to part of the geotechnical engineering for the overall project will vary, of course, and this may or may not be an integral part of the presentation of the slope stability evaluation.

One or more cross sections should be prepared to show the geometry of the slope and immediate foundation area that are being evaluated. Where field investigations have been performed using exploratory borings, test pits, and nonintrusive geophysical methods, soil profile cross sections or an equivalent graphical representation of the subsurface (fence diagrams, three-dimensional solid and surface models, etc.) should be prepared.

A cross-section drawing should be prepared showing all the details that are considered important for the stability analyses, but the cross section may exclude extraneous and minor features if they are known to be unimportant. The cross section should be drawn to scale.

Trial F = 1.613

1	2	3	4	5	6	7	8	9	10	11	12	13	14	15	16
Slice No.	b	h_{shell}	γ_{shell}	h_{core}	γ_{core}	W	α	$W \sin\alpha$	c'	ϕ'	$h_{piez.}$	u	$c'b + (W - ub)\tan\phi'$	m_α	$c'b + (W - ub)\tan\phi' + m_\alpha$
1	20.0	13.1	140	0.0	120	36659	43.2	25110	0	38	0.0	0	28641	1.06	27011
2	35.2	20.8	140	26.6	120	215174	40.2	138831	0	20	23.2	1450	59724	0.91	65660
3	65.0	10.0	140	76.0	120	683527	35.0	391572	0	20	38.3	2392	192187	0.95	202533
4	62.5	29.5	140	84.9	120	894190	28.7	429214	0	20	55.3	3449	247055	0.99	250670
5	89.9	52.5	140	69.1	120	1406585	21.7	519784	0	20	59.9	3741	389600	1.01	384751
6	105.7	82.2	140	34.2	120	1648280	13.1	374514	0	20	34.2	2131	517970	1.03	505280
7	21.2	101.4	140	5.6	120	315386	7.7	42435	0	20	11.5	719	109240	1.02	106965
8	81.8	92.2	140	0.0	120	1055716	3.4	62854	0	38	12.9	805	773349	1.03	752971
9	114.7	54.1	140	0.0	120	867907	-4.8	-72569	0	38	7.2	452	637567	0.96	666912
10	49.8	14.2	140	0.0	120	98993	-11.7	-20099	0	38	0.0	0	77342	0.88	87806
							$\Sigma =$	1891646						$\Sigma =$	3050560
														$F' =$	1.613

Note: Trial F adjusted repeatedly until computed value, F', and trial value are the same.

Figure 15.1   Sample spreadsheet for simplified Bishop procedure.

SOIL PROPERTY EVALUATION

The basis for the soil properties used in a stability evaluation should be described and appropriate laboratory test data should be presented. If properties are estimated based on experience, or using correlations with other soil properties or from data from similar sites, this should be explained. Results of laboratory tests should be summarized to include index properties, water content, and unit weights. For compacted soils, suitable summaries of compaction moisture–density data are useful. A summary of shear strength properties is particularly important and should include both the original data and the shear strength envelopes used for analyses (Mohr–Coulomb diagrams, modified Mohr–Coulomb diagrams, τ_{ff} vs. σ'_{fc} diagrams).

The principal laboratory data that are used in slope stability analyses are the unit weights and shear strength envelopes. If many more extensive laboratory data are available, the information can be presented separately from the stability analyses in other sections, chapters, or separate reports. Only the summary of shear strength and unit weight information need be presented with the stability evaluation in such cases.

PORE WATER PRESSURES

For effective stress analyses the basis for pore water pressures should be described. If pore water pressures are based on measurements of groundwater levels in bore holes or with piezometers, the measured data should be described and summarized in appropriate figures or tables. If seepage analyses are performed to compute the pore water pressures, the method of analysis, including computer software, that was used should be described. Also, for such analyses the soil properties and boundary conditions as well as any assumptions used in the analyses should be described. Soil properties should include the hydraulic conductivities. Appropriate flow nets or contours of pore water pressure, total head, or pressure head should be presented to summarize the results of the analyses.

SPECIAL FEATURES

Slopes sometimes have special features of their makeup and construction that should be included in the presentation of results. For example, for reinforced slopes the type of reinforcement (nails, geogrid, geotextile, tieback anchors, piles, etc.) is important and should be indicated. In the case of reinforcement, the basis for determining the forces in the reinforcement should be described, including what factors of safety were applied to the reinforcement forces. Important details of any nonsoil materials such as geosynthetics and the interfaces between soil and nonsoil materials or between nonsoil and nonsoil materials (e.g., geomem-

Table 15.1 Sample Description of Contents of Spreadsheet Presented in Figure 15.1

Column no.	Description
1	Slice number
2	Horizontal width of slice
3	Height of portion of slice in the *shell* material; measured at midpoint of slice, h_{shell}
4	Total unit weight of shell material, γ_{shell}
5	Height of portion of slice in the *core* material; measured at midpoint of slice, h_{core}
6	Total unit weight of *core* material, γ_{core}
7	Weight of slice, $W = b(\gamma_{shell}h_{shell} + \gamma_{core}h_{core})$
8	Inclination of the bottom of the slice, α, in degrees measured from the horizontal [positive when the bottom of the slice is inclined in the same direction as the slope face (e.g., near the crest of the slope); negative when the base of the slice is inclined in the direction opposite that of the slope face (e.g., near the toe of the slope)]
9	Product of slice weight and sine of the angle, α
10	Cohesion value for soil at bottom of slice
11	Friction angle for soil at bottom of slice
12	Height of piezometric line (h_{piez}) above the center of the base of the slice
13	Pore water pressure at the center of the bottom of the slice, $u = \gamma_{water}h_{piez}$
14	Intermediate term used to compute numerator in equation for factor of safety: $c'b + (W - ub)\tan\phi'$
15	Value of the quantity m_{α} for trial factor of safety; trial factor of safety is shown above column 15 ($m_{\alpha} = \cos\alpha + \sin\alpha\tan\phi'/F$)
16	Column 14 divided by column 15, representing terms in the numerator of the expression for factor of safety; column 16 summed for all slices and divided by the summation of column 9 to compute a new factor of safety, F'

branes and geotextiles) should also be included in the report.

For compacted clay fills, the compaction procedures and construction quality control are important in determining the shear strength values. In some cases the construction sequence will determine the geometries that must be considered and may dictate consideration of special stages in construction that need attention, due either to the shear strengths that will exist at that time or the geometric configuration of the slope. Important details of construction should be addressed in the report.

Operational details can also be important and need to be included in a report. For example, the operating procedures for a dam or reservoir will control the amount of drawdown and thus have a direct bearing on stability. Temporary construction or stockpiling of materials or other heavy items near the crest of a slope will influence its stability and should be duly noted and accounted for in the stability evaluation. Also, for seismic areas the characteristics of the design earthquake should be described.

CALCULATION PROCEDURE

Presentations of slope stability analyses should clearly identify the procedures used to perform the stability calculations. When computer software is used, the software and references to documentation (manuals) should be specified; software without manuals or documentation should not be used. Inclusion of detailed computer output is usually unnecessary if proper related documentation is presented as described elsewhere in this chapter. If computer output is included, it should be organized in appendixes and clearly labeled.

If spreadsheets are included as part of a presentation of results, a complete description of the content of items (rows, columns) in the spreadsheet should be included. An example for this type of documentation is the spreadsheet for the Simplified Bishop procedure shown in Figure 15.1. Table 15.1 contains a description of the contents of the columns in this spreadsheet, including the equations used for computation of the contents of each column, except where the information is self-explanatory.

ANALYSIS SUMMARY FIGURE

Results of stability calculations for each condition (e.g., end-of-construction, long-term, sudden drawdown) should be presented in separate figures showing, at least, the slope and subsurface geometry. The figure should also show the most critical slip surface that gave the minimum factor of safety. The figure need not include all of the details shown for the cross section described earlier in the section "Site Characterization and Representation" but should include sufficient detail that the location of important features relative to the critical slip surface can be readily seen. For example, in an embankment dam, the various zones of pervious and impervious soils should be shown on each cross section. For embankments and excavated slopes in natural soils, the locations of various strata should be delineated.

The summary figure for each stability condition should also include information showing what set of shear strength properties were used and, for effective stress analyses, what pore water pressure conditions were used in the analyses. For simple soil and slope conditions the shear strength properties may be shown directly on the figure, at applicable locations in the cross section (Figure 15.2) or in a separate table included on the figure (Figure 15.3). For more complex shear strength conditions, a separate table with references between the table and drawing of the slope may be needed. If a separate table is used, the material properties and the zones to which they apply should be identified clearly. It is usually appropriate in this

case to number the materials as well as to use descriptive terms (e.g., 1, silty sand; 2, clay foundation; 3, slope protection; 4, filters) (Figure 15.4 and Table 15.2).

If anisotropic shear strengths or nonlinear Mohr failure envelopes are used in the stability calculations, additional figures should be prepared to show the shear strength. For anisotropic shear strengths the variation in shear strength with the orientation of the failure plane (slip surface) should be shown (Figure 15.5). For nonlinear Mohr failure envelopes the envelope should be shown on a Mohr–Coulomb diagram (Figure 15.6).

For reinforced slopes the reinforcement pattern assumed for the analyses should be shown on an appropriate figure (Figure 15.7). The reinforcement may either be shown on the main figure used to summarize the analysis, or if the reinforcement is relatively com-

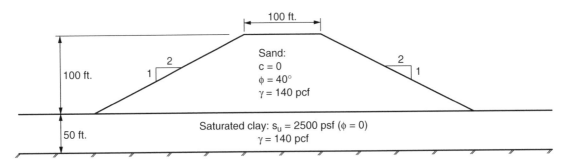

Figure 15.2   Slope with soil properties shown on the cross section.

Mat'l	Description	Unit Wt. (pcf)	c' (psf)	ϕ' (degs)
1	Foundation overburden	134	0	35
2	Core (clay)	130	200	29
3	Transition	135	0	32
4	Shell (gravel)	140	0	37

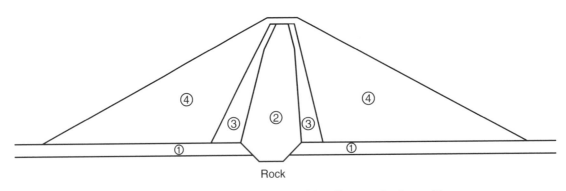

Figure 15.3   Presentation of slope with soil properties in a table.

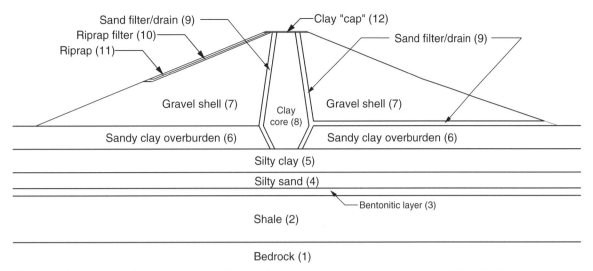

Note: Numbers in parentheses correspond to material numbers shown in accompanying Table 15.2 of soil properties.

Figure 15.4 Slope with material property identification numbers and accompanying Table 15.2 with soil properties.

Table 15.2 Soil Properties for Slope Shown in Figure 15.4

Mat'l	Description	Unit weight (pcf)	Cohesion, c' (psf)	Friction angle, ϕ' (deg)
1	Bedrock	150	10,000	0
2	Shale (foundation)	133	100	21
3	Bentonitic layer (foundation)	127	0	12
4	Silty sand (foundation)	124	0	29
5	Silty clay (foundation)	131	100	26
6	Sandy clay overburden	125	0	32
7	Gravel shell	142	0	38
8	Clay core	135	250	24
9	Sand filter/drain	125	0	35
10	Riprap filter	125	0	33
11	Riprap	130	0	36
12	Clay cap on crest	125	1000	0

plex, a separate figure may be used. The figure should clearly show the spacing of the reinforcement and its length. Appropriate information should also be given for the force in the reinforcement. This may consist of information ranging from a simple note on the drawing to indicate that all reinforcement was assumed to carry a certain, constant force, to a table showing the specific forces in each layer of reinforcement. If the force varies along the length of the reinforcement, appropriate information on the variation should be shown either

graphically (Figure 15.8) or in tabular form. Caution should be exercised in using computer software that automatically assigns forces to reinforcement based on other information (e.g., manufacturers names or product numbers); the values of the forces that are assigned and eventually used to compute stability should be given clearly.

A summary of results of stability calculations for more than one stability condition on a single figure is not desirable. Similarly, summary on one figure of re-

α - deg	s_u - psf
-90	3500
-75	3400
-60	3000
-45	2400
-30	2000
-15	1850
0	1800
15	1900
30	2100
45	2400
60	2800
75	3300
90	3500

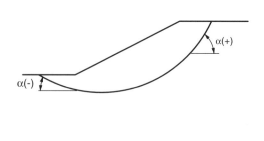

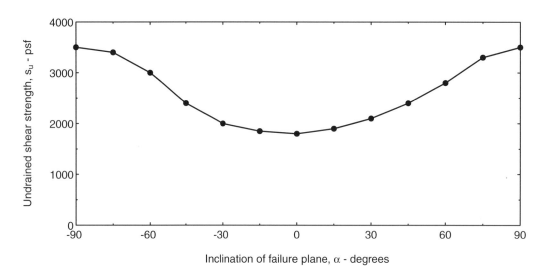

Figure 15.5 Presentation of the variation of undrained shear strength with failure plane orientation for a soil that is anisotropic.

sults of parametric studies in which more than one variable was changed (see the next section) should be avoided.

When analyses are performed using slope stability charts, the summary figure should include the critical slip surface if it can be determined from the chart. The summary figure or accompanying text should designate what charts were used and other relevant information, including appropriate summary calculations. Several examples of summary calculations with slope stability charts are presented in the Appendix.

PARAMETRIC STUDIES

Parametric studies are useful to examine the effect of various assumptions about important quantities, especially when there is significant uncertainty about the values. Parametric studies may be performed by varying the shear strengths, pore water pressures, surface or seismic loads, and slope and subsurface geometry and computing the factor of safety for each set of assumed value. When parametric studies are performed, only one quantity at a time should be varied. For ex-

σ' - psf	τ - psf
0	500
5000	2000
10000	3000
15000	3600
20000	4000

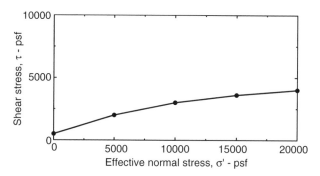

Figure 15.6 Presentation of nonlinear Mohr failure envelope.

ample, the shear strength of a particular stratum might be varied, or different reservoir levels and seepage patterns might be assumed.

When several quantities are varied in parametric studies, separate figures should be presented for each quantity that is varied. For example, a figure might be prepared to show how the factor of safety varies with the shear strength of a particular stratum (Figure 15.9). The figure might show the cross section and, on the same figure, a diagram or table showing how the factor of safety varies with the shear strength. If the critical slip surfaces vary significantly as the value of a quantity is varied, selected individual critical slip surfaces or surfaces representing the extremes might be illustrated. However, there is no need to show critical slip surfaces for each value of each quantity that is varied if there is little effect on the location of the critical slip surface.

DETAILED INPUT DATA

If it is known that someone else will perform additional analyses, and a significant amount of input data (coordinate or pore water pressure data) will be needed, it is helpful to provide such data in tabular

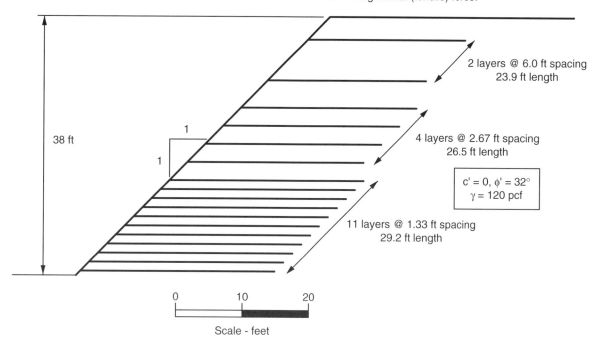

Figure 15.7 Presentation of reinforcement layout.

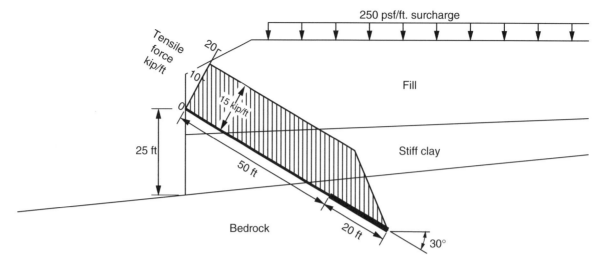

Figure 15.8 Presentation of reinforcement (tie-back anchor) with longitudinal force distribution.

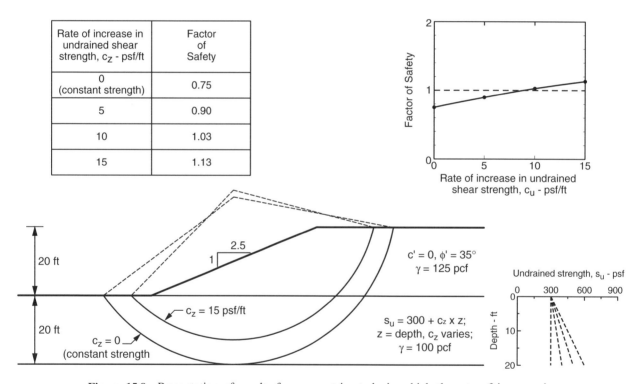

Figure 15.9 Presentation of results for parametric study in which the rate of increase in undrained shear strength (c_z) for the foundation was varied.

form to reduce the amount of effort that will be required later. If such data are included in a presentation they should be neatly organized in appropriate appendixes and clearly labeled and described. Electronic versions of the data on removable storage media may also be provided.

TABLE OF CONTENTS

A sample table of contents for presentation of a stability evaluation is shown in Table 15.3. Details will vary from slope to slope, but most of the items shown should be contained in any presentation of results.

**Table 15.3   Sample Table of Contents for
Presentation of Results of Stability Analyses**

Introduction
Description of site
   Location
   Geology
Soil properties
   Laboratory testing program (or basis for selection)
   Unit weights
   Shear strengths
      Undrained shear strengths
      Drained shear strengths
Groundwater, seepage, and pore water pressure
   conditions
Special features
   Slope reinforcing
   Nonsoil materials and interfaces
   Construction procedures
Stability evaluation procedures
   Computer software
   Slope stability charts
   Empirical correlations and experience with slope in
      area
Summary of results
   End-of-construction stability
   Long-term stability
   Stability for rapid drawdown
   Seismic (earthquake stability)
Discussion and recommendation

CHAPTER 16

Slope Stabilization and Repair

The causes and the nature of a slope failure should be understood before embarking on corrective action. Does the failure involve only the soil above the toe of the slope, or does it extend into the foundation? Was the failure caused by an excessively thick fill on a weak foundation; by an excessively steep slope; by a rise in the groundwater level: by blockage of seepage paths; by erosion at the toe; by loss of soil strength over time due to swelling, creep, and weathering?

When investigating what caused a slope failure, it is well to remember that there may be more than a single cause, as noted by Sowers (1979): "In most cases, several 'causes' exist simultaneously; therefore, attempting to decide which one finally produced failure is not only difficult but also technically incorrect. Often the final factor is nothing more than a trigger that sets a body of earth in motion that was already on the verge of failure. Calling the final factor *the cause* is like calling the match that lit the fuse that detonated the dynamite that destroyed the building *the* cause of the disaster."

Thorough geological study and detailed exploration are the first steps to investigate slope failures. Topographic surveys and measurements on surface markers help to define the area affected and the magnitudes of vertical and horizontal movements. The location of the shear zone can often be determined using test borings, accessible borings, trenches, or slope indicators. Piezometers and observation wells can be used to determine groundwater levels within the slope.

USE OF BACK-ANALYSIS

As discussed in Chapter 12, back-analysis can be used to determine what shear strength would correspond to a factor of safety equal to 1.0 for the conditions at the time of failure. The shear strength determined through back-analysis provides a highly reliable basis for evaluating the factor of safety of the slope after stabilization. Back-analysis not only provides shear strengths that are consistent with the slope failure but results in a complete analytical model (soil stratigraphy, soil unit weights, seepage conditions, etc.) that is consistent with the failure. Such an analytical model, based on the experience gained through the failure, is more reliable than an analytical model based on the results of laboratory tests and idealized estimates of groundwater conditions. When soil strengths and other conditions have been assessed through back-analysis, it is justified to use lower-than-conventional factors of safety for the stabilized slope.

FACTORS GOVERNING SELECTION OF METHOD OF STABILIZATION

Many methods have been used to stabilize slopes, each of them found to be appropriate for a particular set of conditions. In choosing among the methods that are technically feasible, the following factors need to be considered:

1. What is the purpose of stabilizing the slope? Is it only to prevent further large movements, or is it to restore the capacity of the moving ground to provide firm support for structures or pavements? It is more difficult to restore the load-carrying capacity of the ground than merely to stop movements, particularly when the ground has already been disrupted by large movements.
2. How much time is available? Is it essential that the repair be accomplished quickly: for example, to open a blocked highway, railroad, or canal, or is time a less critical element? If time is of the

essence, expeditious methods that can be undertaken without delay are the only ones appropriate. If time is not so critical, it may be possible to fine-tune the fix through thorough study and to devise a less expensive solution for the problem. If it is possible to wait until the dry season before undertaking permanent repair, it may be feasible to use methods, such as excavation of the sliding mass and reconstruction of the slope, that make the slope temporarily steeper.

3. How accessible is the site, and what types of construction equipment can be mobilized there? If the site is reachable only by small roads, or by water, or if steep terrain rules out the use of heavy equipment, considerations of access may limit the methods of stabilization that can be used.

4. What would be the cost of the repair? If the costs exceed the benefits, can less expensive methods be used? Unless political factors dictate otherwise, it is illogical to stabilize a slope when the costs exceed the benefits.

DRAINAGE

Drainage is by far the most frequently used means of stabilizing slopes. Slope failures are very often precipitated by a rise in the groundwater level and increased pore pressures. Therefore, lowering groundwater levels and reducing pore pressures is a logical means of improving stability. In addition, improving drainage is often less expensive than other methods of stabilization, and a large volume of ground can frequently be stabilized at relatively low cost. As a result, drainage is an often-used method, either alone or in conjunction with other methods.

Drainage improves slope stability in two important ways: (1) It reduces pore pressures within the soil, thereby increasing effective stress and shear strength; and (2) it reduces the driving forces of water pressures in cracks, thereby reducing the shear stress required for equilibrium. Once a system of drainage has been established, it must be maintained to keep it functional. Erosion may disrupt surface drains and ditches, and underground drains may become clogged by siltation or bacterial growth. Siltation can be minimized by constructing drains of materials that satisfy filter criteria, and bacterial clogging can be removed by flushing with chemical agents, such as bleach.

Surface Drainage

Preventing water from ponding on the ground surface, and directing surface flow away from the slide area,

will help to reduce groundwater levels and pore pressures within the slide mass. Means to improve surface drainage include:

1. Establishing lined or paved ditches and swales to convey water away from the site
2. Grading to eliminate low spots where water can pond
3. Minimizing infiltration—in the short term by covering the ground with plastic, in the long term through the use of vegetation or paving

Covering the ground with plastic has some drawbacks. Once an area is covered, it is no longer visible, making observation of the condition of the slope and movement of the ground impossible. Undulations in the surface of the plastic tend to collect water, and because individual sheets of plastic are not sealed to each other, water can reach the ground at points of overlap between the sheets. Vegetation increases resistance to erosion by surface runoff and stabilizes the top foot or two of soil at the surface of the slope. In the long term, evapotranspiration helps to lower the groundwater level.

Paving the surface of a slope promotes runoff and impedes infiltration, but it also impedes evaporation and may actually cause water to collect beneath the paved surface. Saleh and Wright (1997), who studied highly plastic clay embankments in Texas, found that paving the slopes helped to reduce the frequency of slides by minimizing the seasonal wetting and drying that can result in gradual degradation in the shear strengths of these clays.

Horizontal Drains

Horizontal drains, sometime called *Hydrauger drains,* after the type of drill first used to install them, are perforated pipes inserted in drilled holes in a slope to provide underground drainage. As shown in Figure 16.1, they usually slope upward into the slope, to permit groundwater to drain by gravity. They are usually 100 to 300 ft long, although longer drains have been

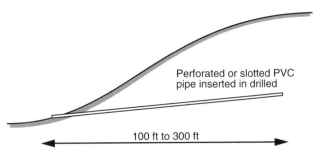

Perforated or slotted PVC pipe inserted in drilled

100 ft to 300 ft

Figure 16.1   Horizontal drains.

used. The drain pipes are commonly perforated or slotted PVC pipe, although steel pipe was used for early applications. The drains are installed by drilling into the slope using a hollow-stem auger, as shown in Figure 16.2, inserting the drain pipe, and withdrawing the auger, leaving the drain in place. The hole is allowed to collapse around the drain pipe. There is no filter between the pipe and the soil.

Horizontal drains are usually installed from points of convenient access for the drill rig, often fanning out as shown in Figure 16.3 to achieve broad coverage of the area. It is commonly found that some drains are very productive and others are nonproductive, but it is very difficult to predict in advance which drains will produce significant flow. Flows usually decline with time after installation and then fluctuate seasonally through wet and dry periods. Rahardjo et al. (2003) found that horizontal drains are most effective when placed low in the slope, provided that the slope does not contain distinct layers of high permeability above the drains.

Drain Wells and Stone Columns

Where soil strata of varying permeability are oriented horizontally, horizontal drains do not provide the most effective means of intercepting seepage. Vertical drains, which cross the layers, are more effective. An example is shown in Figure 16.4. Two-foot diameter holes were drilled in a line along the crest of the slope and were filled with drain rock that satisfied filter criteria for the intercepted soils. The wells were tapped by drilling from the base of the slope to provide drainage by gravity. Vertical wells can be drained using deep pumps, but the requirement for continual power and pump maintenance makes this a less desirable alternative.

Drain wells can be designed using gravity well flow theory (U.S. Army Corps of Engineers, 1986). Between the wells, the phreatic surface rises above the level at the wells, as shown in Figure 16.5. The effective average is about two-thirds of the way up from the head in the drain well to the maximum head between wells. The maximum head between wells decreases as the spacing between wells decreases.

Stone columns provide drainage in much the same way as drain wells, provided that they have a low-level outlet for the water they collect. Because the material in stone columns is compacted as it is placed in the drilled holes, they have the further beneficial effect of increasing the strength of the surrounding soil by densification and increase in lateral stress.

Wellpoints and Deep Wells

Wellpoints are small-diameter vacuum wells that are driven or jetted into place. Vacuum is applied to the tops of the wellpoints through a header—a horizontal pipe that applies vacuum to suck water up the wellpoints. They work best in clean sand and less well in fine-grained soils. Because the water is drawn up the riser pipe by vacuum, their maximum effectiveness is

Figure 16.2 Drilling horizontal drains in the right abutment of La Esperanza Dam in Ecuador.

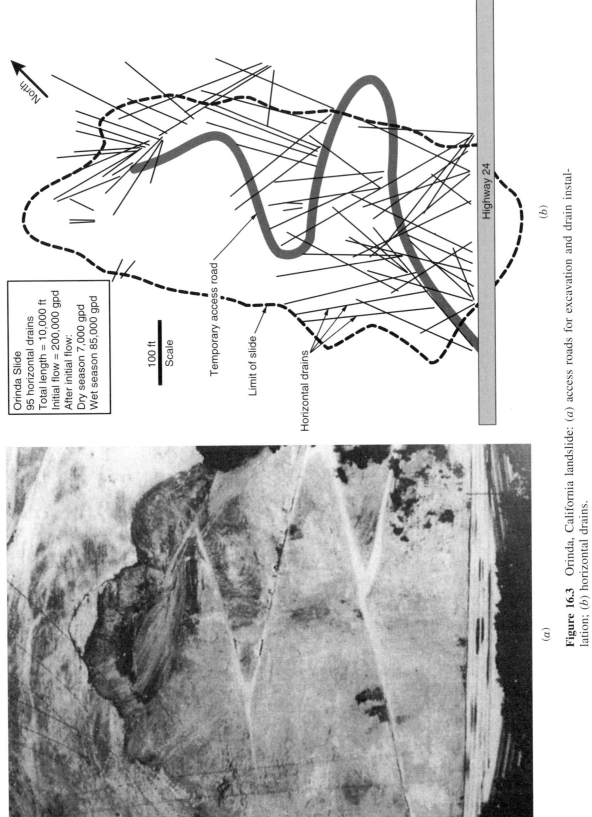

Orinda Slide
95 horizontal drains
Total length = 10,000 ft
Initial flow = 200,000 gpd
After initial flow:
Dry season 7,000 gpd
Wet season 85,000 gpd

North

100 ft
Scale

Temporary access road

Limit of slide

Horizontal drains

Highway 24

(b)

(a)

Figure 16.3 Orinda, California landslide: (a) access roads for excavation and drain installation; (b) horizontal drains.

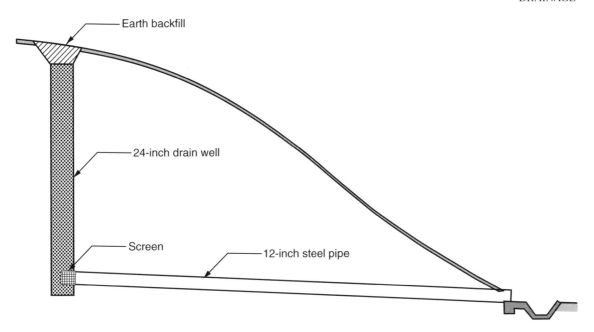

Figure 16.4 Drain wells used to stabilize four landslides near Seattle. (After Palmer et al., 1950.)

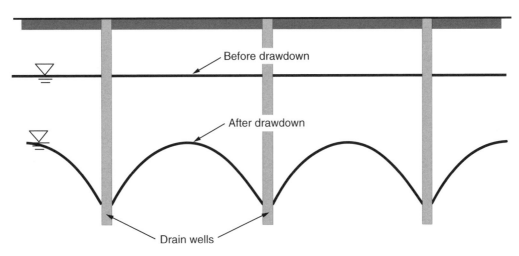

Figure 16.5 Water level between drain wells.

limited to 20 to 25 ft (Mansur and Kaufman, 1962; Sowers, 1979).

Multistage systems can be used to drain deeper excavations. Figure 16.6a shows a 200-ft deep excavation in Australia (Anon., 1981) that was dewatered using eight wellpoint stages and 10 deep wells. Figure 16.6b shows the slope failures that occurred in the sides of the excavation when pumping from the wellpoints was stopped.

Deep wells use submerged pumps to push water to the top of the well and are not limited to a lift of 20

to 25 ft, like suction wells. Each well has its own pump and operates independently. The wells are usually 12 to 24 in. in diameter and have filters surrounding a perforated casing. Like wellpoints, they must be operated continuously to remain effective.

Trench Drains

Trench drains are excavated trenches filled with drain rock that satisfies filter criteria for the surrounding soil, as shown in Figure 16.7. They are sloped to drain by

(a)

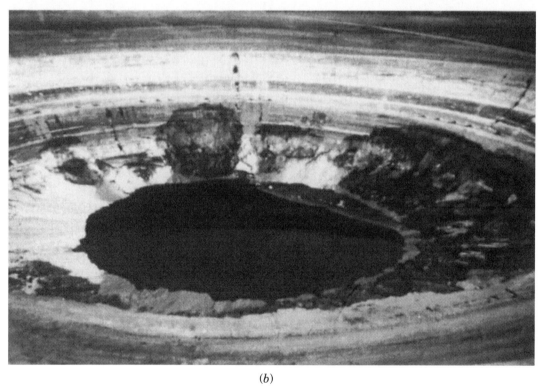

(b)

Figure 16.6 (*a*) Aerial view of Bowman's trial pit during excavation and dewatering; (*b*) view of Bowman's trial pit after dewatering was stopped.

Figure 16.7 Trench drain at Lawrence Berkeley Laboratory.

gravity and may contain a pipe to increase flow capacity. Where pipes are used, manholes are usually provided at intervals for inspection and maintenance. The maximum depth that a trench drain can extend below the ground surface is governed by the requirement that the sides of the trench must remain stable without support until they are backfilled with drain rock.

Drainage Galleries

Where drainage is needed deep within a hillside, a drainage gallery (tunnel) can be used. As shown in Figure 16.8, drains can be drilled outward from the tunnel, extending the drainage through the slope. This technique was used to stabilize the hillside below the Getty Museum in Los Angeles, where improved stability was needed, but environmental considerations made it impossible to flatten the slopes or to construct access roads for the purpose of installing horizontal drains. A deep drainage gallery, with drilled drain holes fanning out from it, was used to stabilize a landslide at the Clyde Power Project in New Zealand (Gil-

lon et al., 1992). At La Esperanza Dam in Ecuador, drainage galleries were tunneled into the abutments, and vertical drains were drilled upward through the roof to drain a permeable layer of brecciated shale (Duncan et al., 1994a).

Finger or Counterfort Drains

Trench drains excavated perpendicular to a slope, as shown in Figure 16.9, are called *finger drains*. Excavating trenches perpendicular to the slope, as shown in Figure 16.9, does not affect the stability of the slope as much as would excavation of a trench drain excavated parallel to the slope. The trench drain that connects the finger drains can be excavated and backfilled in short sections to avoid compromising stability of the slope.

EXCAVATIONS AND BUTTRESS FILLS

A slope can be made more stable by excavation to reduce its height or make it less steep, as illustrated in

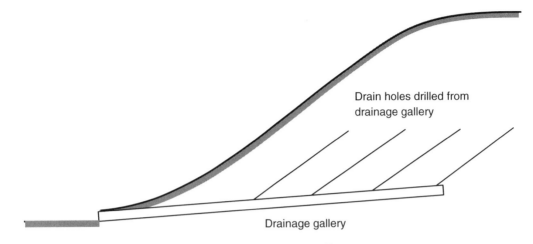

Figure 16.8 Drainage gallery.

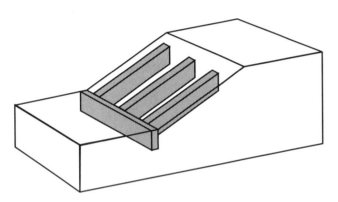

Figure 16.9 Finger drains.

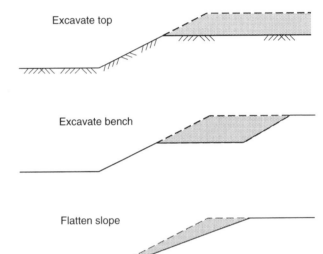

Figure 16.10 Slope repair by excavation.

Figure 16.10. Flattening a slope or reducing its height as shown in Figure 16.10 reduces the shear stresses along potential sliding surfaces and increases the factor of safety. As shown in Figure 16.10, any type of excavation results in a reduction of the useful area at the crest of the slope. Improving stability by excavation requires (1) that an area at the top of the slope can be sacrificed to improve stability, (2) that the site is accessible to construction equipment, and (3) that an area is available for disposal of the excavated material.

Buttress fills are of two types. A buttress of high-strength well-compacted material (see Figure 16.11) provides strength and weight, both of which improve stability. A berm of uncompacted material at the bottom of a slope, sometimes called a *gravity berm,* provides weight and reduces the shear stresses in the slope, even if it consists of weak and compressible soil. The effectiveness of either type of berm is improved if it is placed on a layer of free-draining material that allows drainage of water from the soil beneath.

An example involving both excavation and buttressing is shown in Figure 16.12. Balancing the volume of

cut and fill makes it unnecessary to dispose of material off-site or to import soil for buttress construction. Even soil that has been involved in sliding can be improved and made suitable for berm construction by compaction to high density near optimum water content.

RETAINING STRUCTURES

Retaining structures can be used to improve slope stability by applying stabilizing forces to slopes, thereby reducing the shear stresses on potential slip surfaces.

Prestressed Anchors and Anchored Walls

Prestressed anchors and anchored walls have the advantage that they do not require slope movement be-

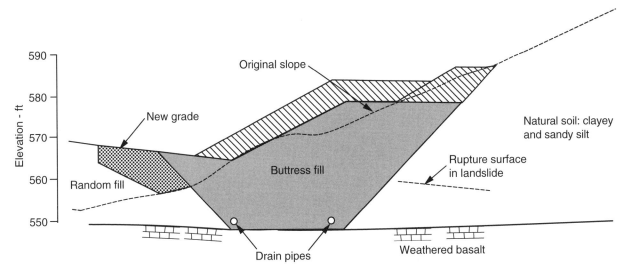

Figure 16.11 Structural buttress for slope stabilization in Portland, Oregon. (After Squier and Versteeg, 1971.)

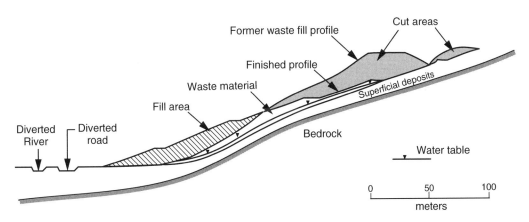

Figure 16.12 Slope stabilization by cut and fill. (After Jones, 1991.)

fore they impose restraining forces. Although anchors can be used without a vertical wall, they do require bearing pads to distribute their loads to the surface of the slope. Figure 16.13 shows an anchored wall constructed to stabilize the Price's Fork landslide, near Blacksburg, Virginia. Soldier piles were driven through the fill at the head of the slide, just behind the slide scarp, and were anchored into rock. After the soil in front of the wall was excavated and wood lagging was fitted between the flanges of the soldier piles, a reinforced concrete footing was constructed in front of the soldier piles, with steel reinforcing bars grouted into rock to restrain the bottoms of the piles, as shown in Figure 16.14. A concrete panel wall was hung in front of the soldier piles to improve the appearance and protect the wood lagging from vandalism.

Figure 16.15 shows anchors used to stabilize a landslide above Tablachaca Dam on the Rio Mantaro River

in Peru (Millet et al., 1992). The slide is on the mountainside above a power plant that provides 40% of the electrical power for Peru. Back-analysis of the slide was performed to assess the shearing resistance of the rock and to provide a means for computing the increase in the factor of safety that could be achieved through the use of anchors. Because the maximum increase in the factor of safety that could be achieved with anchors was less than desired, a drainage tunnel and a berm at the toe of the slope were used to improve stability further.

The improvement in stability afforded by anchors and anchored walls can be evaluated using conventional limit equilibrium slope stability analyses, with the force applied by the anchors included in the analysis as a force of known magnitude and direction, acting at a known location on the slope. The anchor force should be a *working load* (i.e., the ultimate anchor ca-

Figure 16.13 Price's fork wall.

pacity divided by a suitable factor of safety). The factor of safety for the anchor forces should reflect the uncertainties involved in evaluating the anchor capacity and the consequences of anchor failure.

Gravity Walls, MSE Walls, and Soil Nailed Walls

Conventional gravity retaining walls, mechanically stabilized earth (MSE) walls, and soil nailed walls, which are not prestressed, must move before they can develop resistance to stabilize a landslide. Such walls can be designed using the following three steps:

1. Using conventional limit equilibrium slope stability analyses, determine the force required at the location of the wall to stabilize the slope (i.e., to raise the factor of safety of the slope to the desired value). These analyses can be performed using any method in which an external force of specified location, direction, and magnitude can be included. The analyses are performed using repeated trials. The magnitude of the force is varied until the desired factor of safety is achieved. Each of the analyses with a new trial force should search for the location of the critical slip surface. This critical slip surface is not the same as the critical surface with no stabilizing force but is often close to that surface.

   The magnitude of the stabilizing force can be determined with acceptable accuracy using a force equilibrium analysis, with the directions of the side forces between slices assumed to be the average of the slope inclination and the failure surface inclination, or using a method that satisfies all conditions of equilibrium. The position of the stabilizing force can reasonably be assumed to be about $0.4H$ above the bottom of the wall, where H is the height measured from the bottom of the wall to the surface of the slope.

2. Using conventional retaining wall design procedures, determine the external dimensions of the retaining wall, MSE wall, or soil nailed wall required for global wall stability, with the force determined in step 1 applied to the wall. The considerations for external stability of the wall include sliding, overturning, bearing capacity, position of the resultant force on the base, and deep-seated sliding (failure through the foundation beneath the wall).

3. Using conventional design procedures, evaluate the requirements for internal strength. For gravity walls, these include the shear and moment capacity of the footing and stem. For MSE walls, these include the length of reinforcement, strength of reinforcement, and spacing of reinforcement. For soil nailed walls, these include nail capacity, nail length, and nail spacing.

REINFORCING PILES AND DRILLED SHAFTS

Piles or drilled shafts that extend through a sliding mass, into more stable soil beneath, can be used to

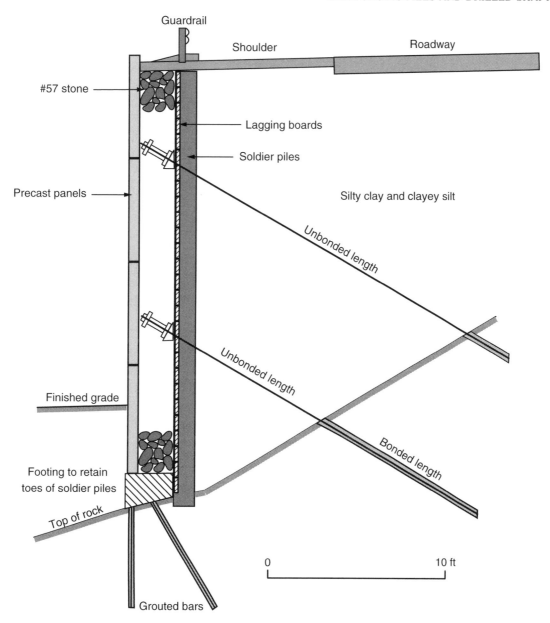

Figure 16.14 Cross section through Price's fork wall.

improve slope stability. Construction of drilled shafts has a smaller adverse effect on slope stability than does driving piles, and drilled shafts are often preferred for this reason. The piles or drilled shafts are installed in one or more lines parallel to the crest of the slope, to provide resistance to down-slope movement. The shafts in each row should be spaced closely enough so that the soil cannot flow between them, rendering them ineffective in stopping slope movements. The usual center-to-center spacing is two to four diameters. Poulos (1995) found that the optimum location of stabilizing shafts is near the center of the potential sliding mass, as opposed to the head or the toe of the slope.

Like retaining walls that are not restrained by prestressed anchors, piles and drilled shafts require movement of the sliding mass before they develop stabilizing forces. The magnitude of the stabilizing force increases as the movement increases, up to the point where the structural capacity of the shafts is reached, or the maximum passive earth pressure is mobilized against the uphill side of the part of the shafts that extend above the sliding surface. The structural capacity of the drilled shafts is controlled by moment rather than shear, and Poulos (1995) indicated that a smaller number of large-diameter drilled shafts results in more effective stabilization than a larger number of

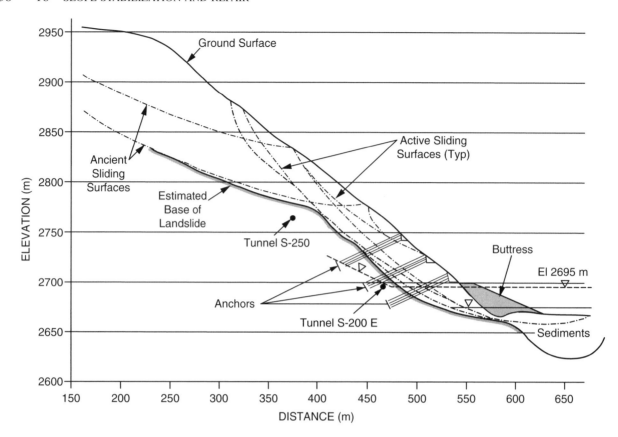

Figure 16.15 Tablachaca slide repair. (Millet et al., 1992.)

small-diameter shafts. The shafts should extend deep enough so that potential slip surfaces passing beneath the shafts have adequate factors of safety.

As in the case of retaining walls, the force required to stabilize the slope can be calculated using any method of slope stability analysis in which an external force of specified magnitude can be included. The magnitude of the required force is determined using repeated trials, varying the magnitude of the force until the desired factor of safety is achieved. Each of the analyses with a new trial force should search for the location of the critical slip surface. This critical slip surface is not the same as the critical surface with no stabilizing force but is often close to that surface.

Methods of evaluating the shear forces and moments in the piles have been proposed by Reese et al. (1992), Poulos (1995), Shmuelyan (1996), Hassiotis et al. (1997), Yamagami et al. (2000), and Reese and Van Impe (2001). Poulos (1995) used boundary element analyses to compute the forces that drilled shafts would apply to the soil above the sliding plane and also the interaction between the shafts and the stable ground beneath the sliding plane, based on assumed patterns and magnitudes of soil movement.

Reese et al. (1992) and Reese and Van Impe (2001) have described methods for evaluating the shear forces and bending moments in piles used to stabilize slopes. These methods use p–y concepts to estimate the stabilizing forces that the piles can exert on the slope, and the shear forces and bending moments in the piles. These procedures can be implemented through the following steps:

1. Estimate the relative movements between the portion of the piles that will extend above the slip surface and the surrounding ground. These estimated movements should be based on the amount of slope movement after pile installation that is considered tolerable, and the estimated amount that the piles will deflect when loaded.
2. Select a trial diameter and center-to-center spacing between piles.
3. Using p–y curves and estimated relative movements between the soil and the section of the pile projecting above the slip surface, determine the value of p at each point along the projecting portion of the pile. An example of such a distribution is shown in Figure 16.16b. The quantity p is the

Design Principles

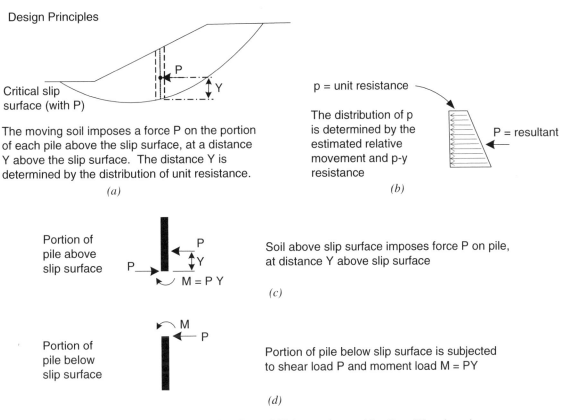

The moving soil imposes a force P on the portion of each pile above the slip surface, at a distance Y above the slip surface. The distance Y is determined by the distribution of unit resistance.

(a)

p = unit resistance

The distribution of p is determined by the estimated relative movement and p-y resistance

P = resultant

(b)

Portion of pile above slip surface

Soil above slip surface imposes force P on pile, at distance Y above slip surface

(c)

Portion of pile below slip surface

Portion of pile below slip surface is subjected to shear load P and moment load M = PY

(d)

Figure 16.16 *(a)* Design principles for stabilizing a slope with piles; *(b)* unit resistance *p* and resultant P_{pile}; *(c)* portion of pile above slip surface; *(d)* portion of pile below slip surface.

soil reaction used in laterally loaded pile analyses and has units of force per unit length, the length being measured along the length of the pile.

4. Calculate the area under the *p*-diagram, denoted here as *P*.

5. Calculate the corresponding stabilizing force per unit length of slope:

$$P_{slope} \text{ (force per unit length)} = \frac{P \text{ (force)}}{S \text{(length)}}$$

(16.1)

where P_{slope} is the stabilizing force per unit length of slope, *P* the force on one pile, and *S* the center-to-center pile spacing, measured parallel to the crest of the slope. If P_{slope} computed from Eq. (16.1) is less than the force required to achieve the desired factor of safety for the slope, increase the pile diameter or reduce the pile spacing, and repeat steps 3 through 5. If P_{slope} is larger than the force required to achieve the desired factor of safety, reduce the pile diameter or increase the pile spacing, and repeat steps 3 through 5. When

pile spacing and diameter have been found that will provide the desired stabilizing force, proceed to step 6.

6. Determine the shear force and bending moment on the part of the pile embedded below the slip surface. The shear force is equal to *P*. The moment is equal to (*P*)(*y*), where *y* is the distance from the slip surface to the *P*, as shown in Figure 16.16*b*.

7. Compute the distributions of shear and bending moment in the part of the pile below the slip surface when subjected to a shear force = *P* and a moment = *Py*. The maximum moment in the lower part of the pile governs the required moment capacity of the pile. The pile should be capable of carrying this moment with a suitable factor of safety against brittle failure or formation of a plastic hinge.

8. Select a pile type, or drilled shaft reinforcing, that is capable of carrying the shear forces and bending moments calculated in step 7. If the moments are too large for the pile, repeat from step 2 with a larger pile diameter and larger spacing between piles.

The key step in this process is the first one: estimating the relative movement between the soil and the part of the pile that projects above the slip surface. A safe assumption, with respect to the loads on the piles, is to assume that the relative movement will be large enough to mobilize the maximum value of p (usually denoted as p_{ult}) along the full length of the part of the pile that projects above the slip surface. However, this extreme loading is unlikely if the pile extends for a large distance above the slip surface, because the flexural deformations of long piles will cause them to deform with the surrounding ground.

The stabilizing force P_{slope} is entered in the slope stability analysis as a known force and is not further reduced during the slope stability analysis. This corresponds to the procedure called method A in Chapter 8.

INJECTION METHODS

Injection methods are attractive because they can be implemented at relatively low cost. Their drawback is that it is difficult to quantify the beneficial effects. In addition, when fluids are injected, the short-term effect may be to make the slope less stable. The beneficial effects may be achieved only later, when the injected material has hardened or has reacted with the soil to alter its properties.

Lime Piles and Lime Slurry Piles

Lime piles are drilled holes filled with lime. Lime slurry piles are drilled holes filled with a slurry of lime and water. Rogers and Glendinning (1993, 1994, and 1997) reviewed the use of lime piles and lime slurry piles to stabilize slopes, and the mechanisms through which they improve soil strength and stability. Handy and Williams (1967) described the use of quicklime placed in drilled holes to stabilize a landslide in Des Moines, Iowa. Six-inch-diameter holes were drilled through a compacted silty clay fill, down to the surface of the underlying shale, where the fill was sliding on the top of the shale. About 50 lb of quicklime was placed in each hole, filling the bottom 3 ft. Water was then added to hydrate the lime, and the holes were backfilled to the surface with soil. Holes were drilled 5 ft apart, stabilizing an area 200 ft by 125 ft using abut 20 tons of quicklime. Physical and chemical tests on the treated soil showed that the lime was reacting with and strengthening the silty clay fill. Movement of the slide essentially stopped within three months after treatment, while movements continued in adjacent untreated areas.

Cement Grout

Stabilizing landslides by injecting cement grout has been used extensively on both American (Smith and Peck, 1955) and British railroads (Purbrick and Ayres, 1956; Ayres, 1959, 1961, 1985). Typical practice involves driving grout points about 5 ft apart in rows parallel to the track, the rows being about 15 ft apart. The tips of the grout points are driven about 3 ft below the estimated depth of the rupture surface, and about 50 ft^3 of grout is injected through each point. Quite high grouting pressures are used for the shallow depths involved: For grouting only 15 ft beneath the surface,

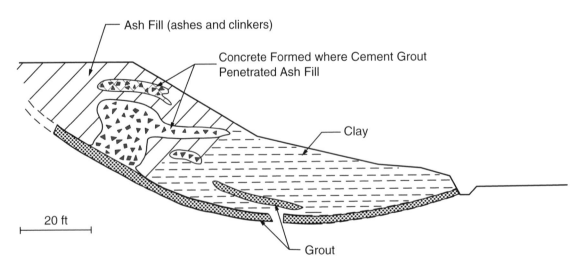

Figure 16.17 Stabilization of landslide at Fenny Compton, England, by injection of neat cement grout. (After Purbrick and Ayres, 1956.)

a grouting pressure of 75 psi might be used for injection of the first 10 ft³, subsequently dropping to 20 psi.

One of the most intriguing aspects of the method is that it is used to stabilize landslides in clay. Cement cannot penetrate the voids of clays because the particles are too large, and the grout pressures cannot cause compaction if the clay is saturated, as it often is. Nevertheless, the method is effective. Trenches excavated into the treated area show how the method works. Figure 16.17 shows a cross section revealed in one such trench. The grout did not penetrate the voids of the clay or the fissures in the clay but did penetrate the voids of the coarser fill called *ash,* which has gravel-sized particles. Within the clay, the grout penetrated along the rupture surface, lifting the mass above, and a solid mass of neat cement concrete was formed along the slip surface when it hardened.

VEGETATION

Vegetation on slopes provides protection against erosion and shallow sliding (Gray and Leiser, 1982; Wu et al., 1994). Roots reinforce or bind the soil and provide cohesion that improves stability against shallow sliding. In addition, plant roots are believed to reduce pore pressures within slopes by intercepting rainfall (reducing infiltration) and by evapotranspiration (Wu et al., 1994). Gray and Sotir (1992) found that living woody plant material (brush), embedded in horizontal layers at the surface of slopes, provided some reinforcement immediately, and more as the plants began to grow and put out new roots. Gray and Sotir (1995) suggested that these brush layers also improve stability by intercepting water flowing within the slope and diverting it to the surface, reducing pore pressures in the process. Use of vegetation in combination with mechanical reinforcement such as geogrids is called *biotechnical stabilization* (Gray and Sotir, 1992).

THERMAL TREATMENT

Thermal treatment has not been widely used to stabilize landslides. One of the few examples, described by Hill (1934), is illustrated in Figure 16.18. The lower part of the slip surface passed through a horizontal clay seam slightly above the toe of the slope. Drainage tunnels driven through the clay, parallel to the crest of the slope, were ineffective because the permeability of the clay was so low that no water drained into the tunnels. To dry out the clay, a gas furnace was constructed, and heated air was blown through a network of interconnected tunnels and drill holes, as shown in Figure 16.18. Hill indicated that the capitalized cost of operating the heating system in perpetuity would be less

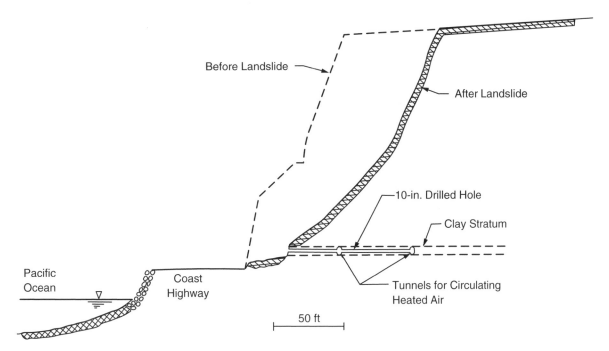

Figure 16.18 Stabilization of landslide near Santa Monica, California, by drying clay stratum. (After Hill, 1934.)

Figure 16.19 Landslide bridge at Lawrence Berkeley Laboratory, California: (*a*) top of bridge; (*b*) excavated area beneath bridge.

than the cost of a restraining structure large enough to stabilize the slide.

Other examples of thermal treatment have been described by Beles and Stanculescu (1958). Landslides in a cut slope, a natural slope, and an embankment were stabilized by drilling down past the slide plane and burning gas in the holes to heat and harden the surrounding soil.

BRIDGING

A landslide on the Lawrence Berkeley Laboratory grounds in California was stabilized using the novel but effective method of building a reinforced concrete bridge on the ground surface near the head of the landslide, and excavating soil from beneath it to unload the slide, as shown in Figure 16.19. The bridge was supported on drilled shafts that were installed before the bridge deck was cast. After the drilled shafts had been constructed and the bridge deck had been cast on the ground surface, about 5000 tons of soil was excavated from beneath the bridge deck, unloading the upper part of the slide. The utilities that ran through the area were hung on supports attached to the bottom of the bridge deck. This ingenious system, conceived and designed by LBL engineer Sherad Talati, halted the movement of the landslide, which had threatened important structures, including the Bevatron building.

REMOVAL AND REPLACEMENT OF THE SLIDING MASS

When a sliding mass has moved a long distance and has become disturbed and softer as the result of the movement, there may be no alternative to removing and replacing the sliding mass if the usefulness of the slide area for supporting structures is to be restored.

Excavating a sliding mass usually makes the slope exposed by excavation even steeper than the slope was before the slide, as shown in Figure 16.20. Therefore, excavation is not undertaken until the stability of the slide has improved (e.g., by drainage). In areas where there are pronounced wet and dry seasons, excavation

can sometimes be accomplished in the dry season. It is important that the excavation be observed carefully, to be sure that it extends below the rupture surface, into undisturbed soil, and that all of the unstable material is removed.

After the sliding mass has been removed, the slope is reconstructed, as illustrated in Figure 16.20. The slope can often be rebuilt using the soil that has been removed, installing drains behind and beneath the compacted fill as it is replaced. Good drainage and well-compacted soil are the keys to improved stability. The process requires an area to store the excavated material temporarily, since it must be removed before reconstruction of the slope can begin. In some cases, where the slide volume is small, the excavated material is wasted, and the slope is reconstructed of free-draining material that requires little or no compaction.

A landslide at the Lawrence Berkeley Laboratory in California was stabilized by removal and replacement, as shown in Figure 16.20. The excavation extended well below the rupture surface, and benches were cut into the stable soils beneath. The material removed was compacted back into place, and drains were constructed behind the fill as it was placed. Horizontal drains, a trench drain, and a drain well were installed as temporary stabilization measures before the sliding mass was excavated. They intercepted considerable

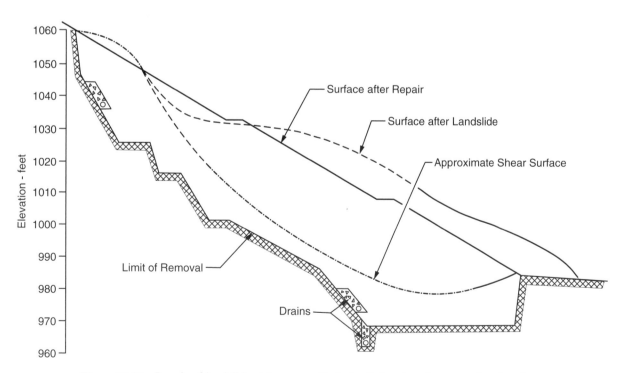

Figure 16.20 Repair of landslide at Lawrence Berkeley Laboratory by removal and replacement of the sliding mass. (After Harding, Miller, Lawson, and Associates, 1970.)

quantities of groundwater at the time they were installed, and continued to flow at a rate of 11,000 gallons per day for some time after the repair was completed (Kimball, 1971).

The cost of this type of repair can be quite large, especially for deep slides. Removal and replacement of a sliding mass 27 ft deep, at a combined cost of $10 per cubic yard for excavation and replacement, would be $10 per square foot, or more than $400,000 per acre. The method is very reliable, however, and can restore full usefulness to the area stabilized.

Recapitulation

- The causes and the nature of a slope failure should be understood before corrective action is undertaken.
- Back-analysis of a slope failure provides a highly reliable basis for designing stabilizing measures and for evaluating the factor of safety after stabilization.
- Drainage is by far the most frequently used means of stabilizing slopes. It can be used alone or in combination with other methods and often provides effective stabilization at relatively low cost.
- Drainage improves slope stability in two ways: (1) it reduces pore pressures within the soil, thereby increasing effective stress and shear strength; and (2) it reduces the driving forces of water pressures in cracks.
- Flattening a slope reduces the shear stresses along potential sliding surfaces, increasing the factor of safety.

- Prestressed anchors and anchored walls do not require slope movement before they impose restraining forces.
- Conventional gravity retaining walls, mechanically stabilized earth (MSE) walls, and soil nailed walls, which are not prestressed, must move before they can develop resistance to stabilize a landslide.
- Piles or drilled shafts that extend through a sliding mass, into more stable soil beneath, can be used to improve slope stability. A combination of limit equilibrium slope stability analyses and $p-y$ analyses can be used to design piles or drilled shafts to achieve the desired increase in factor of safety of the slope.
- Slopes have been stabilized using lime piles, grouting with cement, vegetation, thermal treatment, and construction of a reinforced concrete bridge on the ground surface, followed by excavation of soil from beneath the bridge to unload the head of a landslide.
- When a sliding mass has been disturbed significantly as the result of slope movement, and the slide area must support structures or pavements, it may be necessary to excavate and replace the entire sliding mass. Excavation and replacement, with good compaction and drains beneath the fill, provide a very reliable means of restoring full usefulness to a slide area. The cost of the method is large when the surface of sliding is deep beneath the ground.

APPENDIX

Slope Stability Charts

USE AND APPLICABILITY OF CHARTS FOR ANALYSIS OF SLOPE STABILITY

Slope stability charts provide a means for rapid analysis of slope stability. They can be used for preliminary analyses and for checking detailed analyses. They are especially useful for making comparisons between design alternatives, because they provide answers so quickly. The accuracy of slope stability charts is usually as good as the accuracy with which shear strengths can be evaluated.

In this appendix, chart solutions are presented for four types of slopes:

1. Slopes in soils with $\phi = 0$ and uniform strength throughout the depth of the soil layer
2. Slopes in soils with $\phi > 0$ and $c > 0$ and uniform strength throughout the depth of the soil layer
3. Infinite slopes in soils with $\phi > 0$ and $c = 0$ and soils with $\phi > 0$ and $c > 0$
4. Slopes in soils with $\phi = 0$ and strength increasing linearly with depth

Using approximations in slope geometry and carefully selected soil properties, these chart solutions can be applied to a wide range of nonhomogeneous slopes. This appendix contains the following charts:

- *Figure A-1:* Slope stability charts for $\phi = 0$ soils
- *Figure A-2:* Surcharge adjustment factors for $\phi = 0$ and $\phi > 0$ soils
- *Figure A-3:* Submergence and seepage adjustment factors for $\phi = 0$ and $\phi > 0$ soils
- *Figure A-4:* Tension crack adjustment factors for $\phi = 0$ and $\phi > 0$ soils
- *Figure A-5:* Slope stability charts for $\phi > 0$ soils
- *Figure A-6:* Steady seepage adjustment factor for $\phi > 0$ soils

- *Figure A-7:* Slope stability charts for infinite slopes
- *Figure A-8:* Slope stability charts for $\phi = 0$ soils, with strength increasing with depth

AVERAGING SLOPE INCLINATIONS, UNIT WEIGHTS, AND SHEAR STRENGTHS

For simplicity, charts are developed for simple homogeneous soil conditions. To apply charts to nonhomogeneous conditions, it is necessary to approximate the real conditions with an equivalent homogeneous slope. The most effective method of developing a simple slope profile for chart analysis is to begin with a cross section of the slope drawn to scale. On this cross section, using judgment, draw a geometrically simple slope that approximates the real slope as closely as possible.

To average the shear strengths for chart analysis, it is useful to know the location of the critical slip surface, at least approximately. The charts contained in the following sections provide a means of estimating the position of the critical circle. Average strength values are calculated by drawing the critical circle determined from the charts on the slope. Then the central angle of arc subtended within each layer or zone of soil is measured with a protractor. The central angles are used as weighting factors to calculate weighted average strength parameters, c_{av} and ϕ_{av}:

$$c_{av} = \frac{\sum \delta_i c_i}{\sum \delta_i} \qquad (A\text{-}1)$$

$$\phi_{av} = \frac{\sum \delta_i \phi_i}{\sum \delta_i} \qquad (A\text{-}2)$$

265

where

c_{av} = average cohesion (stress units)

ϕ_{av} = average angle of internal friction (degrees)

δ_i = central angle of arc, measured around the center of the estimated critical circle, within zone i (degrees)

c_i = cohesion in zone i (stress units)

ϕ_i = angle of internal friction in zone i degrees

One condition in which it is preferable not to use these averaging procedures is the case in which an embankment overlies a weak foundation of saturated clay, with $\phi = 0$. Using Eqs. (A-1) and (A-2) to develop average values of c and ϕ in such a case would lead to a small value of ϕ_{av} (perhaps 2 to 5°). With $\phi_{av} > 0$, it would be necessary to use the chart shown in Figure A-5, which is based entirely on circles that pass through the toe of the slope. With weak $\phi = 0$ foundation soils, the critical circle usually goes below the toe into the foundation. In these cases it is better to approximate the embankment as a $\phi = 0$ soil and to use the $\phi = 0$ slope stability charts shown in Figure A-1. The equivalent $\phi = 0$ strength of the embankment soil can be estimated by calculating the average normal stress on the part of the slip surface within the embankment (one-half the average vertical stress is usually a reasonable approximation of the normal stress on this part of the slip surface) and determining the corresponding shear strength at that point on the shear strength envelope for the embankment soil. This value of strength is treated as a value of s_u for the embankment, with $\phi = 0$. The average value of s_u is then calculated for both the embankment and the foundation using the same averaging procedure as described above:

$$(s_u)_{av} = \frac{\sum \delta_i (s_u)_i}{\sum \delta_i} \qquad (A-3)$$

where $(s_u)_{av}$ is the average undrained shear strength (in stress units), δ_i the central angle of arc, measured around the center of the estimated critical circle, within zone i (degrees), and $(s_u)_i$ the s_u in layer i (in stress units). This average value of s_u is then used, with $\phi = 0$, for analysis of the slope.

To average unit weights for use in chart analysis, it is usually sufficient to use layer thickness as a weighting factor, as indicated by the following expression:

$$\gamma_{av} = \frac{\sum \gamma_i h_i}{\sum h_i} \qquad (A-4)$$

where γ_{av} is the average unit weight (force per length cubed), γ_i the unit weight of layer i (force per length cubed), and h_i the thickness of layer i (in length units). Unit weights should be averaged only to the depth of the bottom of the critical circle. If the material below the toe of the slope is a $\phi = 0$ material, the unit weight should be averaged only down to the toe of the slope, since the unit weight of the material below the toe has no effect on stability in this case.

SOILS WITH $\phi = 0$

The slope stability chart for $\phi = 0$ soils developed by Janbu (1968) is shown in Figure A-1. Charts providing adjustment factors for surcharge loading at the top of the slope are shown in Figure A-2. Charts providing adjustment factors for submergence and seepage are shown in Figure A-3. Charts providing adjustment factors to account for tension cracks are shown in Figure A-4.

Steps for the use of $\phi = 0$ charts are:

Step 1. Using judgment, select the range of depths for possible critical circles to be investigated. For uniform soil conditions, the critical circle passes through the toe of the slope if the slope is steeper than about 1 (horizontal) on 1 (vertical). For flatter slopes, the critical circle usually extends below the toe. The chart in Figure A-1 can be used to compute factors of safety for circles extending to any depth, and three or more depths should be analyzed, to be sure that the overall critical circle and overall minimum factor of safety have been found.

Step 2. The following criteria can be used to determine which possibilities should be examined:

a. If there is water outside the slope, a circle passing above the water may be critical.

b. If a soil layer is weaker than the one above it, the critical circle may extend into the lower (weaker) layer. This applies to layers both above and below the toe.

c. If a soil layer is stronger than the one above it, the critical circle may be tangent to the top of the layer.

The following steps are performed for each potential critical circle.

Step 3. Calculate the depth factor, d, using the formula

$$d = \frac{D}{H} \qquad (A-5)$$

where D is the depth from the toe of the slope to the lowest point on the slip circle (L; length) and H

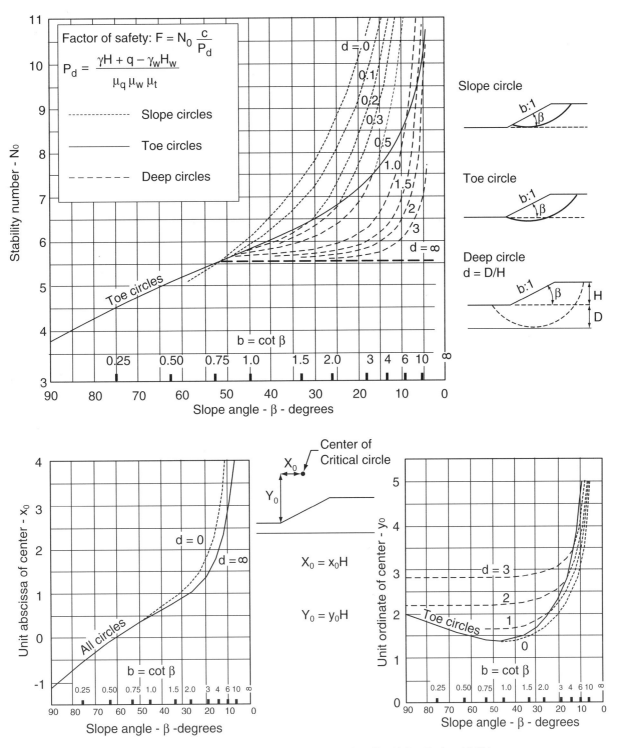

Figure A-1 Slope stability charts for $\phi = 0$ soils. (After Janbu, 1968.)

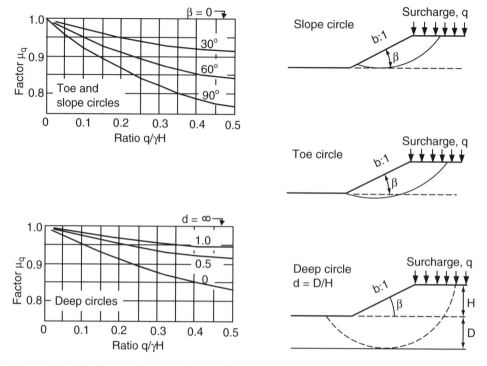

Figure A-2 Surcharge adjustment factors for $\phi = 0$ and $\phi > 0$ soils. (After Janbu, 1968.)

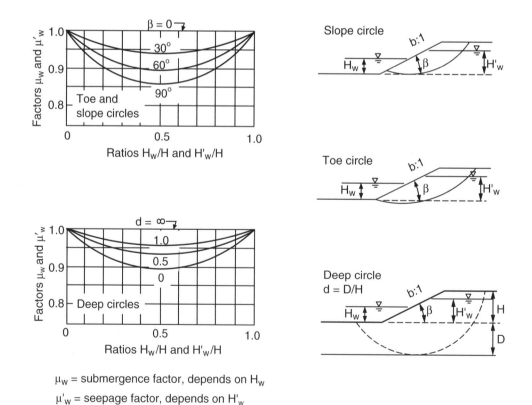

μ_w = submergence factor, depends on H_w

μ'_w = seepage factor, depends on H'_w

Figure A-3 Submergence and seepage adjustment factors for $\phi = 0$ and $\phi > 0$ soils. (After Janbu, 1968.)

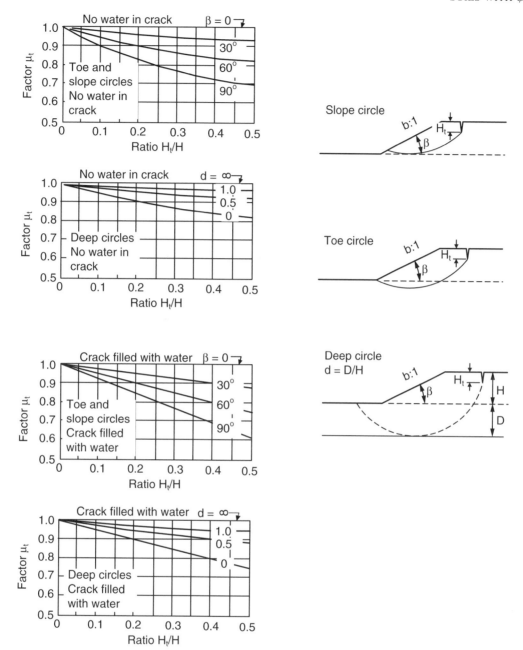

Figure A-4 Tension crack adjustment factors for $\phi = 0$ and $\phi > 0$ soils. (After Janbu, 1968.)

is the slope height above the toe of the slope (L). The value of d is zero if the circle does not pass below the toe of the slope. If the circle being analyzed is entirely above the toe, its point of intersection the slope should be taken as an adjusted toe and all dimensions (e.g., D, H, and H_w) adjusted accordingly in the calculations.

Step 4. Find the center of the critical circle for the trial depth using the charts at the bottom of Figure A-1,

and draw this circle to scale on a cross section of the slope.

Step 5. Determine the average value of the strength, $c = s_u$, for the circle, using Eq. (A-3).

Step 6. Calculate the quantity P_d using Eq. (A-6):

$$P_d = \frac{\gamma H + q - \gamma_w H_w}{\mu_q \mu_w \mu_t} \qquad \text{(A-6)}$$

where

γ = average unit weight of soil (F/L^3)
H = slope height above toe (L)
q = surcharge (F/L^2)
γ_w = unit weight of water (F/L^3)
H_w = height of external water level above toe (L)
μ_q = surcharge adjustment factor (Figure A-2)
μ_w = submergence adjustment factor (Figure A-3)
μ_t = tension crack adjustment factor (Figure A-4)

If there is no surcharge, $\mu_q = 1$; if there is no external water above the toe, $\mu_w = 1$; if there are no tension cracks, $\mu_t = 1$.

Step 7. Using the chart at the top of Figure A-1, determine the value of the stability number, N_o, which depends on the slope angle, β, and the value of d.

Step 8. Calculate the factor of safety, F:

$$F = \frac{N_o c}{P_d} \qquad (A\text{-}7)$$

where N_o is the stability number and c is the average shear strength $= (s_u)_{av}$ (F/L^2).

The example problems in Figures A-9 and A-10 illustrate the use of these methods. Note that both problems involve the same slope, and that the only difference between the two problems is the depth of the circle analyzed.

SOILS WITH $\phi > 0$

The slope stability chart for $\phi > 0$ soils, developed by Janbu (1968), is shown in Figure A-5. Adjustment factors for surcharge are shown in Figure A-2. Adjustment factors for submergence and seepage are shown in Figure A-3. Adjustment factors for tension cracks are shown in Figure A-4. The stability chart in Figure A-5 may be used for analyses in terms of effective stresses. The chart may also be used for total stress analysis for slopes in soils with $\phi > 0$.

Steps for the use of $\phi > 0$ charts are:

Step 1. Estimate the location of the critical circle. For most conditions of slopes in uniform soils with

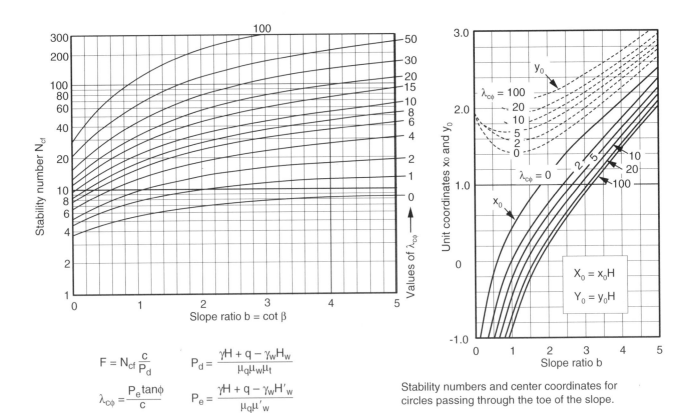

$$F = N_{cf}\frac{c}{P_d} \qquad P_d = \frac{\gamma H + q - \gamma_w H_w}{\mu_q \mu_w \mu_t}$$

$$\lambda_{c\phi} = \frac{P_e \tan\phi}{c} \qquad P_e = \frac{\gamma H + q - \gamma_w H'_w}{\mu_q \mu'_w}$$

Stability numbers and center coordinates for circles passing through the toe of the slope.

Figure A-5 Slope stability charts for $\phi > 0$ soils. (After Janbu, 1968.)

$\phi > 0$, the critical circle passes through the toe of the slope. The stability numbers given in Figure A-5 were developed by analyzing toe circles. When $c = 0$, the critical mechanism is shallow sliding, which can be analyzed as the infinite slope failure mechanism. The stability chart shown in Figure A-7 can be used in this case. If there is water outside the slope, the critical circle may pass above the water.

If conditions are not homogeneous, a circle passing above or below the toe may be more critical than the toe circle. The following criteria can be used to determine which possibilities should be examined:

a. If there is water outside the slope, a circle passing above the water may be critical.
b. If a soil layer is weaker than the one above it, the critical circle may be tangent to the base of the lower (weaker) layer. This applies to layers both above and below the toe.
c. If a soil layer is stronger than the one above it, the critical circle may be tangent to the base of either layer, and both possibilities should be examined. This applies to layers both above and below the toe.

The charts in Figure A-5 can be used for nonuniform conditions provided that the values of c and ϕ used in the calculation represent average values for the circle considered. The following steps are performed for each circle.

Step 2. Calculate P_d:

$$P_d = \frac{\gamma H + q - \gamma_w H_w}{\mu_q \mu_w \mu_t} \qquad (A-8)$$

where

γ = average unit weight of soil (F/L^3)
H = slope height above toe (L)
q = surcharge (F/L^2)
γ_w = unit weight of water (F/L^3)
H_w = height of external water level above toe (L)
μ_q = surcharge reduction factor (Figure A-2)
μ_w = submergence reduction factor (Figure A-3)
μ_t = tension crack reduction factor (Figure A-4)

If there is no surcharge, $\mu_q = 1$; if there is no external water above the toe, $\mu_w = 1$; and if there are no tension cracks, $\mu_t = 1$.

If the circle being studied passes above the toe of the slope, the point where the circle intersects the slope face should be taken as the toe of the slope for the calculation of H and H_w.

Step 3. Calculate P_e:

$$P_e = \frac{\gamma H + q - \gamma_w H'_w}{\mu_q \mu'_w} \qquad (A-9)$$

where H'_w is the height of water within the slope (L) and μ'_w is the seepage correction factor (Figure A-3). The other factors are as defined previously.

H'_w is the average level of the piezometric surface within the slope. For steady seepage conditions this is related to the position of the phreatic surface beneath the crest of the slope as shown in Figure A-6. If the circle being studied passes above the toe of the slope, H'_w is measured relative to the adjusted toe. If there is no seepage, $\mu'_w = 1$, and if there is no surcharge, $\mu_q = 1$. In a total stress analysis, internal pore water pressure is not considered, so $H'_w = 0$ and $\mu'_w = 1$ in the formula for P_e.

Step 4. Calculate the dimensionless parameter, $\lambda_{c\phi}$:

$$\lambda_{c\phi} = \frac{P_e \tan\phi}{c} \qquad (A-10)$$

where ϕ is the average value of ϕ and c is the average value of c (F/L^2). For $c = 0$, $\lambda_{c\phi}$ is infinite. Use the charts for infinite slopes in this case.

Steps 4 and 5 are iterative steps. On the first iteration, average values of $\tan\phi$ and c are estimated using judgment rather than averaging.

Step 5. Using the chart at the top of Figure A-5, determine the center coordinates of the circle being investigated. Plot the critical circle on a scaled cross section of the slope and calculate the weighted average values of ϕ and c using Eqs. (A-1) and (A-2).

Return to step 4 with these average values of the shear strength parameters and repeat this iterative process until the value of $\lambda_{c\phi}$ becomes constant. One iteration is usually sufficient.

Step 6. Using the chart at the left side of Figure A-5, determine the value of the stability number N_{cf}, which depends on the slope angle, β, and the value of $\lambda_{c\phi}$.

Step 7. Calculate the factor of safety:

$$F = N_{cf} \frac{c}{P_d} \qquad (A-11)$$

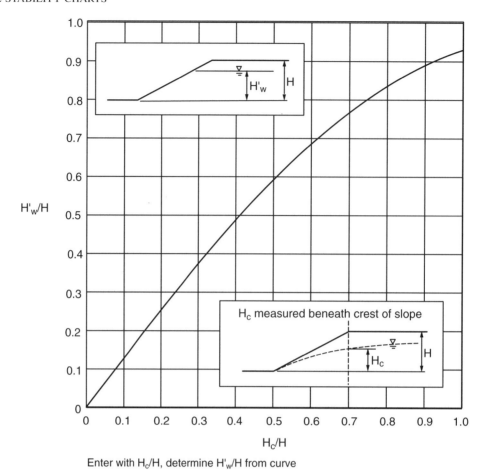

Enter with H$_c$/H, determine H$'_w$/H from curve

Figure A-6   Steady seepage adjustment factor for $\phi > 0$ soils. (After Duncan et al., 1987.)

The example problems in Figures A-11 and A-12 illustrate the use of these methods for total stress and effective stress analyses.

INFINITE SLOPE CHARTS

Two types of conditions can be analyzed using the charts shown in Figure A-7:

1. Slopes in cohesionless materials, where the critical failure mechanism is shallow sliding or surface raveling.
2. Slopes in residual soils, where a relatively thin layer of soil overlies firmer soil or rock, and the critical failure mechanism is sliding along a plane parallel to the slope, at the top of the firm layer.

Steps for use of charts for effective stress analyses are:

Step 1. Determine the pore pressure ratio, r_u, which is defined by

$$r_u = \frac{u}{\gamma H} \qquad (A\text{-}12)$$

where u is the pore pressure (F/L^2), γ the total unit weight of soil (F/L^3), and H the depth corresponding to pore pressure, u (L).

For an existing slope, the pore pressure can be determined from field measurements using piezometers installed at the depth of sliding or estimated for the most adverse anticipated seepage condition. For seepage parallel to the slope, which is a condition frequently used for design, the value of r_u can be calculated using the formula

$$r_u = \frac{X}{T} \frac{\gamma_w}{\gamma} \cos^2\beta \qquad (A\text{-}13)$$

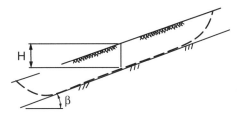

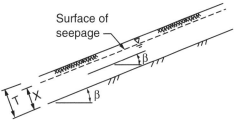

Surface of seepage

Seepage parallel to slope

$$r_u = \frac{X}{T}\frac{\gamma_w}{\gamma}\cos^2\beta$$

γ = total unit weight of soil
γ_w = unit weight of water
c = cohesion intercept
ϕ' = friction angle
r_u = pore pressure ratio = $u/\gamma H$
u = pore pressure at depth H

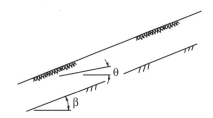

Seepage emerging from slope

$$r_u = \frac{\gamma_w}{\gamma}\frac{1}{1+\tan\beta\tan\theta}$$

Steps:

1. Determine r_u from measured pore pressure or formulas at right.
2. Determine A and B from charts below.
3. Calculate $F = A\frac{\tan\phi'}{\tan\beta} + B\frac{c'}{\gamma H}$

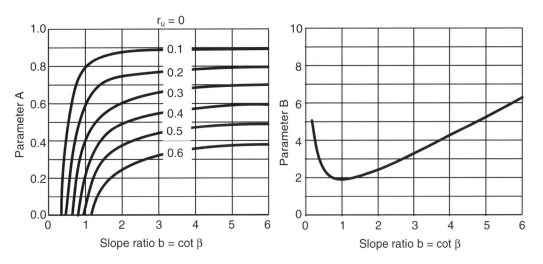

Figure A-7 Slope stability charts for infinite slopes. (After Duncan et al., 1987.)

where

X = distance from the depth of sliding to the surface of seepage, *measured normal to the surface of the slope* (L)

T = distance from the depth of sliding to the surface of the slope, *measured normal to the surface of the slope* (L)

γ_w = unit weight of water (F/L³)
γ = total unit weight of soil (F/L³)
β = slope angle

For seepage emerging from the slope, which is more critical than seepage parallel to the slope, the value or r_u can be calculated using the formula

$$r_u = \frac{\gamma_w}{\gamma} \frac{1}{1 + \tan \beta \tan \theta} \qquad \text{(A-14)}$$

where θ is the angle of seepage measured from the horizontal direction. The other factors are as defined previously. Submerged slopes, with no *excess* pore pressures, can be analyzed using $\gamma = \gamma_b$ (buoyant unit weight) and $r_u = 0$.

Step 2. Determine the values of the dimensionless parameters A and B from the charts at the bottom of Figure A-7.

Step 3. Calculate the factor of safety:

$$F = A \frac{\tan \phi'}{\tan \beta} + B \frac{c'}{\gamma H} \qquad \text{(A-15)}$$

where ϕ' is the angle of internal friction in terms of effective stress, c' the cohesion intercept in terms of effective stress (F/L^2), β the slope angle, H the depth of sliding mass *measured vertically* (L), and the other factors are as defined previously.

Steps for use of charts for total stress analyses are:

Step 1. Determine the value of B from the chart in the lower right corner of Figure A-7.

Step 2. Calculate the factor of safety:

$$F = \frac{\tan \phi}{\tan \beta} + B \frac{c}{\gamma H} \qquad \text{(A-16)}$$

where ϕ is the angle of internal friction in terms of total stress and c is the cohesion intercept in terms of total stress (F/L^2). The other factors are as defined previously.

The example in Figure A-13 illustrates use of the infinite slope stability charts.

SOILS WITH $\phi = 0$ AND STRENGTH INCREASING WITH DEPTH

The chart for slopes in soils with $\phi = 0$ and strength increasing with depth is shown in Figure A-8. Steps for use of the chart are:

Step 1. Select the linear variation of strength with depth that best fits the measured strength data. As shown in Figure A-8, extrapolate this straight line upward to determine H_0, the height at which the straight line intersects zero.

Step 2. Calculate $M = H_0/H$, where H is the slope height.

Step 3. Determine the dimensionless stability number, N, from the chart in the lower right corner of Figure A-8.

Step 4. Determine the value of c_b, the strength at the elevation of the bottom (the toe) of the slope.

Step 5. Calculate the factor of safety:

$$F = N \frac{c_b}{\gamma (H + H_0)} \qquad \text{(A-17)}$$

where γ_{total} is the total unit weight of soil for slopes above water, γ_{buoyant} is the buoyant unit weight for submerged slopes, and γ is the weighted average unit weight for partly submerged slopes. The example shown in Figure A-14 illustrates use of the stability chart shown in Figure A-8.

EXAMPLES

Example A-1. Figure A-9 shows a slope in $\phi = 0$ soil. There are three layers, each with different strength. There is water outside the slope. Two circles were analyzed for this slope: shallow circle tangent to elevation -8 ft and a deep circle tangent to elevation -20 ft.

The shallower circle, tangent to elevation -8 ft, is analyzed first. For this circle:

$$d = \frac{D}{H} = \frac{0}{24} = 0$$

$$\frac{H_w}{H} = \frac{8}{24} = 0.33$$

Using the charts at the top of Figure A-1, with $\beta = 50°$ and $d = 0$:

$$x_0 = 0.35 \quad \text{and} \quad y_0 = 1.4$$

$$X_0 = (H)(x_0) = (24)(0.35) = 8.4 \text{ ft}$$

$$Y_0 = (H)(y_0) = (24)(1.4) = 33.6 \text{ ft}$$

Plot the critical circle on the slope. The circle is shown in Figure A-9.

Measure the central angles of arc in each layer using a protractor. Calculate the weighted average strength parameter c_{av} using Eq. (A-1):

$$c_{\text{av}} = \frac{\sum \delta_i c_i}{\sum \delta_i} = \frac{(22)(600) + (62)(400)}{22 + 62} = 452 \text{ psf}$$

From Figure A-3, with $\beta = 50°$ and $H_w/H = 0.33$, find $\mu_w = 0.93$.

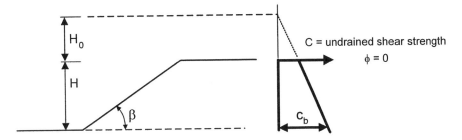

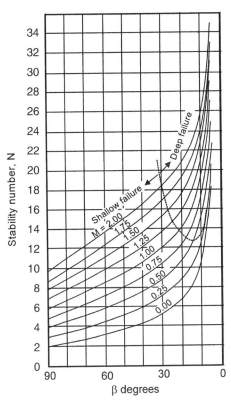

Steps

1. Extrapolate strength profile upward to determine value of H_0, where strength profile intersects zero.

2. Calculate $M = H_0/H$.

3. Determine stability number from chart below.

4. Determine c_b = strength at elevation of toe of slope.

5. Calculate $F = N \dfrac{c_b}{\gamma\left(H + H_0\right)}$

Use $\gamma = \gamma_{\text{buoyant}}$ for submerged slope

Use $\gamma = \gamma_{\text{total}}$ for no water outside slope

Use average γ for partly submerged slope

Figure A-8   Slope stability charts for $\phi = 0$ soils, with strength increasing with depth. (After Hunter and Schuster, 1968.)

Use layer thickness to average the unit weights. Unit weights are averaged only to the bottom of the critical circle.

$$\gamma_{\text{av}} = \frac{\sum \gamma_i h_i}{\sum h_i} = \frac{(120)(12) + (100)(12)}{12 + 12} = 110$$

Calculate the driving force term P_d as follows:

$$P_d = \frac{\gamma H + q - \gamma_w H_w}{\mu_q \mu_w \mu_t}$$

$$= \frac{(110)(24) + 0 - (62.4)(8)}{(1)(0.93)(1)} = 2302$$

From Figure A-1, with $d = 0$ and $\beta = 50°$, find $N_0 = 5.8$.

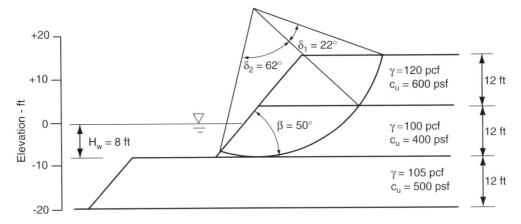

Figure A-9   Circle tangent to elevation −8 ft for cohesive soil with $\phi = 0$.

Calculate the factor of safety using Eq. (A-7):

$$F = \frac{N_0 c}{P_d} = \frac{(5.8)(452)}{2302} = 1.14$$

Example A-2.   Figure A-10 shows the same slope as in Figure A-9. The deeper circle, tangent to elevation −20 ft, is analyzed as follows. For this circle:

$$d = \frac{D}{H} = \frac{12}{24} = 0.5$$

$$\frac{H_w}{H} = \frac{8}{24} = 0.33$$

Using the charts at the bottom of Figure A-1, with $\beta = 50°$ and $d = 0.5$:

$$x_0 = 0.35 \quad \text{and} \quad y_0 = 1.5$$

$$X_0 = (H)(x_0) = (24)(0.35) = 8.4 \text{ ft}$$

$$Y_0 = (H)(y_0) = (24)(1.5) = 36 \text{ ft}$$

Plot the critical circle on the slope as shown in Figure A-10.

Measure the central angles of arc in each layer using a protractor. Calculate the weighted average strength parameter c_{av} using Eq. (A-1).

$$c_{av} = \frac{\sum \delta_i c_i}{\sum \delta_i}$$

$$= \frac{(16)(600) + (17)(400) + (84)(500)}{16 + 17 + 84} = 499 \text{ psf}$$

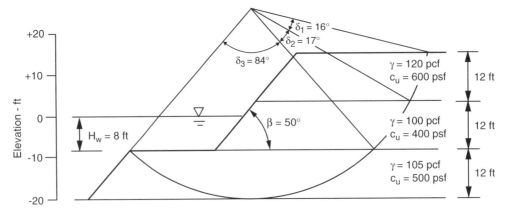

Figure A-10   Circle tangent to elevation −20 ft for cohesive soil with $\phi = 0$.

From Figure A-3, with $d = 0.5$ and $H_w/H = 0.33$, $\mu_w = 0.95$. Use layer thickness to average the unit weights. Since the material below the toe of the slope is a $\phi = 0$ material, the unit weight is averaged only down to the toe of the slope. The unit weight below the toe has no influence on stability if $\phi = 0$.

$$\gamma_{av} = \frac{\sum \gamma_i h_i}{\sum h_i} = \frac{(120)(12) + (100)(12)}{12 + 12} = 110$$

Calculate the driving force term P_d as follows:

$$P_d = \frac{\gamma H + q - \gamma_w H_w}{\mu_q \mu_w \mu_t}$$

$$= \frac{(110)(24) + 0 - (62.4)(8)}{(1)(0.95)(1)} = 2253$$

From Figure A-1, with $d = 0.5$ and $\beta = 50°$, $N_0 = 5.6$. Calculate the factor of safety using Eq. (A-7):

$$F = \frac{N_o c}{P_d} = \frac{(5.6)(499)}{2253} = 1.24$$

This circle is less critical than the circle tangent to elevation -8 ft analyzed previously.

Example A-3. Figure A-11 shows a slope in soils with both c and ϕ. There are three layers, each with different strength. There is no water outside the slope. The factor of safety for a toe circle is calculated as follows.

Use the layer thickness to average the unit weights. Unit weights are averaged down to the toe of the slope, since the unit weight of the material below the toe has little effect on stability.

$$\gamma_{av} = \frac{\sum \gamma_i h_i}{\sum h_i} = \frac{(115)(20) + (110)(20)}{20 + 20} = 112.5$$

Since there is no surcharge, $\mu_q = 1$; since there is no external water above the toe, $\mu_w = 1$; since there is no seepage, $\mu_w' = 1$; since there are no tension cracks, $\mu_t = 1$.

Calculate the driving force term:

$$P_d = \frac{\gamma H + q - \gamma_w H_w}{\mu_q \mu_w \mu_t} = \frac{(112.5)(40)}{(1)(1)(1)} = 4500 \text{ psf}$$

Calculate P_e as follows:

$$P_e = \frac{\gamma H + q - \gamma_w H_w'}{\mu_q \mu_w'} = \frac{(112.5)(40)}{(1)(1)} = 4500 \text{ psf}$$

Estimate $c_{av} = 700$ psf and $\phi_{av} = 7°$, and calculate $\lambda_{c\phi}$ as follows:

$$\lambda_{c\phi} = \frac{P_e \tan\phi}{c} = \frac{(4500)(0.122)}{700} = 0.8$$

From Figure A-5, with $b = 1.5$ and $\lambda_{c\phi} = 0.8$:

$$x_0 = 0.6 \quad \text{and} \quad y_0 = 1.5$$

$$X_0 = (H)(x_0) = (40)(0.6) = 24 \text{ ft}$$

$$Y_0 = (H)(y_0) = (40)(1.5) = 60 \text{ ft}$$

Plot the critical circle on the given slope, as shown in Figure A-11.

Calculate c_{av}, $\tan \phi_{av}$, and $\lambda_{c\phi}$ as follows:

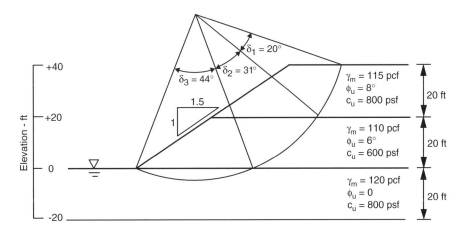

Figure A-11 Total stress analysis of a toe circle in soils with both c and ϕ.

$$c_{av} = \frac{\sum \delta_i c_i}{\sum \delta_i}$$

$$= \frac{(20)(800) + (31)(600) + (44)(800)}{20 + 31 + 44}$$

$$= 735 \text{ psf}$$

$$\tan \phi_{av} = \frac{\sum \delta_i \tan \phi_i}{\sum \delta_i}$$

$$= \frac{(20)(\tan 8°) + (31)(\tan 6°) + (44)(\tan 0°)}{20 + 31 + 44}$$

$$= 0.064$$

$$\lambda_{c\phi} = \frac{P_e \tan \phi}{c} = \frac{(4500)(0.064)}{735} = 0.4$$

From Figure A-5, with $b = 1.5$ and $\lambda_{c\phi} = 0.4$:

$$x_0 = 0.65 \quad \text{and} \quad y_0 = 1.45$$

$$X_0 = (H)(x_0) = (40)(0.65) = 26 \text{ ft}$$

$$Y_0 = (H)(y_0) = (40)(1.45) = 58 \text{ ft}$$

This circle is close to the previous iteration, so keep $\lambda_{c\phi} = 0.4$ and $c_{av} = 735$ psf. From Figure A-5, with $b = 1.5$ and $\lambda_{c\phi} = 0.4$, $N_{cf} = 6.0$. Calculate the factor of safety:

$$F = N_{cf} \frac{c}{P_d} = 6.0 \left(\frac{735}{4500} \right) = 1.0$$

According to this calculation, the slope is on the verge of instability.

Example A-4. Figure A-12 shows the same slope as shown in Figure A-11. Effective stress strength parameters are shown in the figure, and the analysis is performed using effective stresses. There is water outside the slope and seepage within the slope.

Use layer thickness to average the unit weights. Unit weights are averaged only down to the toe of the slope.

$$\gamma_{av} = \frac{\sum \gamma_i h_i}{\sum h_i} = \frac{(115)(20) + (115)(20)}{20 + 20} = 115$$

For this slope:

$$\frac{H_w}{H} = \frac{10}{40} = 0.25$$

$$\frac{H'_w}{H} = \frac{30}{40} = 0.75$$

Since there is no surcharge, $\mu_q = 1$. Using Figure A-3 for toe circles, with $H_w/H = 0.25$ and $\beta = 33.7°$, find $\mu_w = 0.96$. Using Figure A-3 for toe circles, with $H'_w/H = 0.75$ and $\beta = 33.7°$, find $\mu'_w = 0.95$. Since

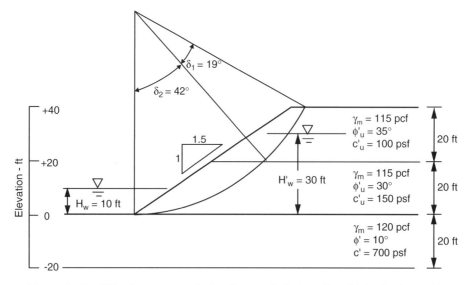

Figure A-12 Effective stress analysis of a toe circle in soils with both c' and ϕ'.

there are no tension cracks, $\mu_t = 1$. Calculate the driving force term:

$$P_d = \frac{\gamma H + q - \gamma_w H_w}{\mu_q \mu_w \mu_t}$$

$$= \frac{(115)(40) + 0 - (62.4)(10)}{(1)(0.96)(1)} = 4141 \text{ psf}$$

Calculate P_e:

$$P_e = \frac{\gamma H + q - \gamma_w H'_w}{\mu_q \mu'_w}$$

$$= \frac{(115)(40) + 0 - (62.4)(30)}{(1)(0.95)} = 2870 \text{ psf}$$

Estimate $c_{av} = 120$ psf and $\phi_{av} = 33°$.

$$\lambda_{c\phi} = \frac{P_e \tan\phi}{c} = \frac{(2870)(0.64)}{120} = 15.3$$

From Figure A-5, with $b = 1.5$ and $\lambda_{c\phi} = 15.3$:

$$x_0 = 0 \quad \text{and} \quad y_0 = 1.9$$

$$X_0 = (H)(x_0) = (40)(0) = 0 \text{ ft}$$

$$Y_0 = (H)(y_0) = (40)(1.9) = 76 \text{ ft}$$

Plot the critical circle on the given slope as shown in Figure A-12. Calculate c_{av}, $\tan \phi_{av}$, and $\lambda_{c\phi}$ as follows:

$$c_{av} = \frac{\sum \delta_i c_i}{\sum \delta_i} = \frac{(19)(100) + (42)(150)}{19 + 42} = 134 \text{ psf}$$

$$\tan \phi_{av} = \frac{\sum \delta_i \tan \phi_i}{\sum \delta_i}$$

$$= \frac{(19)(\tan 35°) + (42)(\tan 30°)}{19 + 42} = 0.62$$

$$\lambda_{c\phi} = \frac{(2870)(0.62)}{134} = 13.3$$

From Figure A-5, with $b = 1.5$ and $\lambda_{c\phi} = 13.3$:

$$x_0 = 0.02 \quad \text{and} \quad y_0 = 1.85$$

$$X_0 = (H)(x_0) = (40)(0.02) = 0.8 \text{ ft}$$

$$Y_0 = (H)(y_0) = (40)(1.85) = 74 \text{ ft}$$

This circle is close to the previous iteration, so keep $\lambda_{c\phi} = 13.3$ and $c_{av} = 134$ psf. From Figure A-5, with $b = 1.5$ and $\lambda_{c\phi} = 13.3$, $N_{cf} = 35$. Calculate the factor of safety:

$$F = N_{cf} \frac{c}{P_d} = 35 \left(\frac{134}{4141} \right) = 1.13$$

With $F = 1.13$, the slope would be very close to failure.

Example A-5. Figure A-13 shows a slope where a relatively thin layer of soil overlies firm soil. The critical failure mechanism for this example is sliding along

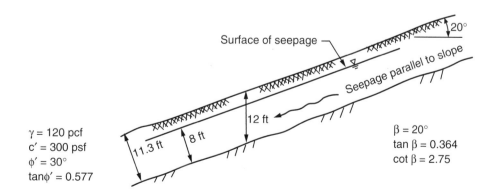

$\gamma = 120$ pcf
$c' = 300$ psf
$\phi' = 30°$
$\tan\phi' = 0.577$

$\beta = 20°$
$\tan \beta = 0.364$
$\cot \beta = 2.75$

Figure A-13 Infinite slope analysis.

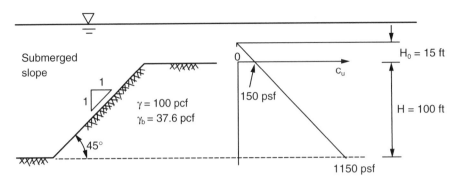

Figure A-14 $\phi = 0$, and strength increasing with depth.

a plane parallel to the slope, at the top of the firm layer. This slope can be analyzed using the infinite slope stability chart shown in Figure A-7. Calculate the factor of safety for seepage parallel to the slope and for horizontal seepage emerging from the slope.

For seepage parallel to the slope:

$$X = 8 \text{ ft} \quad \text{and} \quad T = 11.3 \text{ ft}$$

$$r_u = \frac{X}{T} \frac{\gamma_w}{\gamma} \cos^2 \beta = \frac{8}{11.3} \left(\frac{62.4}{120}\right) (0.94)^2 = 0.325$$

From Figure A-7, with $r_u = 0.325$ and $\cot \beta = 2.75$, $A = 0.62$ and $B = 3.1$. Calculate the factor of safety:

$$F = A \frac{\tan \phi'}{\tan \beta} + B \frac{c'}{\gamma H} = 0.62 \left(\frac{0.577}{0.364}\right)$$

$$+ 3.1 \left[\frac{300}{(120)(12)}\right] = 0.98 + 0.65 = 1.63$$

For horizontal seepage emerging from slope, $\theta = 0°$:

$$r_u = \frac{\gamma_w}{\gamma} \frac{1}{1 + \tan \beta \tan \theta}$$

$$= \frac{62.4}{120} \left[\frac{1}{1 + (0.364)(0)}\right] = 0.52$$

From Figure A-7, with $r_u = 0.52$ and $\cot \beta = 2.75$, $A = 0.41$ and $B = 3.1$. Calculate the factor of safety:

$$F = A \frac{\tan \phi'}{\tan \beta} + B \frac{c'}{\gamma H}$$

$$= 0.41 \left(\frac{0.577}{0.364}\right) + 3.1 \left[\frac{300}{(120)(12)}\right]$$

$$= 0.65 + 0.65 = 1.30$$

Note that the factor of safety for seepage emerging from the slope is smaller than the factor of safety for seepage parallel to the slope.

Example A-6. Figure A-14 shows a submerged clay slope with $\phi = 0$ and strength increasing linearly with depth. The factor of safety is calculated using the slope stability chart shown in Figure A-8. Extrapolating the strength profile up to zero gives $H_0 = 15$ ft. Calculate M as follows:

$$M = \frac{H_0}{H} = \frac{15}{100} = 0.15$$

From Figure A-8, with $M = 0.15$ and $\beta = 45°$, $N = 5.1$. From the soil strength profile, $c_b = 1150$ psf. Calculate the factor of safety:

$$F = N \frac{c_b}{\gamma(H + H_0)} = 5.1 \left[\frac{1150}{(37.6)(115)}\right] = 1.36$$

References

Abrams, T. G., and Wright, S. G. (1972). *A Survey of Earth Slope Failures and Remedial Measures in Texas,* Research Report 161-1, Center for Highway Research, University of Texas, Austin, TX, December.

Abramson, L. W., Lee, T. S., Sharma, S., and Boyce, G. M. (2002). *Slope Stability and Stabilization Methods,* 2nd ed., Wiley, Hoboken, NJ.

Al-Hussaini, M., and Perry, E. B. (1978). Analysis of rubber membrane strip reinforced earth wall, *Proceedings of the Symposium on Soil Reinforcing and Stabilizing Techniques,* pp. 59–72.

Anon. (1981). 18 million litres a day! That's the massive dewatering task involving Hanson Sykes for this trial coal pit. *Earthmover and Civil Contractor,* Peter Attwater, Australia, July, 1981.

Arai, K., and Tagyo, K. (1985). Determination of noncircular slip surface giving the minimum factor of safety in slope stability analysis, *Soils and Foundations,* 25(1), 43–51.

Ausilio, E., Conte, E., and Dente, G. (2000). Seismic stability analysis of reinforced slopes, *Soil Dynamics and Earthquake Engineering,* 19, 159–172.

Ausilio, E., Conte, E., and Dente, G. (2001). Stability analysis of slopes reinforced with piles, *Computers and Geotechnics,* 28, 591–611.

Ayres, D. J. (1959). Grouting and the civil engineer, *Transactions of the Society of Engineers,* 114–124.

Ayres, D. J. (1961). The treatment of unstable slopes and railway track formations, *Journal of the Society of Engineers,* 52, 111–138.

Ayres, D. J. (1985). Stabilization of slips in cohesive soils by grouting, in *Failures in Earthworks,* Thomas Telford, London.

Azzouz, A. S., Baligh, M. M., and Ladd, C. C. (1981). Three-dimensional stability analysis of four embankment failures, *Proceedings of the 10th International Conference on Soil Mechanics and Foundation Engineering,* Stockholm, June, Vol. 3, pp. 343–346.

Baker, R. (1980). Determination of the critical slip surface in slope stability computations, *International Journal for Numerical and Analytical Methods in Geomechanics,* 4(4), 333–359.

Becker, E., Chan, C. K., and Seed, H. B. (1972). *Strength and Deformation Characteristics of Rockfill Materials in Plane Strain and Triaxial Compression Tests,* Report TE-72-3, Office of Research Services, University of California, Berkeley, CA.

Beles, A. A., and Stanculescu, I. I. (1958). Thermal treatments as a means of improving the stability of earth masses, *Geotechnique,* 8, 156–165.

Bishop, A. W. (1954). The use of pore pressure coefficients in practice, *Geotechnique,* 4(4), 148–152.

Bishop, A. W. (1955). The use of slip circles in the stability analysis of earth slopes, *Geotechnique,* 5(1), 7–17.

Bishop, A. W., and Bjerrum, L. (1960). The relevance of the triaxial test to the solution of stability problems, *Proceedings of the ASCE Research Conference on the Shear Strength of Cohesive Soils,* Boulder, CO.

Bishop, A. W., and Morgenstern, N. (1960). Stability coefficients for earth slopes, *Geotechnique,* 10(4), 129–150.

Bjerrum, L. (1967). Progressive failure in slopes of overconsolidated plastic clay and clay shales, *ASCE, Journal of the Soil Mechanics and Foundation Division,* 93(5), 1–49.

Bjerrum, L. (1972). Embankments on soft ground, *Proceedings of the ASCE Specialty Conference on Performance of Earth and Earth-Supported Structures,* Purdue University, Vol. II, pp. 1–54.

Bjerrum, L. (1973). Problems of soil mechanics and construction on soft clays, *Proceedings of the 8th International Conference on Soil Mechanics and Foundation Engineering,* Moscow, Vol. 3, pp. 11–159.

Bjerrum, L. and Simons, N. E. (1960). Comparisons of shear strength characteristics of normally consolidated clays, *Proceedings of the ASCE Research Conference on the Shear Strength of Cohesive Soils,* Boulder, CO, pp. 711–726.

Blacklock, J. R., and Wright, P. J. (1986). Injection stabilization of failed highway embankments, presented at the

65th Annual Meeting of the Transportation Research Board, Washington, DC.

Bonaparte, R. and Christopher, B. R. (1987). Design and construction of reinforced embankments over weak foundations, *Transportation Research Record* No. 1153, Transportation Research Board, National Research Council, pp. 26–39.

Bonarparte, R., Holtz, R. D. and Giroud, J. P. (1987). Soil reinforcement design using geotextiles and geograids, *Geotextile Testing and the Design Engineer, ASTM, Special Technical Publication 952.*

Bouazzack, A., and Kavazanjian, E. (2001). Construction on old landfills, *Environmental Geotechnics: Proceedings of the 2nd Australia and New Zealand Conference on Environmental Geotechnics–Geoenvironment,* Australian Geomechanics Society, Newcastle Chapter, pp. 467–482.

Boutrup, E., Lovell, C. W. (1980), Searching techniques in slope stability Analyses, *Engineering Geology,* Vol. 16, No. 1/2, pp. 51–61.

Brahma, S. P., and Harr, M. E. (1963). Transient development of the free surface in a homogeneous earth dam, *Geotechnique,* 12(4), 183–302.

Brandl, H. (1981). Stabilization of slippage-prone slopes by lime piles, *Proceedings of the 8th International Conference on Soil Mechanics and Foundation Engineering,* pp. 738–740.

Brandon, T. L., Ed. (2001). *Foundations and Ground Improvement,* Geotechnical Special Publication 113, ASCE, Reston, VA.

Brandon, T. L., Duncan, J. M., and Huffman, J. T. (1990). *Classification and Engineering Behavior of Silt,* final report to the U.S. Army Corps of Engineers, Lower Mississippi Valley Division.

Bray, J. D., Rathje, E. M., Augello, A. J., and Merry, S. M. (1998). Simplified seismic design procedure for geosynthetic-lined, solid-waste landfills. *Geosynthetics International,* 5(1–2), 203–235.

Bromhead, E. N. (1992). *The Stability of Slopes,* 2nd ed., Blackie, New York.

Bromwell, L. G., and Carrier, W. D. I. (1979). Consolidation of fine-grained mining wastes, *Proceedings of the 6th Pan-American Conference on Soil Mechanics and Foundation Engineering,* Lima, Peru, Vol. 1, pp. 293–304.

Bromwell, L. G., and Carrier, W. D. I. (1983). Reclamation alternatives for phosphatic clay disposal areas, *Proceedings of the Symposium on Surface Mining, Hydrology, Sedimentology and Reclamation,* University of Kentucky, Lexington, KY, pp. 371–376.

Browzin, B. S. (1961). Non-steady flow in homogeneous earth dams after rapid drawdown, *Proceedings of the 5th International Conference on Soil Mechanics and Foundation Engineering,* Paris, Vol. 2, pp. 551–554.

Busbridge, J. R., Chan, P., Milligan, V., La Rochelle, P., and Lefebvre, L. D. (1985). The effect of geogrid reinforcement on the stability of embankments on a soft sensitive Champlain clay deposit, draft report prepared for the Transportation Development Center, Montreal, Quebec, by Golder Associates and Laval University.

Byrne, J. (2003). Personal communication.

Byrne, R. J., Kendall, J., and Brown, S. (1992). Cause and mechanism of failure, Kettlemen Hills landfill B-19, unit IA, *Proceedings of the ASCE Specialty Conference on Performance and Stability of Slopes and Embankments: II,* Vol. 2, pp. 1188–1215.

Carter, M., and Bentley, S. P. (1985). The geometry of slip surfaces beneath landslides: predictions from surface measurements, *Canadian Geotechnical Journal,* 22(2), 234–238.

Casagrande, A. (1937). Seepage through dams, *Journal of the New England Water Works Association,* 51(2). Reprinted in *Contributions to Soil Mechanics, 1925–1940,* Boston Society of Civil Engineers, Boston, 1940, pp. 295–336.

Celestino, T. B., and Duncan, J. M. (1981). Simplified search for noncircular slip surfaces, *Proceedings of the 10th International Conference on Soil Mechanics and Foundation Engineering,* Stockholm, June, Vol. 3, pp. 391–394.

Chandler, R. J. (1977). Back analysis techniques for slope stabilization works: a case record, *Geotechnique,* 27(4), 479–495.

Chandler, R. J., Ed. (1991). Slope stability engineering: developments and applications, *Proceedings of the International Conference on Slope Stability,* Isle of Wight, England, Thomas Telford, London.

Chang, C. Y., and Duncan, J. M. (1970). Analysis of soil movements around a deep excavation, *ASCE, Journal of the Soil Mechanics and Foundation Division,* 96(5), 1655–1681.

Chen, L. T., and Poulos, H. G. (1997). Piles subjected to lateral soil movements, *ASCE, Journal of Geotechnical and Geoenvironmental Engineering,* 123(9), 802–811.

Chen, Z.-Y., and Morgenstern, N. R. (1983). Extensions to the generalized method of slices for stability analysis, *Canadian Geotechnical Journal,* 20(1), 104–119.

Chen, Z.-Y., and Shao, C. M. (1988). Evaluation of minimum factor of safety in slope stability analysis, *Canadian Geotechnical Journal,* 25(4), 735–748.

Chirapuntu, S., and Duncan, J. M. (1975). *The Role of Fill Strength in the Stability of Embankments on Soft Clay Foundations,* Geotechnical Engineering Report TE 75-3, University of California, Berkeley, CA.

Chirapuntu, S., and Duncan, J. M. (1977). Cracking and progressive failure of embankments on soft clay foundations, *Proceedings of the International Symposium on Soft Clay,* Bangkok, Thailand, pp. 453–470.

Christian, J. T. (1996). Reliability methods for stability of existing slopes, *Uncertainty in the Geologic Environment: From Theory to Practice, Proceedings of Uncertainty '96,* Madison, WI, ASCE Geotechnical Special Publication 58, pp. 409–418.

Christian, J. T., and Alfredo, U. (1998). Probabilistic evaluation of earthquake-induced slope failure, *ASCE, Journal of Geotechnical and Geoenvironmental Engineering,* 124(11), 1140–1143.

Christian, J. T., and Baecher, G. B. (2001). Discussion on "Factors of safety and reliability in geotechnical engineering," by J. M. Duncan, *ASCE, Journal of Geotechnical and Geoenvironmental Engineering,* 127(8), 700–702.

Christian, J. T., Ladd, C. C., and Baecher, G. B. (1994). Reliability applied to slope stability analysis, *ASCE, Journal of Geotechnical Engineering,* 120(12), 2187–2207.

Christopher, B R. and Holtz, R. D. (1985). *Geotextiles Engineering Manual,* National Highway Institute, FHWA, Washington, DC, Contract DTFH61-80-C-00094.

Chugh, A. K. (1981). Pore water pressures in natural slopes, *International Journal for Numerical Methods in Geomechanics,* 5(4), 449–454.

Chugh, A. K. (1982). Procedure for design of restraining structures for slope stabilization problems, *Geotechnical Engineering,* 13, 223–234.

Collins, S. A., Rogers, W., and Sowers, G. F. (1982). Report of embankment reanalysis—Mohicanville dikes, *Report for Department of the Army, Huntington District Corps of Engineers,* Huntington, West Virginia, by Law Engineering Testing Company, July.

Cooper, M. R. (1984). The application of back-analysis to the design of remedial works for failed slopes, *Proceedings of the 4th International Symposium on Landslides,* Toronto, Vol. 2, pp. 387–392.

Cruden, D. M. (1986). The geometry of slip surfaces beneath landslides: predictions from surface measurements: discussion, *Canadian Geotechnical Journal,* 23(1), 94.

Dai, S.-H., and Wang, M.-O. (1992). *Reliability Analysis in Engineering Applications.* Van Nostrand Reinhold, New York.

Das, B. M. (2002). *Principles of Geotechnical Engineering,* 5th ed., Brooks/Cole, Pacific Grove, CA.

de Mello, V. (1971). The standard penetration test—a state-of-the-art report, *Proceedings of the 4th Panamerican Conf. on Soil Mechanids and Found. Engineering,* Puerto Rico, pp. 1–86.

Desai, C. S. (1972). Seepage analysis of earth banks under drawdown, *ASCE, Journal of the Soil Mechanics and Foundations Division,* 98(11), 1143–1162.

Desai, C. S. (1977). Drawdown analysis of slopes by numerical method, *ASCE, Journal of the Geotechnical Engineering Division,* 103(7), 667–676.

Desai, C. S., and Sherman, W. S. (1971). Unconfined transient seepage in sloping banks, *ASCE, Journal of the Soil Mechanics and Foundations Division,* 97(2), 357–373.

Donald, I., and Giam, P. (1989). *Soil Slope Stability Programs Review,* ACADS Publication U255, Association of Computer Aided Design, Melbourne, Australia, April.

Duncan, J. M. (1970). *Strength and Stress–Strain Characteristics of Atchafalaya Levee Foundation Soils,* Report TE 70-1, Office of Research Services, University of California, Berkeley, CA.

Duncan, J. M. (1971). Prevention and correction of landslides, *Proceedings of the 6th Annual Nevada Street and Highway Conference.*

Duncan, J. M. (1972). Finite element analyses of stresses and movements in dams, excavations, and slopes, State-of-the-Art Report, *Proceedings of the Symposium on Applications of the Finite Element Method in Geotechnical Engineering,* U.S. Army Corps of Engineers Waterways Experiment Station, Vicksburg, MS, pp. 267–324.

Duncan, J. M. (1974). Finite element analyses of slopes and excavations, State-of-the-Art Report, *Proceedings of the First Brazilian Seminar on the Application of the Finite Element Method in Soil Mechanics,* COPPE, Rio de Janeiro, Brazil, pp. 195–208.

Duncan, J. M. (1986). "Methods of analyzing the stability of natural slopes, notes for a lecture to be presented at the 17th Annual Ohio River Valley Soil Seminar, Louisville, KY.

Duncan, J. M. (1988). Prediction, design and performance in geotechnical engineering, keynote paper for the 5th Australia–New Zealand Conference on Geomechanics, Sydney, Australia.

Duncan, J. M. (1992). State-of-the-art static stability and deformation analysis, *Stability and Performance of Slopes and Embankments: II,* Geotechnical Special Publication 31, ASCE, Reston, VA, pp. 222–266.

Duncan, J. M. (1993). Limitations of conventional analysis of consolidation settlement, *ASCE, Journal of Geotechnical Engineering,* 119(9), 1333–1359.

Duncan, J. M. (1996a). *Landslides: Investigation and Mitigation,* Transportation Research Board, National Research Council, National Academy Press, Washington, DC, pp. 337–371.

Duncan, J. M. (1996b). State of the art: limit equilibrium and finite element analysis of slopes, *ASCE, Journal of Geotechnical Engineering,* 122(7), 577–596.

Duncan, J. M. (1997). Geotechnical solutions to construction problems at La Esperanza dam, *Proceedings of the Central Pennsylvania Conference on Excellence in Geotechnical Engineering.*

Duncan, J. M. (1999). The use of back analysis to reduce slope failure risk, *Civil Engineering Practice, Journal of the Boston Society of Civil Engineers,* 14(1), 75–91.

Duncan, J. M. (2000). Factors of safety and reliability in geotechnical engineering, *ASCE, Journal of Geotechnical and Geoenvironmental Engineering,* 126(4), 307–316.

Duncan, J. M. (2001). Closure to discussion on "Factors of safety and reliability in geotechnical engineering, by J. M. Duncan, *ASCE, Journal of Geotechnical and Geoenvironmental Engineering,* 127(8), 717–721.

Duncan, J. M., and Buchignani, A. L. (1973). Failure of underwater slope in San Francisco bay, *ASCE, Journal of the Soil Mechanics and Foundation Division,* 99(9), 687–703.

Duncan, J. M., and Chang, C. Y. (1970). Nonlinear analysis of stress and strain in soils, *ASCE, Journal of the Soil Mechanics and Foundation Division,* 96(5), 1629–1653.

Duncan, J. M., and Dunlop, P. (1969). Slopes in stiff-fissured clays and shales, *ASCE, Journal of the Soil Mechanics and Foundation Division,* 95(2), 467–492.

Duncan, J. M., and Houston, W. N. (1983). Estimating failure probabilities for California levees, *ASCE, Journal of Geotechnical Engineering,* 109(2), 260–268.

Duncan, J. M., and Schaefer V. R. (1988). Finite element consolidation analysis of embankments, *Computers and Geotechnics,* Special Issue on Embankment Dams.

Duncan, J. M., and Seed, H. B. (1965). *The Effect of Anisotropy and Reorientation of Principal Stresses on the Shear Strength of Saturated Clay,* Report TE 65-3, Office of Research Services, University of California, Berkeley, CA.

Duncan, J. M., and Seed, H. B. (1966a). Anisotropy and stress reorientation in clay, *ASCE, Journal of the Soil Mechanics and Foundation Division,* 92(5), 21–50.

Duncan, J. M., and Seed, H. B. (1966b). Strength variation along failure surfaces in clay, *ASCE, Journal of the Soil Mechanics and Foundation Division,* 92(6), 81–103.

Duncan, J. M., and Stark, T. D. (1989). The causes of the 1981 slide in San Luis dam, Henry M. Shaw Lecture, Department of Civil Engineering, North Carolina State University, Raleigh, NC.

Duncan, J. M., and Stark, T. D. (1992). Soil strengths from back analysis of slope failures, *Stability and Performance of Slopes and Embankments: II,* Geotechnical Special Publication 31, University of California, Berkeley, CA, pp. 890–904.

Duncan, J. M., and Wong, K. S. (1983). Use and mis-use of the consolidated-undrained triaxial test for analysis of slope stability during rapid drawdown, paper prepared for the 25th Anniversary Conference on Soil Mechanics, Venezuela.

Duncan, J. M., and Wong, K. S. (1999). *SAGE User's Guide,* Vol. II, *Soil Properties Manual,* report of the Center for Geotechnical Practice and Research, Virginia Polytechnic Institute and State University, Blacksburg, VA.

Duncan, J. M., and Wright, S. G. (1980). The accuracy of equilibrium methods of slope stability analysis, *Engineering Geology,* 16(1), 5–17. Also, *Proceedings of the International Symposium on Landslides,* New Delhi, India, June 1980.)

Duncan, J. M., Byrne, P., Wong, K. S., and Mabry, P. (1980a). *Strength, Stress–Strain and Bulk Modulus Parameters for Finite Element Analyses of Stresses and Movements in Soil Masses,* Geotechnical Engineering Report UCB/GT/80-01, University of California, Berkeley, CA.

Duncan, J. M., Lefebvre, G., and Lade, P. V. (1980b). *The Landslide at Tuve,* near Göteborg, *Sweden, on November 30, 1977,* National Academy Press, Washington, DC.

Duncan, J. M., Low, B. K., and Schaefer V. R. (1985). *STABGM: A Computer Program for Slope Stability Analysis of Reinforced Embankments and Slopes,* Geotechnical Engineering Report, Department of Civil Engineering, Virginia Polytechnic Institute and State University, Blacksburg, VA.

Duncan, J. M., Buchignani, A. L., and De Wet, M. (1987). *Engineering Manual for Slope Stability Studies,* Charles E. Via, Jr. Department of Civil Engineering, Virginia Polytechnic Institute and State University, Blacksburg, VA.

Duncan, J. M., Schaefer V. R., Franks L. W., and Collins S. A. (1988). *Design and Performance of a Reinforced Embankment for Mohicanville Dike No. 2 in Ohio,* Transportation Research Record 1153, Transportation Research Board, National Research Council, National Academy Press, Washington, DC.

Duncan, J. M., Horz, R. C., and Yang, T. L. (1989). *Shear Strength Correlations for Geotechnical Engineering,* Charles E. Via, Jr. Department of Civil Engineering, Virginia Polytechnic Institute and State University, Blacksburg, VA.

Duncan, J. M., Wright, S. G., and Wong, K. S. (1990). Slope stability during rapid drawdown, *Proceedings of the H. Bolton Seed Memorial Symposium,* May, Vol. 2, pp. 253–272.

Duncan, J. M., Bolinaga, F., and Morrison, C. S. (1994a). Analysis and treatment of landslides on the abutments of La Esperanza dam, *Proceedings of the First Pan-American Symposium on Landslides,* Guayaquil, Ecuador, pp. 319–330.

Duncan, J. M., Evans, L. T. Jr., and Ooi, P. S. K. (1994b). Lateral load analysis of single piles and drilled shafts, *ASCE, Journal of Geotechnical Engineering,* 120(5), 1018–1033.

Duncan, J. M., Low, B. K., Schaefer, V. R., and Bentler, D. J. (1998). *STABGM 2.0: A Computer Program for Slope Stability Analysis of Reinforced and Unreinforced Embankments and Slopes,* Department of Civil Engineering, Virginia Polytechnic Institute and State University, Blacksburg, VA, April.

Duncan, J. M., Navin, M., and Patterson, K. (1999). *Manual for Geotechnical Engineering Reliability Calculations,* report of a study sponsored by the Center for Geotechnical Practice and Research, Virginia Polytechnic Institute and State University, Blacksburg, VA.

Dunlop, P. and Duncan, J. M. (1970), Development of failure around excavated slopes, *ASCE, Journal of the Soil Mechanics and Foundations Division,* 96(2), 471–493.

Dunlop, P., Duncan, J. M., and Seed, H. B. (1968). *Finite Element Analyses of Slopes in Soil,* report to the U.S. Army Corps of Engineers, Waterways Experiment Station, Report TE 68-3.

Dunn, I. S., Anderson, L. R., and Kiefer, F. W. (1980). *Fundamentals of Geotechnical Analysis,* Wiley, Hoboken, NJ.

Edris, E. V., Jr., and Wright, S. G. (1987). *User's Guide: UTEXAS2 Slope Stability Package,* Vol. 1, *User's Manual,* Instruction Report GL-87-1, Geotechnical Laboratory, Department of the Army, Waterways Experiment Station, U.S. Army Corps of Engineers, Vicksburg, MS, August.

Eid, H. T., Stark, T. D., Evans, W. D., and Sherry, P. E. (2000). Municipal solid waste slope failure: waste and foundation soil properties, *ASCE, Journal of Geotechnical and Geoenvironmental Engineering,* 126(5), 391–407.

Elias, V., and Christopher, B. R. (1997). *Mechanically Stabilized Earth Walls and Reinforced Soil Slopes: Design and Construction Guidelines,* Report FHWA-SA-96-071, FHWA Demonstration Project 82, U.S. Department of Transportation, Washington, DC.

EMRL (2001). *GMS 3.1 Online User Guide,* Environmental Modeling Research Laboratory, Brigham Young University, Provo, UT.

ENR (1982). Fast fix: San Luis dam up and filling, *Engineering News Record,* Vol. 28, April 1, pp. 26–28.

Evans, M. (1987), *Undrained Cyclic Triaxial Testing of Gravels—The Effects of Membrane Compliance,* Ph.D. Dissertation, University of California, Berkeley, California.

Fellenius, W. (1922). *Statens Jarnjvagars Geoteknniska Commission,* Stockholm, Sweden.

Fellenius, W. (1936). Calculation of the stability of earth dams, *Transactions of the 2nd Congress on Large Dams,* International Commission on Large Dams of the World Power Conference, Vol. 4, pp. 445–462.

FHWA (2000). *Mechanically Stabilized Earth Walls and Reinforced Soil Slopes: Design and Construction Guidelines,* Report FHWA-NHI-00-043, Federal Highways Administration, U.S. Department of Transportation, Washington, DC; available at no cost through the U.S. Department of Transportation Web site at *http://isddc.dot.gov.* (Select "Get a document," then enter the publication number.)

Filz, G. M., Brandon, T. L., and Duncan, J. M. (1992). *Back Analysis of Olmsted Landslide Using Anistropic Strengths,* Transportation Research Record 1343, Transportation Research Board, National Research Council, National Academy Press, Washington, DC, pp. 72–78.

Filz, G. M., Esterhuizen, J. J. B., and Duncan, J. M. (2001). Progressive failure of lined waste impoundments, *ASCE, Journal of Geotechnical and Geoenvironmental Engineering,* 127(10), 841–848.

Fleming, L. N., and Duncan, J. M. (1990). Stress-deformation characteristics of Alaskan silt, *ASCE, Journal of Geotechnical Engineering,* 116(3), 377–393.

Folayan, J. I., Hoeg, K., and Benjamin, J. R. (1970). Decision theory applied to settlement prediction, *ASCE, Journal of the Soil Mechanics and Foundation Division,* 96(4), 1127–1141.

Forester, T., and Morrison, P. (1994). *Computer Ethics: Cautionary Tales and Ethical Dilemmas in Computing,* 2nd ed., MIT Press, Cambridge, MA.

Fowler, J. (1982). Theoretical design considerations for fabric-reinforced embankments, *Proceedings, Second International Conference on Geotextiles,* Vol. 3, Las Vegas, Nevada, August 1–6, pp. 665–670.

Fowler, J., Leach, R. E., Peters, J. F., and Horz, R. C. (1983). *Mohicanville Reinforced Dike No. 2—design memorandum,* Geotechnical Laboratory, U.S. Army Waterways Experiment Station, Vicksburg, Mississippi, September.

Franks, L. W., Duncan, J. M., Collins S. A., Fowler, J., Peters, J. F., and Schaefer V. R. (1988). Use of reinforcement at Mohicanville Dike No. 2, *Proceedings of the 2nd International Conference on Case Histories in Geotechnical Engineering,* St. Louis, MO.

Franks, L. W., Duncan, J. M., and Collins S. A. (1991). Design and construction of Mohicanville Dike No. 2, *Proceedings of the 11th Annual U.S. Committee on Large Dams Lecture Series, Use of Geosynthetics in Dams,* White Plains, NY.

Fredlund, D. G., and Krahn, J. (1977). Comparison of slope stability methods of analysis, *Canadian Geotechnical Journal,* 14(3), 429–439.

Frohlich, O. K. (1953). The factor of safety with respect to sliding of a mass of soil along the arc of a logarithmic spiral, *Proceedings of the 3rd International Conference on Soil Mechanics and Foundation Engineering,* Switzerland, Vol. 2, pp. 230–233.

Fukuoka, M. (1977). The effects of horizontal loads on piles due to landslides, *Proceedings of the 10th Special Session, 9th International Conference on Soil Mechanics and Foundation Engineering,* pp. 27–42.

Geo-Slope (2002). *SLOPE/W for Slope Stability Analysis, Version 5: User's Guide,* Geo-Slope International, Calgary, Alberta, Canada.

Gibbs, H. J. and Holtz, W. G. (1957). Research on determining the density of sands by spoon penetration testing, *Proceedings of the 4th International Conf. Soil Mech. Fund. Eng.,* (London), vol. I.

Gilbert, R. B., Long, J. H., and Moses, B. E. (1996a). Analytical model of progressive slope failure in waste containment systems, *International Journal for Numerical and Analytical Methods in Geomechanics,* 20(1), 35–56.

Gilbert, R. B., Wright, S. G., and Liedtke, E. (1996b). Uncertainty in back analysis of slopes, *Uncertainty in the Geologic Environment: From Theory to Practice,* Geotechnical Special Publication 58, ASCE, Reston, VA, Vol. 1, pp. 494–517.

Gilbert, R. B., Wright, S. G., and Liedtke, E. (1998). Uncertainty in back analysis of slopes: Kettleman Hills case history, *ASCE, Journal of Geotechnical and Geoenvironmental Engineering,* 124(12), 1167–1176.

Gillon, M. D., Graham, C. J., and Grocott, G. G. (1992). Low level drainage works at the Brewery Creek slide, *Uncertainty in the Geologic Environment: From Theory to Practice,* Geotechnical Special Publication 58, ASCE, Reston, VA, Vol. 1, pp. 715–720.

Glendinning, S. (1995). Deep stabilization of slopes using lime piles, Ph.D. dissertation, Loughborough University, Leicestershire, England.

Golder Associates. (1991). *Cause and Mechanism of the March, 1988 Failure in Landfill B-19, Phase IA Kettleman Hill Facility, Kettleman City, California,* report to Chemical Waste Management, Inc., prepared by Golder Associates, Inc. Redmond, WA.

Gray, D. H., and Leiser, A. T. (1982). *Biotechnical Slope Protection and Erosion Control,* Van Nostrand Reinhold, New York.

Gray, D., and Sotir, R. (1992). Biotechnical stabilization of cut and fill slopes, *Stability and Performance of Slopes and Embankments: II,* Geotechnical Special Publication 31, ASCE, Reston, VA.

Gray, D., and Sotir, R. (1995). *Biotechnical Stabilization of Steepened Slopes,* Transportation Research Record 1474, Transportation Research Board, National Research Council, National Academy Press, Washington, DC.

Gucma, P. R., and Kehle, R. O. (1978). Bearpaw mountains rockslide, Montana, U.S.A., in *Natural Phenomena,* Vol. 1, *Rockslides and Avalanches,* Barry Voight, Ed., Elsevier Scientific, Amsterdam, pp. 393–421.

Haliburton, T. A., Anglin, C. C., and Lawmaster, J. D. (1978). Testing of geotechnical fabric for use as reinforcement, *Geotechnical Testing Journal,* vol. 1, Dec. 1978, ASTM, pp. 203–212.

Haliburton T. A., Lawmaster, J. D., and McGuffey, V. E. (1982). *Use of Engineering Fabric in Transportation Related Applications,* final report under Contract NO. DTFH-80-C-0094, FHWA, National Highway Institute, Washington, D.C.

Handy, R. L., and Williams, N. W. (1967). Chemical stabilization of an active landslide, *Civil Engineering,* 37(8), 62–65.

Hansbo, S. (1981). Consolidation of fine-grained soils by prefabricated drains, *Proceedings of the 10th International Conference on Soil Mechanics and Foundation Engineering,* Vol. 3, pp. 677–682.

Harding, Miller, Lawson Associates (1970). *Geotechnic Engineering Report on a Landslide at the Lawrence Berkeley Laboratory,* engineering report to the University of California, Berkeley.

Harr, M. E. (1987). *Reliability-Based Design in Civil Engineering,* McGraw-Hill, New York.

Hassiotis, S., Chameau, J. L., and Gunaratne, M. (1997). Design method for stabilization of slopes with piles, *ASCE, Journal of Geotechnical and Geoenvironmental Engineering,* 123(4), 314–323.

Henkel, D. J. (1957). Investigation of two long-term failures in London clay slopes at Wood Green, *Proceedings of the 4th International Conference on Soil Mechanics and Foundation Engineering,* Butterworth Scientific, London, Vol. 2, pp. 315–320.

Hill, R. A. (1934). Clay stratum dried out to prevent landslips, *Civil Engineering,* 4, 403–407.

Holtz, R. D., and Kovacs, W. D. (1981). *An Introduction to Geotechnical Engineering,* Prentice Hall, Englewood Cliffs, NJ.

Holtz, R. D., Lancellotta, R., Jamiolkowski, M., and Pedrona, S. (1991). *Prefabricated Vertical Drains: Design and Performance,* Construction Industry Research and Information Association, Butterworth-Heinemann.

Hong, W. P., and Han, J. G. (1996). The behavior of stabilizing piles installed in slopes, *Proceedings of the 7th International Symposium on Landslides,* Rotterdam, pp. 1709–1714.

Hong, W. P., Han, J. G., and Nam, J. M. (1997). Stability of a cut slope reinforced by stabilizing piles, *Proceedings of the International Conference on Soil Mechanics and Foundation Engineering,* pp. 1319–1322.

Hull, T. S., and Poulos, H. G. (1999). Design method for stabilization of slopes with piles (discussion), *ASCE, Journal of Geotechnical and Geoenvironmental Engineering,* 125(10), 911–913.

Hull, T. S., Lee, C. Y., and Poulos, H. G. (1991). *Mechanics of Pile Reinforcement for Unstable Slopes,* Report 636, School of Civil and Mining Engineering, University of Sydney, Sydney, Australia.

Hunter, J. H., and Schuster, R. L. (1968). Stability of simple cuttings in normally consolidated clays, *Geotechnique,* 18(3), 372–378.

Hvorslev, M. J. (1949). *Subsurface Exploration and Sampling of Soils for Civil Engineering Purposes,* U.S. Army Corps of Engineers, Waterways Experiment Station, Vicksburg, MS.

Hutchinson, J. N., and Sarma, S. K. (1985), Discussion of three-dimensional limit equilibrium analysis of slopes by R. H. Chen and J. L. Chameau, *Geotechnique,* Institution of Civil Engineers, Great Britain, Vol. 35, No. 2, pp. 215–216.

Hynes-Griffin, M. E., and Franklin, A. G. (1984). *Rationalizing the Seismic Coefficient Method,* Final Report, Miscellaneous Paper GL-84-13, Department of the Army, U.S. Army Corps of Engineers, Waterways Experiment Station, Vicksburg, MS, July.

Hynes, M. E. (1988), *Pore Pressure Generation Characteristics of Gravel Under Undrained Cyclic Loading,* Ph.D. Dissertation, University of California, Berkeley, California.

Idriss, I. M., and Duncan, J. M. (1988). Earthquake analysis of embankments, in *Advanced Dam Engineering,* R. J. Jansen, Ed., Van Nostrand Reinhold, New York, pp. 239–255.

Idriss, I. M., and Seed, H. B. (1967). Response of earth banks during earthquakes, *ASCE, Journal of the Soil Mechanics and Foundation Division,* 93(3), 61–82.

Ingold, T. S. (1982). *Reinforced Earth,* Thomas Telford, London.

Ito, T., and Matsui, T. (1975). Methods to estimate lateral force acting on stabilizing piles, *Soils and Foundations,* 15(4), 43–60.

Ito, T., Matsui, T., and Hong, W. P. (1981). Design method for stabilizing piles against landslide: one row of piles, *Soils and Foundations,* 21(1), 21–37.

Jamiolkowski, M., Ladd, C. C., Germaine, J. T., and Lancellotta, R. (1985). New developments in field and laboratory testing of soils, *Proceedings of the 11th International Conference on Soil Mechanics and Foundation Engineering,* San Francisco, Vol. 1, pp. 57–153.

Janbu, N. (1954a). Application of composite slip surface for stability analysis, *Proceedings of the European Conference on Stability of Earth Slopes,* Stockholm, Vol. 3, pp. 43–49.

Janbu, N. (1954b). *Stability Analysis of Slopes with Dimensionless Parameters,* Harvard Soil Mechanics Series 46, Harvard University Press, Cambridge MA.

Janbu, N. (1968). Slope stability computations, *Soil Mechanics and Foundation Engineering Report,* The Technical University of Norway, Trondheim.

Janbu, N. (1973). Slope stability computations, in *Embankment-Dam Engineering: Casagrande Volume,* R. C. Hirschfeld and S. J. Pouros, Eds., Wiley, Hoboken, NJ, pp. 47–86.

Janbu, N., Bjerrum, L., and Kjærnsli, B. (1956). *Veiledning ved Løsning av Fundamenteringsoppgaver* (Soil Mechanics Applied to Some Engineering Problems), Publication 16, Norwegian Geotechnical Institute, Oslo.

Jewel, R. A. (1990). Strength and deformation in reinforced soil design, *Proceedings of the 4th International Conference on Geotextiles, Geomembranes and Related Products,* Balkema, Rotterdam, The Hague, The Netherlands, pp. 913–946.

Jewel, R. A. (1996). *Soil Reinforcement with Geotextiles,* CIRIA Special Publication 123:45-6.

Jibson, R. W. (1993). *Predicting Earthquake Induced Landslide Displacements Using Newmark's Sliding Block Analysis,* Transportation Research Record 1411, Transportation Research Board, National Research Council, National Academy Press, Washington, DC, pp. 9–17.

Jones, D. B. (1991). Slope stabilization experience in South Wales, UK, in *Slope Stability Engineering,* Thomas Telford, London.

Jones, N. L. (1990). Solid modeling of earth masses for applications in geotechnical engineering, Dissertation presented to the faculty of the graduate school of the University of Texas at Austin in partial fulfillment of the requirements for the degree of Doctor of Philosophy, University of Texas, Austin, TX.

Jumikis, A. R. (1962). *Active and Passive Earth Pressure Coefficient Tables,* Engineering Research Publication 43, College of Engineering, Bureau of Engineering Research, Rutgers University, New Brunswick, NJ.

Kavazanjian, E. Jr., (1999). Seismic design of solid waste containment facilities, *Proceedings of the 8th Canadian Conference on Earthquake Engineering,* pp. 51–68.

Kavazanjian, E., Jr. (2001). Mechanical properties of municipal solid waste, *Proceedings of Sardinia 2001: 8th International Waste Management and Landfilling Symposium,* Cagliari (Sardinia), Italy, pp. 415–424.

Kavazanjian, E., Jr., and Matasovic, N. (1995). Seismic analysis of solid waste landfills, Geoenvironment 2000, Geotechnical Special Publication 46, ASCE, Reston, VA, pp. 1066–1080.

Kavazanjian, E., Jr., Matasovic, N., Bonaparte, R., and Schmertmann, G. R. (1995). Evaluation of MSW properties for seismic analysis, *Geoenvironment 2000,* Geotechnical Special Publication 46, ASCE, Reston, VA, pp. 1126–1141.

Kavazanjian, E., Jr., Matasovic, N., Hadj-Hamou, T., Sabatini, P. J. (1997). *Design Guidance: Geotechnical Earthquake Engineering for Highways,* Vol. 1, *Design Principles,* Geotechnical Engineering Circular 3, Publication FHWA-SA-97-076, Federal Highways Administration, U.S. Department of Transportation, Washington, DC, May.

Kayyal, M. K., and Wright, S. G. (1991). Investigation of long-term strength properties of Paris and Beaumont clays in earth embankments, *Research Rep. 1195-2F,* Center for Transportation Research, Univ. of Texas at Austin, Austin, Tex.

Kimball, G. H. (1971). Personal communication.

Koerner, R. M. (1998). *Designing with Geosynthetics,* 4th ed., Prentice Hall, Upper Saddle River, NJ.

Kramer, S. L. (1996). *Geotechnical Earthquake Engineering,* Prentice Hall, Upper Saddle River, NJ.

Kramer, S. L., and Smith, M. W. (1997). Modified Newmark model for seismic displacements of compliant slopes, *Journal of Geotechnical and Geoenvironmental Engineering,* 123(7), pp. 635–644.

Kulhawy, F. H. (1992). *On the Evaluation of Soil Properties,* Geotechnical Special Publication 31, ASCE, Reston, VA, pp. 95–115.

Kulhawy, F. H., and Duncan, J. M. (1970). *Nonlinear Finite Element Analysis of Stresses and Movements in Oroville Dam,* Report TE-70-2, Geotechnical Engineering, Department of Civil Engineering, University of California, Berkeley, CA.

Kulhawy, F. H., and Duncan, J. M. (1972). Stresses and movements in Oroville dam, *ASCE, Journal of the Soil Mechanics and Foundation Division,* 98(7), 653–665.

Kulhawy, F. H., Duncan, J. M. and Seed, H. B. (1969). *Finite Element Analyses of Stresses and Movements in Embankments During Construction,* Office of Research Services, Report No. TE 69-4, University of California, 169 pages.

Lacasse, S., and Nadim, F. (1997). *Uncertainties in Characterizing Soil Properties,* Publication 21, Norwegian Geotechnical Institute, Oslo, Norway, pp. 49–75.

Ladd, C. C. (1991). Stability evaluation during staged construction, *ASCE, Journal of Geotechnical Engineering,* 117, 540–615.

Ladd, C. C., and Foott, R. (1974). New design procedure for stability of soft clays, *ASCE, Journal of Geotechnical Engineering,* 100(7), 763–786.

Ladd, C. C., Foott, R., Ishihara, K., Schlosser, F., and Poulos, H. G. (1977). Stress-deformation and strength characteristics, *Proceedings of the 9th International Conference on Soil Mechanics and Foundation Engineering,* Tokyo, pp. 421–494.

Lee, C. Y., Hull, T. S., and Poulos, H. G. (1995). Simplified pile-slope stability analysis, *Computers and Geotechnics,* 17, 1–16.

Lee, D. T., and Schachter, B. J. (1980). Two algorithms for constructing a Delaunay triangulation, *International Journal of Computer and Information Sciences,* 9(3), 219–242.

Lee, I. K., White, W., and Ingles, O. G. (1983). *Geotechnical Engineering,* Pitman Publishing, Boston.

Lee, K. L., and Duncan, J. M. (1975). *Landslide of April 25, 1974 on the Mantaro River, Peru,* report of inspection submitted to the Committee on Natural Disasters, National Research Council, National Academy of Sciences, Washington, DC.

Lee, K. L. and Seed, H. B. (1967). Drained strength characteristics of sands, *ASCE Soil Mechanics and Foundations Division Journal,* vol 93, no SM6, pp. 117–141.

Lefebvre, G., and Duncan, J. M. (1973). *Finite Element Analyses of Traverse Cracking in Low Embankment Dams,* Geotechnical Engineering Report TE 73-3, University of California, Berkeley, CA.

Leshchinsky, D. (1997). *Design Procedure for Geosynthetic Reinforced Steep Slopes,* Technical Report REMR-GT-23, U.S. Army Corps of Engineers, Waterways Experiment Station, Vicksburg, MS, January.

Leshchinsky, D. (1999). Stability of geosynthetic reinforced steep slopes, in *Proceedings of the International Conference on Slope Stability Engineering,* vol. 1, IS-Shikoku '99, Matsuyama, Japan, Nov. 1999, pp. 49–66.

Leshchinsky, D. (2001). Design dilemma: use peak or residual strength of soil, *Geotextiles and Geomembranes,* 19, 111–125.

Leshchinsky, D., and Baker, R. (1986). Three-dimensional slope stability: end effects, *Soils and Foundations,* 26(4), 98–110.

Leshchinsky, D., and Boedeker, R. H. (1989). Geosynthetic reinforced soil structures, *ASCE, Journal of Geotechnical Engineering,* 115(10), 1459–1478.

Leshchinsky, D., and Huang, C. C. (1992). Generalized three-dimensional slope-stability analysis, *ASCE, Journal of Geotechnical Engineering,* 118(11), 1748–1764.

Leshchinsky, D., and San, K. C. (1994). Pseudostatic seismic stability of slopes: design charts, *ASCE, Journal of Geotechnical Engineering,* 120(9), 1514–1532.

Leshchinsky, D., and Volk, J. C. (1985). *Stability Charts for Geotextile-Reinforced Walls,* Transportation Research Record 1031, Transportation Research Board, National Research Council, National Academy Press, Washington, DC, pp. 5–16.

Leshchinsky, D., Ling, H. I., and Hanks, G. (1995). Unified design approach to geosynthetics reinforced slopes and segmental walls, *Geosynthetics International,* 2(4), 845–881.

Liu, C., and Evett, J. B. (2001). *Soils and Foundations,* 5th ed., Prentice Hall, Upper Saddle River, NJ.

Li, K. S., and White, W. (1987), Rapid evaluation of the critical slip surface in slope stability problems, *International Journal for Numerical and Analytical Methods in Geomechanics,* John Wiley and Sons, Vol. 11, No. 5, pp. 449–473.

Low, B. K. and Duncan, J. M. (1985). *Analysis of the Behavior of Reinforced Embankments on Weak Foundations,* Geotechnical Engineering Report, Department of Civil Engineering, Virginia Polytechnic Institute and State University, Blacksburg, VA.

Low, B. K., Gilbert, R. B., and Wright, S. G. (1998). Slope reliability analysis using generalized method of slices, *ASCE, Journal of Geotechnical and Geoenvironmental Engineering,* 124(4), 350–362.

Lowe, J., and Karafiath, L. (1959). Stability of earth dams upon drawdown, *Proceedings of the First PanAmerican Conference on Soil Mechanics and Foundation Engineering,* Mexico City, Vol. 2, pp. 537–552.

Lowe, J., III, and Karafiath, L. (1960). Effect of anisotropic consolidation on the undrained shear strength of compacted clays, *Proceedings of the ASCE Research Conference on Shear Strength of Cohesive Soils,* Boulder, CO, pp. 837–858.

Lunne, T., and Kleven, A. (1982). *Role of CPT in North Sea Foundation Engineering,* Norwegian Geotechnical Institute Publication #139.

Makdisi, F. I., and Seed, H. B. (1977). *A Simplified Procedure for Estimating Earthquake-Induced Deformation in Dams and Embankments,* Report UCB/EERC-77/19, Earthquake Engineering Research Center, University of California, Berkeley, CA.

Makdisi, F. I., and Seed, H. B. (1978). A simplified procedure for estimating dam and embankment earthquake-induced deformations, *ASCE, Journal of the Geotechnical Engineering Division,* 104(7), 849–867.

Mansur, C. I., and Kaufman, R. I. (1962). Dewatering, Chapter 3 in *Foundation Engineering,* G. A. Leonards, Ed., McGraw-Hill Civil Engineering Series, McGraw-Hill, New York.

Marachi, N. D., Chan, C. K., Seed, H. B., and Duncan, J. M. (1969). *Strength and Deformation Characteristics of Rockfill Materials,* Report TE 69-5, Office of Research Services, University of California, Berkeley, CA.

Marcuson, W. F., Hynes, M. E., and Franklin, A. G. (1990). Evaluation and use of residual strength in seismic safety analysis of embankments, *Earthquake Spectra,* 6(3), 529–572.

Matasovic, N., and Kavazanjian, E. Jr., (1998). Cyclic characterization of OIL landfill solid waste, *ASCE, Journal of Geotechnical and Geoenvironmental Engineering,* 124(3), 197–210.

Maugeri, M., and Motta, E. (1992). Stresses on piles used to stabilize landslides, *Proceedings of the 6th International Symposium on Landslides,* Christchurch, New Zealand, pp. 785–790.

McCarty, D. F. (1993). *Essentials of Soil Mechanics and Foundations: Basic Geotechnics,* 4th ed., Prentice Hall, Upper Saddle River, NJ.

McCullough, D. (1999). *Path Between the Seas: The Creation of the Panama Canal, 1870–1914,* Touchstone Books, New York.

McGregor, J. A., and Duncan, J. M. (1998). *Performance and Use of the Standard Penetration Test in Geotechnical Engineering Practice,* Center for Geotechnical Practice and Research, Charles E. Via, Jr. Department of Civil and Environmental Engineering, Virginia Polytechnic Institute and State University, Blacksburg, VA.

Mesri, G. (1989). A reevaluation of $S_{u(mob)} = 0.22\sigma'_p$ using laboratory shear tests, *Canadian Geotechnical Journal,* 26, 162–164.

Meyerhof, G. G. (1956). Penetration tests and bearing capacity of piles, *ASCE, Journal of the Soil Mechanics and Foundation Division,* 82(1), paper 886, 19 pp.

Meyerhof, G. G. (1976). Bearing capacity and settlement of pile foundations, *ASCE, Journal of Geotechnical Engineering,* 102(3), 195–228.

Millet, R. A., Lawton, G. M., Repetto, P. C., and Garga, V. K., (1992). Stabilization of Tablachaca Dam landslide, *Stability and Performance of Slopes and Enbankments-II,* vol. 2, ASCE Special Technical Publication No. 31, pp. 1365–1381.

Mirante, A., and Weingarten, N. (1982). The radial sweep algorithm for constructing triangulated irregular networks, *IEEE Computer Graphics and Applications,* May, pp. 11–21.

Mitchell, J. K. (1993). *Fundamentals of Soil Behavior,* 2nd ed., Wiley, Hoboken, NJ.

Mitchell, J. K., Seed, R. B., and Seed, H. B. (1990). Kettleman Hills waste landfill slope failure: I. Linear system properties, *ASCE, Journal of Geotechnical Engineering,* 116(4), 647–668.

Morgenstern, N. R. (1963). Stability charts for earth slopes during rapid drawdown, *Geotechnique,* 13(2), 121–131.

Morgenstern, N. R., and Price, V. E. (1965). The analysis of the stability of general slip surfaces, *Geotechnique,* 15(1), 79–93.

Morgenstern, N. R., and Price, V. E. (1967). A numerical method for solving the equations of stability of general slip surfaces, *Computer Journal,* 9(4), 388–393.

NAVFAC (1986) *Foundations and Earth Structures, Design Manual 7.02,* U.S. Naval Facilities Command, Alexandria, Virginia.

Newlin, C. W., and Rossier, S. C. (1967). Embankment drainage after instantaneous drawdown, *ASCE, Journal of the Soil Mechanics and Foundations Division,* 93(6), 79–96.

Newmark, N. M. (1965), Effects of earthquakes on dams and embankments, *Geotechnique,* Vol. 15, No. 2, June, pp. 139–160.

Nguyen, V. (1985). Determination of critical slope failure surfaces, *ASCE, Journal of Geotechnical Engineering,* 111(2), 238–250.

Oakland, M. W., and Chameau, J.-L. A. (1984). Finite element analysis of drilled piers used for slope stabilization, in *Laterally Loaded Deep Foundations: Analysis and Performance,* ASTM STP 835, J. A. Langer, E. T. Mosley, and C. D. Thompson, Eds., ASTM, West Conshohocken, PA, 182–193.

Olson, S. M., and Stark, T. D. (2002). Liquefied strength ratio from liquefaction flow failure case histories, *Canadian Geotechnical Journal,* 39, 629–647.

Ooi, P. S. K., and Duncan, J. M. (1994). Lateral load analysis of groups of piles and drilled shafts, *ASCE, Journal of Geotechnical Engineering,* 120(5), 1034–1050.

Palmer, L. A., Thompson, J. B., and Yeomans, C. M. (1950). The control of a landslide by subsurface drainage, *Proceedings of the Highway Research Board,* Vol. 30, Washington, DC, pp. 503–508.

Peck, R. B. (1969). A man of judgment, 2nd R. P. Davis Lecture on the Practice of Engineering, *West Virginia University Bulletin.*

Peck, R. B., Hanson, W. E., and Thornburn, T. H. (1974). *Foundation Engineering,* 2nd ed., Wiley, New York.

Peterson, R., Iverson, N. L., and Rivard, P. J. (1957). Studies of several dam failures clay foundations, *Proceedings of the 4th International Conference on Soil Mechanics and Foundation Engineering,* Vol. 2, Butterworth Scientific, London, pp. 348–352.

Petterson, K. E. (1955). The early history of circular sliding surfaces, *Geotechnique,* 5(4), 275–296.

Pockoski, M., and Duncan, J. M. (2000). *Comparison of Computer Programs for Analysis of Reinforced Slopes,* report of a study sponsored by the Virginia Tech Center for Geotechnical Practice and Research.

Poulos, H. G. (1973). Analysis of piles in soil undergoing lateral movement, *ASCE, Journal of the Soil Mechanics and Foundation Division,* 99(5), 391–406.

Poulos, H. G. (1995). Design of reinforcing piles to increase slope stability, *Canadian Geotechnical Journal,* 32(5), 808–818.

Poulos, H. G. (1999). Design of slope stabilizing piles, *Proceedings of the International Conference on Slope Stability Engineering,* IS-Shikoku, Matsuyama, Japan, Vol. 1, pp. 67–81.

Poulos, S. J., Castro, G., and France, J. W. (1985). Liquefaction evaluation procedure, *ASCE, Journal of Geotechnical Engineering,* 111(6), 772–792.

Purbrick, M. C., and Ayres, D. J. (1956). Uses of aerated cement grout and mortar in stabilizing of slips in embankments, large scale tunnel works and other works, *Proceedings of the Institution of Civil Engineers,* Part II, Vol. 5, No. 1.

Rahardjo, H., Hritzuk, K. J., Leong, E. C., and Rezaur, R. B. (2003). Effectiveness of horizontal drains for slope stability, *Engineering Geology,* 2154, 1–14.

Reese, L. C., and Van Impe, W. F. (2001). *Single Piles and Pile Groups Under Lateral Loading,* A. A. Balkema, Brookfield, VT.

Reese, L. C., Wang, S.-T., and Fouse J. L. (1992). Use of drilled shafts in stabilizing a slope, *Stability and Performance of Slopes and Embankments: II,* Geotechnical Special Publication 31, ASCE, Reston, VA, Vol. 2, pp. 1318–1322.

Reuss, R. F., and Schattenberg, J. W. (1972). Internal piping and shear deformation, Vistor Braunig Dam, San Antonio, Texas, *Proceedings of the ASCE Specialty Conference on Performance of Earth and Earth-Supported Structures,* Vol. 1, Part 1, pp. 627–651.

Robertson, P. K. and Campanella, R. G. (1983). Interpretation of cone penetration tests, 1: sand, *Canadian Geotechnical Journal,* vol 20, No. 4, pp. 718–733.

RocScience (2002). Slide 4.0, *2D Limit Equilibrium for Soil and Rock Slopes: Online User Guide,* RocScience, Toronto, Ontario, Canada.

Rogers, C. D. F., and Glendinning, S. (1993). Stabilization of embankment clay fills using lime piles, *Proceedings of the International Conference on Engineered Fills,* Thomas Telford, London, pp. 226–238.

Rogers, C. D. F., and Glendinning, S. (1994). *Deep Slope Stabilization Using Lime,* Transportation Research Record 1440, Transportation Research Board, National Research Council, National Academy Press, Washington, DC, pp. 63–70.

Rogers, C. D. F., and Glendinning, S. (1997). Improvement of clay soils in situ using lime piles in the UK, *Engineering Geology,* 47(3), 243–257.

Rose, A. T. (1994). The undrained behavior of saturated dilatant silts, Ph.D. dissertation, Virginia Polytechnic Institute and State University, Blacksburg, VA.

Rose, A. T., Brandon, T. L., and Duncan, J. M. (1993). *Classification and Engineering Behavior of Silts,* submitted to the U.S. Army Corps of Engineers.

Rowe, R. K., Gnanendran, C. T., Landva, A. O., and Valsangkar, A. J. (1996). Calculated and observed behavior of a reinforced embankment over soft compressible soil, *Canadian Geotechnical Journal,* vol. 33, pp. 324–338.

Rowe, R. K. and Mylleville, B. L. J. (1996). A geogrid reinforced embankment on peat over organic silt: a case history, *Canadian Geotechnical Journal,* vol. 33, pp. 106–122.

Rowe, R. K. and Soderman, K. L. (1985). An approximate method for estimating the stability of geotextile-reinforced embankments, *Canadian Geotechnical Journal,* vol. 22, pp. 392–398.

Ruenkrairergsa, T., and Pimsarn, T. (1982). Deep hole lime stabilization for unstable clay shale embankment, *Proceedings of the 7th Southeast Asia Geotechnics Conference,* Hong Kong, pp. 631–645.

Sabatini, P. J., Griffin, L. M., Bonaparte, R., Espinoza, R. D., and Giroud, J. P. (2002). Reliability of state of practice for selection of shear strength parameters for waste containment system stability analyses, *Geotextiles and Geomembranes,* 20, 241–262.

Saleh, A. A., and Wright, S. G. (1997). *Shear Strength Correlations and Remedial Measure Guidelines for Long-Term Stability of Slopes Constructed of Highly Plastic Clay Soils,* Research Report 1435-2F, Center for Transportation Research, Bureau of Engineering Research, University of Texas, Austin, TX.

Sarma, S. K. (1973). Stability analysis of embankments and slopes, *Geotechnique,* 23(3), 423–433.

Schaefer, V. R., and Duncan, J. M. (1986). *Evaluation of the Behavior of Mohicanville Dike No. 2,* report of research conducted for the Huntington District, Corps of Engineers, Department of Civil Engineering, Virginia Polytechnic Institute and State University, Blacksburg, VA.

Schaefer, V. R., and Duncan, J. M. (1987). *Analysis of Reinforced Embankments and Foundations Overlying Soft Soils,* Geotechnical Engineering Report, Department of Civil Engineering, Virginia Polytechnic Institute and State University, Blacksburg, VA.

Schmertmann, J. H. (1975). Measurement of in-situ shear strength—state of the art review, *Proceedings of a conference on in-situ measurement of soil properties,* vol. 2, North Carolina State University, pp. 57–138.

Schmertmann, G. R., Chouery-Curtis, V. E., Johnson, R. D., and Bonaparte, R. (1987). Design charts for geogrid-reinforced soil slopes, *Proceedings from Geosynthetics '87,* vol. 1, New Orleans, pp. 108–120.

Seed, H. B. (1966). A method for earthquake resistant design of earth dams, *ASCE, Journal of the Soil Mechanics and Foundations Division,* 92(1), 13–41.

Seed, H. B. (1979). Considerations in the earthquake-resistant design of earth and rockfill dams, Nineteenth Rankine Lecture, *Geotechnique,* 29(3), 215–263.

Seed, R. B., and Harder, L. F., Jr. (1990). SPT-based analysis of cyclic pore pressure generation and undrained residual strength, *Proceedings of the H. Bolton Seed Memorial Symposium,* May, Vol. 2, pp. 351–376.

Seed, R. B., Mitchell, J. K., and Seed, H. B. (1990). Kettleman Hills waste landfill slope failure: II. Stability analysis, *ASCE, Journal of Geotechnical Engineering,* 116(4), 669–690.

Seed, H. B., Lee, K. L., and Idriss, I. M. (1969). Analysis of Sheffield dam failure, *ASCE, Journal of the Soil Mechanics and Foundations Division,* 95(6), 1453–1490.

Seed, H. B., Seed, R. B., Harder, L. F., Jr., and Jong, H.-L. (1988). *Re-evaluation of the Slide in the Lower San Fernando Dam in the Earthquakes of February 9, 1971,*

Report UCB/EERC-88/04, University of California, Berkeley, CA.

Shmuelyan, A. (1996). Piled stabilization of slopes, *Proceedings of the International Symposium on Landslides,* Vol. 3, pp. 1799–1804.

Siegel, R. A., Kovacs, W. D., and Lovell, C. W. (1981). Random surface generation in stability analysis, Technical Note, *ASCE, Journal of the Geotechnical Engineering Division,* 107(7) 996–1002.

Skempton, A. W. (1948). The $\phi = 0$ analysis of stability and its theoretical basis, *Proceedings of the 2nd International Conference on Soil Mechanics and Foundation Engineering,* Rotterdam, Vol. 1, pp. 72–78.

Skempton, A. W. (1954). The pore pressure coefficients A and B, *Geotechnique,* 4(4), 143–147.

Skempton, A. W. (1964). Longterm stability of clay slopes, *Geotechnique,* 14(2), 77–102.

Skempton, A. W. (1970). First-time slides in overconsolidated clays, *Geotechnique,* 20(3), 320–324.

Skempton, A. W. (1977). Slope stability of cuttings in brown London clay, *Proceedings of the 9th International Conference on Soil Mechanics,* Tokyo, Vol. 3, pp. 261–270.

Skempton, A. W. (1985). Residual strength of clays in landslides, flooded strata and the laboratory, *Geotechnique,* 35(1), 3–18.

Skempton, A. W., and LaRochelle, P. (1965). The Bradwell slip: a short-term failure in London clay, *Geotechnique,* 15(3), 221–242.

Smith, R., and Peck, R. B. (1955). Stabilization by pressure grouting on American railroads, *Geotechnique,* 5, 243–252.

Sowers, G. F. (1979). *Introductory Soil Mechanics and Foundations: Geotechnical Engineering,* 4th ed., Macmillan, New York.

Spencer, E. (1967). A method of analysis of the stability of embankments assuming parallel inter-slice forces, *Geotechnique,* 17(1), 11–26.

Squier, L. R., and Versteeg, J. H. (1971). History and correction of the OMSI-zoo landslide, *Proceedings of the 9th Engineering Geology and Soils Engineering Symposium,* pp. 237–256.

Srbulov, M. (2001). Analyses of stability of geogrid reinforced steep slopes and retaining walls, *Computers and Geotechnics,* 28, 255–268.

Stark, T. D., and Duncan, J. M. (1987). *Mechanisms of Strength Loss in Stiff Clays,* report to the U.S. Bureau of Reclamation, Geotechnical Engineering Report, Department of Civil Engineering, Virginia Polytechnic Institute and State University, Blacksburg, VA.

Stark, T. D., and Duncan, J. M. (1991). Mechanisms of strength loss in stiff clays, *ASCE, Journal of Geotechnical Engineering,* 117(1), 139–154.

Stark, T. D., and Eid, H. T. (1993). Modified Bromhead ring shear apparatus, *ASTM, Geotechnical Testing Journal,* 16(1), 100–107.

Stark, T. D., and Eid, H. T. (1994). Drained residual strength of cohesive soils, *ASCE, Journal of Geotechnical Engineering,* 120, 856–871.

Stark, T. D., and Eid, H. T. (1997). Slope stability analyses in stiff fissured clay, *ASCE, Journal of Geotechnical and Geoenvironmental Engineering,* 123(4), 335–343.

Stark, T. D., and Mesri, G. (1992). Undrained shear strength of liquefied sands for stability analysis, *ASCE, Journal of Geotechnical Engineering,* 121(11), 1727–1747.

Stark, T. D., and Poeppel, A. R. (1994). Landfill line interface strengths from torsional ring shear tests, *ASCE, Journal of Geotechnical Engineering,* 120(3), 597–615.

Stauffer, P. A., and Wright, S. G. (1984). *An Examination of Earth Slope Failures in Texas,* Research Report 353-3F, Center for Transportation Research, University of Texas, Austin, TX, November.

Svano, G., and Nordal, S. (1987). Undrained effective stress stability analysis, *Proceedings of the 9th European Conference on Soil Mechanics and Foundation Engineering,* Dublin. Also, Bulletin 22 of the Geotechnical Division, Norwegian Institute of Technology, University of Trondheim, 1989.)

Tang, W. H. (1984). Principles of probabilistic characterization of soil properties, *Proceedings of the Symposium on Probabilistic Characterization of Soil Properties,* ASCE, Reston, VA, pp. 74–89.

Tang, W. H., Stark, T. D., and Angulo, M. (1999). Reliability in back analysis of slope failures, *Journal of Soil Mechanics and Foundations,* October.

Tavenas, F., Mieussens, C., and Bourges, F. (1979). Lateral displacements in clay foundations under embankments, *Canadian Geotechnical Journal,* vol. 16, No. 3, pp. 532–550.

Taylor, D. W. (1937). Stability of earth slopes, *Journal of the Boston Society of Civil Engineers,* 24(3). Reprinted in *Contributions to Soil Mechanics, 1925–1940,* Boston Society of Civil Engineers, Boston, 1940, pp. 337–386.

Taylor, D. W. (1948). *Fundamentals of Soil Mechanics,* Wiley, Hoboken, NJ.

Tervans, F., Mieussens, C., and Bourges, F. (1979). Lateral displacements in clay foundations under embankments, *Canadian Geotechnical Journal,* 16(3), 532–550.

Terzaghi, K. V. (1936). Stability of slopes of natural clay, *Proceedings of the First International Conference on Soil Mechanics and Foundation Engineering,* Cambridge, MA, Vol. 1, pp. 161–165.

Terzaghi, K. (1943). *Theoretical Soil Mechanics,* Wiley, Hoboken, NJ.

Terzaghi, K. (1950). *Mechanisms of Landslides,* Engineering Geology (Berkeley) Volume, Geological Society of America, Boulder, CO, November, pp. 83–123.

Terzaghi, K., and Peck, R. B. (1967). *Soil Mechanics in Engineering Practice,* 2nd ed., Wiley, Hoboken, NJ.

Terzaghi, K. Peck, R. B., and Mesri, G. (1996). *Soil Mechanics in Engineering Practice,* 3rd ed., Wiley, Hoboken, NJ, 549 pages.

Tokimatsu, K. and Yoshimi, Y. (19893), Empirical correlation of soil liquefaction based on SPT N-value and fines content, *Soils and Foundations,* Vol. 23, No. 4, pp. 56–74.

Tracy, F. T. (1991). *Application of Finite Element, Grid Generation, and Scientific Visualization Techniques to 2-D and 3-D Seepage and Groundwater Modeling,* Technical Report ITL-91-3, Department of the Army, U.S. Army Corps of Engineers, Waterways Experiment Station, Vicksburg, MS, September.

Tschebotarioff, G. P. (1973). *Foundations, Retaining and Earth Structures,* 2nd ed., McGraw-Hill, New York.

Turnbull, W. J., and Hvorslev, M. J. (1967). Special problems in slope stability, *ASCE, Journal of the Soil Mechanics and Foundations Division,* 93(4), 499–528. Also, in *Stability and Performance of Slopes and Embankments, Proceedings of an ASCE Specialty Conference,* Berkeley, CA, August 22–26, 1966, pp. 549–578.

U.S. Army Corps of Engineers (1970). *Engineering and Design: Stability of Earth and Rock-Fill Dams,* Engineer Manual EM 1110-2-1902, Department of the Army, Corps of Engineers, Office of the Chief of Engineers, Washington, DC, April.

U.S. Army Corps of Engineers (1986). *Seepage Analysis and Control for Dams,* Engineering Manual EM 1110-2-1901, Department of the Army, Washington, DC.

U.S. Army Corps of Engineers (1998). *Risk-Based Analysis in Geotechnical Engineering for Support of Planning Studies.* Engineering Circular 1110-2-554, Department of the Army, Washington, DC.

U.S. Department of the Interior, Bureau of Reclamation (1973). *Design of Small Dams,* A Water Resources Technical Publication, 2nd ed., Government Printing Office, Washington, DC.

United States Department of the Interior, Bureau of Reclamation (1974). *Earth Manual,* 2nd edition, Denver, Colorado.

Vanmarcke, E. H. (1977). Reliability of earth slopes, *ASCE, Journal of Geotechnical Engineering,* 103(11), 1247–1265.

Vaughan, P. R., and Walbancke, H. J. (1973). Pore pressure changes and the delayed failure of cutting slopes in over-consolidated clay, *Geotechnique,* 23, 531–539.

Viggiani, C. (1981). Ultimate lateral load on piles used to stabilize landslides, *Proceedings of the 10th International Conference on Soil Mechanics and Foundation Engineering,* Stockholm, Vol. 3, pp. 555–560.

Wang, J. L., and Vivatrat, V. (1982). Geotechnical properties of Alaska OC marine silts, *Proceedings of the 14th Annual Offshore Technology Conference,* Houston, TX.

Wang, M. C., Wu, A. H., and Scheessele, D. J. (1979). Stress and deformation in single piles due to lateral movement of surrounding soils, in *Behavior of Deep Foundations,* R. Lunggren, Ed., ASTM 670, ASTM, West Conshohocken, PA, pp. 578–591.

Watson, D. F., and Philip, G. M. (1984). Survey: systematic triangulations, *Computer Vision, Graphics, and Image Processing,* 26(2), 217–223.

Whitman, R. V., and Bailey, W. A. (1967). Use of computers for slope stability analyses, *ASCE, Journal of the Soil Mechanics and Foundations Division,* 93(4), 475–498. Also, in *Stability and Performance of Slopes and Embankments, Proceedings of an ASCE Specialty Conference,* Berkeley, CA, August 22–26, 1966, pp. 519–542.

Wolff, T. F. (1994). *Evaluating the Reliability of Existing Levees,* Report on Research Project: Reliability of Existing Levees, prepared for U.S. Army Corps of Engineers, Waterways Experiment Station Geotechnical Laboratory, Vicksburg, MI.

Wolff, T. F. (1996). Probabilistic slope stability in theory and practice, *Uncertainty in the Geologic Environment, ASCE Proceedings of Special Conference,* Madison, WI, pp. 419–433. Also, ASCE Geotechnical Special Publication 58.

Wolski, W., Szymanski, A., Mirecki, J., Lechowicz, Z., Larsson, R., Hartlen, J., Garbulewski, K., and Bergdahl, U. (1988). *Two Stage-Constructed Embankments on Organic Soils: Field and Laboratory Investigations, Instrumentation, Prediction and Observation of Behavior,* SGI Report 32, 63(+/3/s), Statens Geotekniska Institut, Linköping, Sweden.

Wolski, W., Szymanski, A., Lechowicz, Z., Larsson, R., Hartlen, J., and Bergdahl, U. (1989). *Full-Scale Failure Test on Stage-Constructed Test Fill on Organic Soil,* SGI Report 36, 87, Linköping, Sweden. Statens Geotekniska Institut.

Wong, K. S., and Duncan, J. M. (1974). *Hyperbolic Stress–Strain Parameters for Nonlinear Finite Element Analyses of Stresses and Movements in soil Masses,* Geotechnical Engineering Report TE 74-3 to the National Science Foundation, Office of Research Services, University of California, Berkeley, CA.

Wong, K. S., Duncan, J. M., and Seed, H. B. (1983). *Comparisons of Methods of Rapid Drawdown Stability Analysis,* Report UCB/GT/82-05, Department of Civil Engineering, University of California, Berkeley, CA, December, 1982. Revised July 1983.

Wright, S. G. (1969). A study of slope stability and the undrained shear strength of clay shales, dissertation submitted in partial satisfaction of the requirements for the degree of Doctor of Philosophy in engineering, University of California, Berkeley, CA.

Wright, S. G. (1974). *SSTAB1: A General Computer Program for Slope Stability Analyses,* Department of Civil Engineering, University of Texas, Austin, TX.

Wright, S. G. (1999). *UTEXAS4: A Computer Program for Slope Stability Calculations,* Shinoak Software, Austin, TX.

Wright, S. G. (2002). Long-term slope stability computation for earth dams using finite element seepage analyses, ASDSO paper reference.

Wright, S. G., and Duncan, J. M. (1969). Anisotropy of clay shales, Specialty Session 10 on Engineering Properties and Behavior of Clay Shales, 7th International Conference on Soil Mechanics and Foundation Engineering, Mexico City, Mexico.

Wright, S. G., and Duncan, J. M. (1972). Analysis of Waco dam slide, *ASCE, Journal of the Soil Mechanics and Foundation Division,* 98(9), 869–877.

Wright, S. G., and Duncan, J. M. (1987). *An Examination of Slope Stability Computation Procedures for Sudden Drawdown,* Miscellaneous Paper GL-87-25, Geotechnical Laboratory, U.S. Army Corps of Engineers, Waterways Experiment Station, Vicksburg, MS, September.

Wright, S. G., and Duncan, J. M. (1991). Limit equilibrium stability analyses for reinforced slopes, *Proceedings of the 70th Annual Meeting of the Transportation Research Board,* Washington, DC.

Wright, S. G., Kulhawy, F. H., and Duncan, J. M. (1973). Accuracy of equilibrium slope stability analyses, *ASCE, Journal of the Soil Mechanics and Foundation Division,* 99(10), 783–791.

Wu, T. H., (1976). *Soil Mechanics,* 2nd ed., Allyn & Bacon, Boston.

Wu, T. H., Riestenberg, M. M., and Flege, A. (1994). Root properties for design of slope stabilization, *Proceedings of the International Conference on Vegetation and Slopes: Stabilization, Protection and Ecology,* Oxford.

Yamagami, T., Jiang, J., and Ueno, K. (2000). A limit equilibrium stability analysis of slopes with stabilizing piles, *Slope Stability 2000,* Geotechnical Special Publication 101, ASCE, Reston, VA, pp. 343–354.

Youd, T. L., Idris, I. M., Andrus, R. D., Arango, I., Castro, C. Christian, J. T., Dobry, R., Finn, W. D. Liam, Harder, J. T., Leslie F., Hynes, M. E., Ishihara, K., Koester, J. P., Liao, S. S. C. Marcuson, W. F., III, Martin, G. R., Mitchell, J. K., Moriwaki, Y., Power, M. S., Robertson, P. K., Seed, R. B., Stokoe, K. H., II, (2001). Liquefaction resistance of soils: summary report from the 1996 NCEER and 1998 NCEER/NSF workshops on evaluation of liquefaction resistance of soils, *Journal of Geotechnical and Geoenvironmental Engineering,* 127(10), 817–833.

Zeller, J., and Wullimann, R. (1957). The shear strength of the shell materials for the Goschenenalp Dam, Switzerland, *Proceedings of the 4th International Conference on Soil Mechanics and Foundation Engineering,* Vol. 2, Butterworth Scientific, London, pp. 399–404.

Zeng, S., and Liang, R. (2002). Stability analysis of drilled shafts reinforced slope, *Soils and Foundations,* 42(2), 93–102.

Zornberg, J. G., Sitar, N., and Mitchell, J. K. (1998a). Limit equilibrium as basis for design of geosynthetic reinforced slopes, *ASCE, Journal of Geotechnical and Geoenvironmental Engineering,* 124(8), 684–698.

Zornberg, J. G., Sitar, N., and Mitchell, J. K. (1998b). Performance of geosynthetics reinforced slopes at failure, *ASCE, Journal of Geotechnical and Geoenvironmental Engineering,* 124(8), 670–683.

INDEX